EVOLUTION'S WEDGE

ORGANISMS AND ENVIRONMENTS

Harry W. Greene, Consulting Editor

1. *The View from Bald Hill: Thirty Years in an Arizona Grassland,* by Carl E. Bock and Jane H. Bock

2. *Tupai: A Field Study of Bornean Treeshrews,* by Louise H. Emmons

3. *Singing the Turtles to Sea: The Comcáac (Seri) Art and Science of Reptiles,* by Gary Paul Nabhan

4. *Amphibians and Reptiles of Baja California, Including Its Pacific Islands and the Islands in the Sea of Cortés,* by L. Lee Grismer

5. *Lizards: Windows to the Evolution of Diversity,* by Eric R. Pianka and Laurie J. Vitt

6. *American Bison: A Natural History,* by Dale F. Lott

7. *A Bat Man in the Tropics: Chasing El Duende,* by Theodore H. Fleming

8. *Twilight of the Mammoths: Ice Age Extinctions and the Rewilding of America,* by Paul S. Martin

9. *Biology of Gila Monsters and Beaded Lizards,* by Daniel D. Beck

10. *Lizards in the Evolutionary Tree,* by Jonathan B. Losos

11. *Grass: In Search of Human Habitat,* by Joe C. Truett, with a foreword by Harry W. Greene

12. *Evolution's Wedge: Competition and the Origins of Diversity,* by David W. Pfennig and Karin S. Pfennig

EVOLUTION'S WEDGE

Competition and the Origins of Diversity

David W. Pfennig
Karin S. Pfennig

WITHDRAWN

UNIVERSITY OF CALIFORNIA PRESS
Berkeley   Los Angeles   London

THE PUBLISHER GRATEFULLY ACKNOWLEDGES THE
GENEROUS SUPPORT OF THE GENERAL ENDOWMENT FUND
OF THE UNIVERSITY OF CALIFORNIA PRESS FOUNDATION.

University of California Press, one of the most distinguished university presses in the United States, enriches lives around the world by advancing scholarship in the humanities, social sciences, and natural sciences. Its activities are supported by the UC Press Foundation and by philanthropic contributions from individuals and institutions. For more information, visit www.ucpress.edu.

Organisms and Environments, No. 12

University of California Press
Berkeley and Los Angeles, California

University of California Press, Ltd.
London, England

© 2012 by The Regents of the University of California

Library of Congress Cataloging-in-Publication Data

Pfennig, David W. (David William), 1955–
   Evolution's wedge : competition and the origins of diversity / David W. Pfennig, Karin S. Pfennig.
      p. cm. — (Organisms and environments ; no. 12)
   Includes bibliographical references and index.
   ISBN 978-0-520-27418-1 (cloth : alk. paper)
   1. Divergence (Biology) 2. Competition (Biology) 3. Animal diversity.
I. Pfennig, Karin S. (Karin Susan), 1969– II. Title.
   QH408.P44 2012
   577.8'3—dc23
                                                  2012009564

Manufactured in the United States of America

19 18 17 16 15 14 13 12
10 9 8 7 6 5 4 3 2 1

The paper used in this publication meets the minimum requirements of ANSI/NISO Z39.48-1992 (R 2002) (*Permanence of Paper*). ♾

Cover illustration: Pencil and watercolor sketches of beetles by Henry Walter Bates. © The Natural History Museum, London.

CONTENTS

Preface · ix

1 Discovery of a Unifying Principle · 1
   A Brief History · 4
   Detecting Character Displacement · 10
   Phenomena Mistaken for Character Displacement · 14
   *What Constitutes Character Displacement? Conflation of Process
     and Pattern* · 19
   Reproductive Character Displacement versus Reinforcement · 20
   Terminology · 21
   Box 1.1: *Alternative Manifestations of Character Displacement* · 22
   A Unifying Principle · 25
   Box 1.2: *Suggestions for Future Research* · 26
   Summary · 27
   Further Reading · 27

2 Why Character Displacement Occurs · 29
   Why Ecological Character Displacement Occurs · 30
   Box 2.1: *Alternative Models of Species Coexistence* · 36
   Box 2.2: *Is Competitively Induced Plasticity Character Displacement?* · 44
   Why Reproductive Character Displacement Occurs · 48
   Box 2.3: *Suggestions for Future Research* · 53

Summary · 54
Further Reading · 54

3   When Character Displacement Occurs · 57
Facilitators of Character Displacement · 58
Variation in the Expression of Character Displacement · 65
How Ecological and Reproductive Character Displacement Facilitate Each Other · 69
How Ecological and Reproductive Character Displacement Can Impede Each Other · 77
Box 3.1: Suggestions for Future Research · 78
Summary · 78
Further Reading · 79

4   How Character Displacement Unfolds · 81
Mechanisms of Divergence · 82
Tempo and Mode of Character Displacement · 93
Summary · 102
Box 4.1: Suggestions for Future Research · 103
Further Reading · 104

5   Diversity and Novelty Within Species · 105
How Intraspecific Character Displacement Works · 106
Intraspecific Character Displacement: Observational Evidence · 108
Intraspecific Character Displacement: Experimental Evidence · 110
Evolution of Alternative Phenotypes · 111
Intraspecific Character Displacement and Species Diversity · 126
Character Displacement Within Versus Between Species · 129
Summary · 130
Box 5.1: Suggestions for Future Research · 131
Further Reading · 131

6   Ecological Consequences · 133
Evolution of the Niche · 134
Partitioning of Resources and Reproduction: A Reprise · 136
Box 6.1: Individual Variation and the Coexistence of Species · 139
Community Organization · 143
Character Displacement and Darwinian Extinction · 146
Species Distributions and Geographic Mosaics · 150
Character Displacement and Species Ranges · 151
Summary · 153
Box 6.2: Suggestions for Future Research · 154
Further Reading · 155

7 Sexual Selection · 157
How Sexual Selection Works · 159
How Character Displacement Affects Sexual Selection · 160
Implications of the Effects of Character Displacement on Sexual Selection · 168
How Sexual Selection Affects Character Displacement · 174
A Cautionary Note: Process Versus Pattern · 175
Box 7.1: *Suggestions for Future Research* · 176
Summary · 176
Further Reading · 177

8 Speciation · 179
What Are Species? · 179
How Are Species' Boundaries Maintained? · 182
The Evolution of Isolating Mechanisms · 183
Box 8.1: *Selection and the Evolution of Reproductive Isolation* · 186
Character Displacement's Role in Speciation · 189
Summary · 202
Box 8.2: *Suggestions for Future Research* · 203
Further Reading · 203

9 Macroevolution · 205
Competition in the Fossil Record · 207
Methods for Studying Macroevolution: Replaying the Tape of Life · 211
Adaptive Radiation · 213
Evolutionary Escalation · 224
Macroevolution: Red Queen or Court Jester? · 228
Summary · 229
Box 9.1: *Suggestions for Future Research* · 230
Further Reading · 230

10 Major Themes and Unsolved Problems · 233
Major Themes of the Book · 234
Some Unsolved Problems · 238
Summary · 242

References · 243
Index · 291

PREFACE

This book examines how a pervasive feature of living systems—competition—drives evolution and generates diversity. Competitive interactions take place whenever any two individuals, populations, or species affect each other adversely. These interactions might involve direct contests for limiting resources or reproductive opportunities. Alternatively, they might assume more subtle forms, such as depleting a shared resource or interfering with each other's ability to identify high-quality conspecific mates.

Regardless of how it is expressed, competition is central to the Darwinian theory of evolution by means of natural selection. Indeed, Darwin recounts in his autobiography (Barlow 1959, p. 120) how he developed his theory only after reading an essay on population growth by Malthus (1797 [1990]). In this essay, Malthus argued that, if unchecked, the rate of human population growth would exceed the rate of increase in the food supply, resulting in a "struggle for existence." Malthus' emphasis on overpopulation and competition immediately inspired Darwin's thinking regarding the mechanism that drove evolution—natural selection.

Upon reading Malthus, Darwin's critical insight was to realize that *all* organisms tend to produce far more offspring than could be supported by the available food supply and habitat, thereby fueling perpetual competition among organisms for these crucial resources. Such competition, Darwin (1859 [2009]) held, favors individuals that are least like their competitors in how they obtain resources. In this way, competitively mediated selection may cause interacting species and populations to evolve different resource-acquisition traits (for example, morphologies and behaviors). Darwin argued that this process might even cause interacting conspecific populations to diverge from one another to such a degree that they might eventually become entirely new species. Thus, according to Darwin (1859 [2009]), competitively mediated selection could explain two important features of the living world: why there are so many different species, and why even closely related species typically differ in ecologically relevant traits.

Although these ideas about how competition generates biodiversity were crucial to Darwin, they are often misunderstood and under-appreciated today. Moreover, Darwin's ideas have equally profound implications for ecology, which are sometimes overlooked. Indeed, competition potentially explains why species are found where they are, why they specialize on the particular resources that they do, and why they (ultimately) go extinct. In short, competition is crucial to any general theory concerning the origins, maintenance, abundance, and distribution of biodiversity.

Trait evolution that arises as an adaptive response to resource competition or deleterious reproductive interactions between species is now known as "character displacement" (*sensu* Brown and Wilson 1956). This book explores the causes and consequences of this process. A central premise of our book is that, by understanding these causes and consequences, we can gain crucial insights into some of the most fundamental issues in evolutionary biology and ecology, including how new traits and new species arise, why species diversify, and how they are able to coexist with each other. Indeed, character displacement is not only central to ecology and evolutionary biology, it can also unify these sometimes disparate fields.

Before proceeding, however, we need to answer an important question: why do we need a book on character displacement? After all, the basic idea has been around for 150 years. Moreover, several books have appeared recently that highlight character displacement to at least some extent (for example, Schluter 2000; Coyne and Orr 2004; Grant and Grant 2008; Nosil 2012). What is the justification for an entire book on character displacement?

Our motivation in writing this book was threefold. First, most prior discussion of character displacement has concentrated on how to detect it, and on whether or not it even occurs. Given that ample evidence now indicates that character displacement does indeed occur (chapter 1), an integrated discussion of the factors that fuel character displacement and its consequences beyond mere trait evolution is needed. Therefore, in the first half of the book, we discuss how competitively mediated selection often (but not always) promotes character displacement (chapter 2), the factors that determine when character displacement occurs (chapter 3), and the proximate (that is, genetic and developmental) mechanisms of character displacement (chapter 4). Some of the topics covered in this section have received scant attention. For example, in chapter 4, we consider whether phenotypic plasticity (the capacity of a single genotype to produce multiple phenotypes in response to variation in the organism's environment) plays a role in character displacement. In fact, facultative niche shifts induced by competitors might often be a significant factor in initiating character displacement.

In the second half of the book, our focus is on character displacement's myriad downstream effects. In particular, we discuss how character displacement can generate novel, complex traits (chapter 5); foster species coexistence within ecological communities (chapter 6); influence the outcome of sexual selection (chapter 7); promote the formation of new species (chapter 8); and trigger adaptive radiation and even propel certain major evolutionary trends (chapter 9). A central theme of this half of the book is that selection acting to minimize costly competitive interactions among organisms—and the resulting evolutionary change occurring within populations (that is, microevolution)—might ultimately bring about large-scale evolutionary change (that is, macroevolution). Indeed, the study of character displacement potentially helps unite microevolution and macroevolution.

A second motivation for writing this book is to bridge two slightly different perspectives in the field of character displacement. Since Darwin's time, researchers have recognized that trait evolution can occur as an adaptive response to *resource* competition. This evolution, involving traits associated with resource use, is referred to as "ecological character displacement." Yet resource competition is not the only form of competition that promotes trait evolution. As we noted above, selection can also act to lessen costly *reproductive* interactions between species, such as when the members of one species interfere with the members of another species in their ability to obtain high-quality mates, or when separate species engage in costly mismatings with each other (that is, hybridization). Selection to minimize such deleterious reproductive interactions operates on traits associated with reproduction, and this selection can lead to a form of trait evolution known as "reproductive character displacement" (as described in chapter 1, the precise definition of "reproductive character displacement" is controversial; however, we define this term broadly to include any trait evolution stemming from selection that lessens deleterious reproductive interactions between species).

Historically, ecological character displacement has been studied largely independently of reproductive character displacement. Consequently, there has been relatively little cross-fertilization of ideas between researchers who study these two forms of character displacement. Yet the causes and consequences of these two forms of character displacement are similar in many respects. Moreover, these two forms are often likely to occur simultaneously: organisms that compete for resources will frequently also interact during reproduction, and vice versa. Therefore, in chapter 3, we consider how one form of character displacement (be it ecological or reproductive) can influence the other form in ways that promote the unfolding of both. Indeed, the interaction between these two related processes can leave an important imprint on a species' ecology and evolution.

A third motivation for writing this book was to broaden interest in, and deepen understanding of, a process that potentially has vast but under-appreciated implications for evolution and ecology. Most researchers probably currently view character displacement simply as a mechanism for promoting differences between co-occurring species. Yet character displacement potentially encompasses a broad spectrum of phenomena, from the evolution of specialization to the evolution of increased competitive ability. These phenomena, in turn, have additional ramifications. For instance, once a lineage evolves increased competitive ability, it exerts selection on any lineages with which it is interacting to also become better competitors, which in turn imposes selection on the original lineage to further enhance its competitive ability, and so on. Such reciprocal selection could thereby engender a co-evolutionary arms race generating an escalation in traits associated with competitive ability (*sensu* Vermeij 1987). The tendency for many evolutionary lineages to increase in body size over time might be a common outgrowth of such escalation. Indeed, the constant struggle to keep up with their competitors (among other enemies) can even explain why living things have, throughout the history of life, generally evolved to become more complex. In other words, character displacement potentially has had far-reaching impacts on evolution, ranging from promoting diversity to generating complexity. We explore these ideas in greater detail in chapter 9.

Related to the previous point—but at the opposite end of the spectrum of biological organization—character displacement might also explain why species harbor so much trait variation within them (in some cases, variation within species can be as pronounced as that normally seen between species). In chapter 5, we examine one possible explanation for this diversity when we discuss how character displacement can also operate *within* species. We describe how selection to lessen resource competition or reproductive interactions among conspecifics can promote trait evolution through "intraspecific character displacement" (*sensu* West-Eberhard 2003). We further discuss how within-species diversity resulting from intraspecific character displacement can rival speciation in providing an important setting for evolutionary innovation and diversification. Indeed, we describe how intraspecific character displacement might play a key role in instigating the appearance of major phenotypic novelties.

We have written this book to appeal to a broad array of readers, especially those interested in the interplay between ecology and evolution. We have therefore attempted to make the book as accessible as possible. Moreover, we provide a list of suggested readings at the end of each chapter for those who want to learn more about the subject of that particular chapter. Additionally, at the end of each chapter, we provide a text box in which we list some key challenges for future research.

Finally, because we believe that "nothing in evolutionary biology makes sense except in the light of ecology" (Grant and Grant 2008, p. 167), throughout the book we highlight the natural history of character displacement. To illustrate important concepts, we draw as much as possible on case studies from natural populations to emphasize how the study of character displacement can help shed light on how organisms are built, function, evolve, and interact with each other in their natural environment. Our hope is that this book will stimulate new thinking and new research into the causes and consequences of character displacement. Indeed, our overarching goal in writing this book is to illustrate how character displacement is central in the origins, maintenance, abundance, and distribution of biodiversity.

In preparing this book, we have benefited greatly from the advice and assistance of many colleagues. Foremost among these are our former graduate students—Amber Rice, Ryan Martin, Cris Ledón-Rettig, and Aaron Leichty—from whom we learned much about the ideas in this book. We also owe a huge debt of gratitude to the members of a graduate seminar group at the University of North Carolina at Chapel Hill who, in spring 2010, critiqued an entire first draft of the book. We thank this group for their many constructive comments, but the following individuals were particularly helpful in clarifying and sharpening our arguments: Chris Willett, Allen Hurlbert, Ryan Martin, David Kikuchi, Sumit Dhole, Aaron Leichty, and Maria Servedio. The book owes much to their efforts.

We gratefully acknowledge the University of California Press's reviewers—Ryan Calsbeek, Greg Grether, and Art Shapiro—who read and commented on the final draft of the book. Their comments and encouragement were much appreciated. We also thank several anonymous readers who provided an assessment of the project when we first began circulating a prospectus of the book.

We owe a special debt to Trevor Price, who provided thorough and insightful comments on three chapters (chapters 3, 6, 8). Additionally, for providing feedback on individual chapters, we thank Dean Adams (chapter 3), Craig Benkman (chapter 6), Lauren Buckley (chapter 6), Carl Gerhardt (chapter 2), Peter Grant (chapter 1), Rosemary Grant (chapter 3), Mark Kirkpatrick (chapter 8), Jonathan Levine (chapters 2 and 6), Jeff McKinnon (chapter 9), Armin Moczek (chapter 5), Dan McShea (chapter 9), Patrik Nosil (chapter 8), Jeff Podos (chapter 3), Rick Relyea (chapter 2), Doug Schemske (chapter 8), Maria Servedio (chapter 8), Skúli Skúlason (chapter 5), Mary Jane West-Eberhard (chapter 4), Haven Wiley (chapter 7), Chris Willett (chapters 4 and 8), and Matt Wund (chapters 4 and 9). Jeff Conner and his lab group provided comments on chapter 7, as did the combined lab groups of Alex Basolo, Eileen Hebets, and Bill Wagner, along with Emilie Snell-Rood. Lisa Bono, Michael Foote, David Jablonski, Cris Ledón-Rettig, Ethan Temeles, Mike Travisano, and Mark Webster graciously answered questions about specific material in the book.

We hasten to add that not all reviewers agreed with everything in the book, and, of course, any errors that remain are entirely our responsibility.

We also thank Kern Reeve and Mohamed Noor for their friendship, advice, and encouragement over the years; our thesis and postdoctoral advisors—Jeff Brawn, Jim Bull, Jim Collins, Jeff Conner, Mark Kirkpatrick, Mike Ryan, Paul Sherman, and Richard Tinsley—for their support, guidance, and insights; Alan Feduccia for pointers on writing and publishing a book; Harry Greene for urging us to write a book, *any* book; and the staff at the University of California Press and at Bookmatters—especially Blake Edgar, Kate Marshall, David Peattie, and Francisco Reinking—for guiding us through the publication process and for producing a beautiful book. Additionally, we thank Hope Steele for her thorough copyediting of the manuscript, Louise Doucette for proofreading the entire book, and Leonard Rosenbaum for preparing the index.

Our research on character displacement has been supported by the National Science Foundation, a National Institutes of Health Office of the Director New Innovator Award (to Karin Pfennig), and the University of North Carolina at Chapel Hill, for which we are grateful. We also thank the American Museum of Natural History's Southwestern Research Station (Portal, AZ), and its directors, Wade Sherbrooke and Dawn Wilson, for providing a supportive environment for field research over the past 25 years.

Last, and most importantly, we thank our parents, family members, teachers/mentors, and friends for nurturing, encouraging, and often sharing our lifelong passion for science and natural history, and especially our daughters, Katrina and Elsa, for their patience and understanding as we wrote this book.

Chapel Hill, NC
December 27, 2011

1

DISCOVERY OF A UNIFYING PRINCIPLE

In a frequently heard—and perhaps apocryphal—story, the evolutionary biologist J. B. S. Haldane, when asked to comment on what could be inferred about the Creator based on the creation, is reported to have said, "He must have had an inordinate fondness of beetles" (Farrell 1998). Why *are* there so many species of beetles (Figure 1.1A)? For that matter, why are there so many different kinds of living things generally?

As it turns out, living things are amazingly diverse. As one measure of this diversity, conservative estimates place the number of species alive today at 8 to 10 million (Stork 1993; Hamilton et al. 2010; Wilson 2010, p. x; Mora et al. 2011). Yet, as staggering as these numbers are, they vastly underestimate life's true diversity: all species consist of individual organisms that are themselves unique. Indeed, each species comprises a bewildering array of morphological, physiological, ecological, and behavioral traits (for example, see Figure 1.1B, C). In short, the total inherited variation of all organisms—"biodiversity," as coined by Wilson and Peter (1988)—is truly astounding.

Biodiversity demands explanation. Why are there so many different species? Why do species typically express different traits, especially when these traits influence resource acquisition or reproduction (for example, see Figure 1.1B)? Why do species harbor so much trait variation within them, which, in some cases, is as pronounced as the variation normally seen between different species (for example, see Figure 1.1C)? What factors trigger the formation of new traits and new species in the first place? Finally, once new species do arise, why do they sometimes coexist with other species and sometimes not?

One hundred and fifty years ago, Charles Darwin (1859 [2009]) offered a scien-

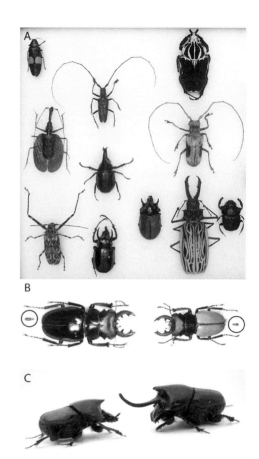

FIGURE 1.1. Among biology's most persistent challenges is to explain why there are so many different kinds of living things. Consider, for example, the beetles. (A) Although scientists do not know how many beetle species there are (about 350,000 species have been described), they are thought to be among the most diverse group of organisms. (B) Even closely related species typically differ in traits associated with resource use and reproduction. For example, male stag beetles from Southeast Asia, *Odontolabis cuvera* (left) and *O. mouhoti* (right), differ in body size, pigmentation, and genitalia length (circled), which are features that are important in obtaining reproduction. (C) Pronounced trait variation can even be found within species. For instance, horned beetles, *Onthophagus nigriventris*, consist of minor males (left) and larger, major males (right), which possess a distinctive thoracic horn. Photo credits: (A): The Natural History Museum, London; (B): Kawano (2003); (C): Alex Wild. Reproduced with the kind permission of the photographers and publishers.

tific explanation for biodiversity—natural selection. Although natural selection is an evolutionary process, Darwin argued that at natural selection's core is an ecological process: competition. (Throughout this book, the term "competition" refers to any direct or indirect interaction between species or populations that reduces access to vital resources or successful reproductive opportunities and that is therefore deleterious—on average—to both parties; see Table 1.1.) According to Darwin (1859 [2009]), all organisms face recurring competition for scarce resources, and this competition favors individuals that are least like their competitors in resource use and associated traits. Consequently, groups of organisms that compete should become increasingly different over time. Selection driven by competition, Darwin held, is the primary engine of diversification.

This book explores how competitively mediated selection generates biodiversity. In particular, we examine how such selection—whether stemming from competition for resources or access to successful reproduction—can promote evolutionary diversification through a process known as "character displacement" (*sensu* Brown and Wilson

TABLE 1.1. Summary of the fitness effects of various ecological interactions between two species or two populations

Ecological interaction	Effect of interaction on Species 1 or Population 1[a]	Effect of interaction on Species 2 or Population 2[a]
Competition[b]	–	–
Amensalism[b]	–	0
Predation/parasitism	+	–
Mutualism	+	+
Commensalism	+	0
Neutralism	0	0

NOTE: Modified from Odum (1959) and Pianka (2000).

[a]Ecological interactions may have deleterious (–), positive (+), or no (0) fitness consequences for members of each of two interacting species/populations. Note that those reproductive interactions that have negative consequences for both parties would constitute competitive interactions under this scheme.

[b]We define "competition" as an ecological interaction in which each species or population affects the other adversely (that is, a minus-minus interaction). A fine line exists, however, between competition and amensalism, which is an ecological interaction in which one participant pays a cost but the other pays no cost (that is, a minus-zero interaction). In particular, when there is extreme asymmetry in competitive ability, the superior competitor's impact on the weaker competitor will be negative, whereas the weaker competitor may seem to have little, if any, immediate impact on the superior competitor. Yet, the superior competitor would likely pay fitness costs in terms of having to produce and maintain phenotypic attributes associated with being a stronger competitor (e.g., large body size, weaponry, toxins). Additionally, superior competitive ability may reflect past selection for escalating competitive ability. Thus, some authors define "amensalism" as asymmetric competition (e.g., see Stiling 2012).

1956). Although evolutionary biologists have long recognized that numerous factors can act as agents of selection and thereby potentially promote diversification (these factors are reviewed in MacColl 2011), competition may be the most common of all selective agents (Vermeij 1987; Amarasekare 2009). Moreover, because competition (unlike, for example, predation) is uniquely mutually costly to both parties involved (Table 1.1), it is a particularly potent agent of divergent selection, which can serve as a wedge that drives competitors apart ecologically and phenotypically.

Our aim in this book is to evaluate character displacement's role in the origins, maintenance, abundance, and distribution of biodiversity. In this opening chapter, we summarize the history of the discovery of character displacement and describe how the process of character displacement is often conflated with the patterns that are predicted to arise from it. We also provide a formal definition of character displacement that avoids such confusion. We conclude the chapter with a discussion of how character displacement serves to unify evolutionary biology and ecology.

A BRIEF HISTORY

Divergence of character . . . is of high importance on my theory, and explains, as I believe, several important facts.

(DARWIN 1859 [2009], p. 111)

Natural selection, also, leads to divergence of character; for more living beings can be supported on the same area the more they diverge in structure, habits, and constitution, of which we see proof by looking at the inhabitants of any small spot or at naturalised productions. Therefore during the modification of the descendants of any one species, and during the incessant struggle of all species to increase in numbers, the more diversified these descendants become, the better will be their chance of succeeding in the battle of life. Thus the small differences distinguishing varieties of the same species, will steadily tend to increase till they come to equal the greater differences between species of the same genus, or even of distinct genera.

(DARWIN 1859 [2009], pp. 127–128)

. . . it is the most closely-allied forms,—varieties of the same species, and species of the same genus or of related genera,—which, from having nearly the same structure, constitution, and habits, generally come into the severest competition with each other. Consequently, each new variety or species, during the progress of its formation, will generally press hardest on its nearest kindred, and tend to exterminate them.

(DARWIN 1859 [2009], p. 110)

With these words, Darwin first proposed that competition acts as a ubiquitous and potent agent of divergent selection. Darwin's ideas were groundbreaking: none of his predecessors had viewed interactions among organisms as being significant in evolution (Ridley 2005). The crux of Darwin's idea is that when organisms compete for scarce resources, natural selection should favor those individuals that are least like their competitors. Consequently, groups of organisms that compete should become more dissimilar over time.

Darwin considered this process, which he dubbed "divergence of character," to be "of high importance" for two reasons. First, he held that this process was crucial to the origin of species. According to Darwin, selection that minimizes competition between "varieties" could drive divergence between them until they became separate species (see the second quote above).

Second, Darwin maintained that his principle of divergence of character could explain why species tend to differ phenotypically and also—perhaps even more radically—why evolution has produced a distinctive "tree-like" typology (reviewed in Ridley 2005). Indeed, Darwin changed the way that we view the *shape* of evolution. Before

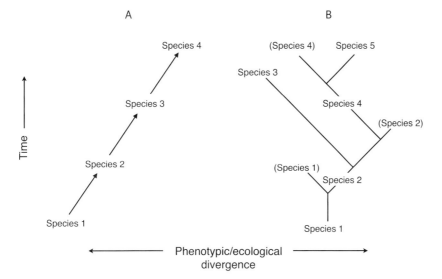

FIGURE 1.2.
Two views of evolution. (A) Before Darwin, the shape of evolution was thought to resemble a ladder, in which ancestral species graded into descendant species, resulting in non-overlapping straight-line evolution. (B) Darwin's view was that evolution produces a distinctive tree-like pattern. Here, lineages split apart, giving rise to two new lineages that (potentially) coexist and that tend to diverge from one another. Redrawn from MacFadden (1992).

Darwin, the image of evolution (attributed mostly to the early evolutionists Jean-Baptiste Lamarck and Geoffrey Saint-Hilaire) was that of a ladder, with ancient, simpler life forms at the base of the ladder and more complex life forms at the top (Gould 1989). In this system, lineages did not branch. Instead, evolutionary lineages persisted indefinitely, gradually accumulating changes over many generations, until they became new species (Figure 1.2A).

Darwin's view of evolution was entirely different. His imagery was that of a tree, in which ancestors and descendants split and coexisted. Moreover, according to Darwin, when evolutionary lineages split, they tended to produce two new lineages that diverged from one another, thereby accentuating the tree-like pattern (Figure 1.2B). Darwin held that evolution's divergent nature could be traced to the tendency for the strength of competition to decrease with increasing divergence between competitors (see the third quote above). Thus, according to Darwin, by continually eliminating intermediate forms, competitively mediated selection has caused species to differ and the history of life to resemble a tree, with numerous, diverging branches.

As the above discussion indicates, the concept of divergence of character was crucial to Darwin's thinking on the origin and diversity of species. Indeed, Darwin devoted as much space in *The Origin* to discussing divergence of character as he did to discussing the idea for which he is most widely known—natural selection (Ridley 2005). Yet, despite the importance that Darwin attached to his principle of divergence of character, he failed to provide any actual examples of competitively mediated divergence in contemporary species. Moreover, although some have questioned whether Darwin's principle of divergence of character is actually synonymous with character displacement (this is reviewed in Pfennig and Pfennig 2010), Darwin's statements that "more living things can be supported on the same area the more they diverge" and that "each new variety or species . . . will generally press hardest on its nearest kindred" suggest that he envisioned the modern process of character displacement, in which sympatric species diverge owing to the action of competitively mediated selection.

As it turns out, for Darwin to come up with examples of competitively mediated divergence in contemporary species would have been no trivial exercise. Indeed, competitively mediated divergence can be notoriously difficult to detect. On the one hand, if two species are phenotypically similar enough to compete, then they probably have not undergone much divergence. On the other hand, if two species have already undergone competitively mediated divergence, then they are probably no longer similar enough to experience much competition with each other for an investigator to detect.

Lack (1947) was the first to propose a way around this conundrum. Lack was strongly influenced by Gause's (1934) hypothesis that, because of competition, no two species could persist in the same locality without possessing ecological differences (we expand on Gause's hypothesis in chapter 2). From this idea, Lack (1947) developed a powerful approach for detecting competition's evolutionary signature in natural populations. In particular, he devised the method of comparing different populations of the same species—those in sympatry with a heterospecific versus those in allopatry—to test the evolutionary effects of interspecific competition. The logic behind this approach is clear: selection to lessen competition between a pair of species will act only in areas where the two species actually co-occur. Thus, if competition plays an important role in evolution, Lack reasoned we should observe a distinctive pattern in which species pairs are more dissimilar where they occur together than where each occurs alone (Figure 1.3).

In developing these ideas, Lack drew on his detailed studies of finches from the Galápagos Islands. This archipelago contains over a dozen endemic species of finches, many of which co-occur on some islands but not on other islands (Grant 1986; Grant and Grant 2008). Interestingly, Darwin visited the Galápagos Islands as a young man in 1835, and these same finches—now referred to as "Darwin's finches" (Lack 1947)—were crucial in sparking Darwin's thinking about evolution and natural selection (Browne 1995). Indeed, according to Sulloway, Darwin was convinced that competitively mediated selection could explain why these species differ in the size and shape of their beaks. However, Darwin lacked evidence demonstrating that different beak morphologies were effective at reducing competition (Sulloway 1982).

Unlike Darwin, Lack was able to demonstrate that competition likely explained the observed differences between species in beak morphology. Specifically, he described

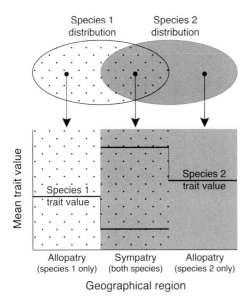

FIGURE 1.3.
When two species occur in both sympatry with each other and in allopatry, competitively mediated selection may produce a distinctive pattern of divergence in which the two species are more dissimilar in sympatry (where such selection should favor divergence) than in allopatry (where there is no such selection). Moreover, within each species, populations in sympatry with the heterospecific would diverge from conspecific populations in allopatry.

several cases in which different species of finches differed in beak morphology more where they were sympatric with each other than where they were allopatric (Figure 1.4). He argued that such divergence between sympatric populations was a signature of past selection that had minimized resource competition. Specifically, Lack stated:

> The significance of these marked beak differences between species otherwise similar has excited speculation from all who have discussed Darwin's finches.... If two species of birds occur together in the same habitat in the same region, eat the same types of food and have the same other ecological requirements, then they should compete with each other, and since the chance of their being equally well adapted is negligible, one of them should eliminate the other completely.... There must be some factor which prevents these species from effectively competing.... I consider that the marked difference in the size of their beaks is an adaptation for taking food of different size ... to enable [different species] to live in the same habitat without effectively competing. (Lack 1947, pp. 61–64)

Although Lack made a convincing case for competition having played a role in the adaptive radiation of finches on the Galápagos Islands, an important question remained: how general was this phenomenon (Mayr 1947)? The answer to this question awaited publication of a landmark paper by Brown and Wilson (1956). This paper was important for three reasons.

First, Brown and Wilson established that Galápagos finches were not unique. They sifted through the literature and applied the method of comparing sympatric and allopatric populations to taxa as diverse as insects, crabs, fish, amphibians, and birds. They presented several compelling cases in which species pairs were recognizably different

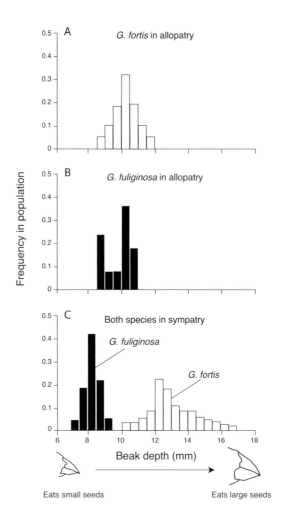

FIGURE 1.4.
Lack first used patterns of geographical variation in beak size of seed-eating ground finches (genus *Geospiza*) from the Galápagos Islands to infer that competitively mediated selection had promoted divergence between different species. In these finches, beak depth correlates positively with the size and hardness of the seeds that an individual eats. Thus, because individuals that are more similar in beak size would be expected to compete more for food, Lack reasoned that competitively mediated selection should promote divergence between species in beak size as a means of minimizing such competition. However, because this selection should only act in areas where species are actually sympatric, a distinctive pattern should emerge in which species exhibit exaggerated divergence in sympatry only (as shown in Fig. 1.3). Lack developed these ideas when studying *Geospiza* finches. He found, for example, that on islands where *G. fortis* and *G. fuliginosa* each occur in the absence of the other species (that is, allopatry), they are similar in beak depth (panels A and B, respectively). By contrast, on islands where the two species occur together, they differ in beak depth (panel C). As an aside, although this particular example is often portrayed as the classic illustration of interspecific character displacement, Boag and Grant (1984) and Schluter et al. (1985) have argued that this case actually reflects "character release" resulting from "*intra*specific character displacement" (see chapter 5). Data based on Lack (1947).

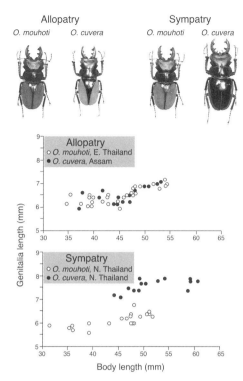

FIGURE 1.5.
Evidence of reproductive character displacement, as revealed by exaggerated divergence in sympatry between two species of stag beetles from Southeast Asia (genus *Odontolabis*). In allopatry, *O. mouhoti* and *O. cuvera* are similar in body size, genitalia length, and coloration. In sympatry, these two species show exaggerated divergence in these characters, all of which have been implicated in mate acquisition. Data from Kawano (2003). Beetle pictures reproduced, with permission, from Kawano (2003).

in sympatry but not in allopatry, which suggested that competition may play a general role in promoting divergence between species.

Second, Brown and Wilson attached a name to the phenomenon. In their paper, they coined the term "character displacement" to describe the pattern of exaggerated divergence in sympatry. In particular, Brown and Wilson (1956, p. 63) defined character displacement as "the situation in which, when two species of animals overlap geographically, the differences between them are accentuated in the zone of sympatry and weakened or lost entirely in the parts of their ranges outside this zone."

Third, Brown and Wilson emphasized that a pattern of exaggerated divergence in sympatry may reflect selection to minimize both resource competition ("ecological" character displacement) and mismatings between species ("reproductive" character displacement). In an earlier version of his book, Lack (1945) too had actually suggested that differences in beak size in Galápagos finches also mediate species recognition signals and might thereby reduce reproductive interactions between species, a conclusion that has been confirmed subsequently by Ratcliffe and Grant (1983).

As a consequence of Brown and Wilson's paper, ecologists and evolutionary biologists became more aware of the phenomenon of character displacement. They also began to consider how both resource competition and reproductive interactions may impose selection favoring divergence between species. Indeed, in the half-century since the publication of Brown and Wilson's influential paper, numerous instances of exaggerated divergence between sympatric species—in characters associated with both resource use *and reproduction* (for example, see Figure 1.5)—have been documented in taxa ranging from plants and protozoa to snails and shrews (see reviews by Howard 1993; Schluter 2000; Dayan and Simberloff 2005). In the next section, we describe some of the approaches that have been used to detect character displacement.

DETECTING CHARACTER DISPLACEMENT

The most compelling evidence for the occurrence of character displacement is to observe it actually taking place. However, because an evolutionary response to selection often takes many generations, direct corroboration is uncommon. In one such direct demonstration of character displacement, Grant and Grant (2006) documented the evolution of character displacement in Galápagos finches (Figure 1.6). Moreover, an evolutionary response to competitors has been demonstrated in laboratory populations of organisms that have short generation times (Barrett and Bell 2006; Tyerman et al. 2008).

Such direct demonstrations of character displacement are not feasible in many species. Nevertheless, most systems can be used to test key predictions of the hypothesis that competitively mediated selection promotes divergence between species. These experiments are conducted within a single generation and focus on species for which prior observational evidence exists to suggest that they have undergone character displacement (we describe how such observational evidence is gathered later in this section). For example, experiments have been used to test a crucial prediction of character displacement theory: that those members of a focal species that are the most similar to a heterospecific competitor in resource use or reproductive traits suffer the most from competition with that heterospecific (for example, see Schluter 1994; Pfennig et al. 2007; Smith and Rausher 2008).

An alternative way to evaluate whether competition promotes divergence is to study species that use phenotypic plasticity to respond adaptively to heterospecific competitors (we discuss such responses in detail in chapters 2 and 4). In these species, individuals respond to competition by changing their phenotype in *developmental time,* rather than over evolutionary time. Therefore, a causal link can be established between the presence of a heterospecific and character change if, in the heterospecific's presence, a member of the focal species facultatively expresses an alternative phenotype less like its heterospecific competitor's phenotype. Such studies are particularly compelling when these experimentally demonstrated niche shifts mirror niche shifts observed in natural populations (for example, see Figure 1.7).

Most of the evidence for character displacement, however, comes from observations of patterns in nature, as first proposed by Lack (reviewed in Howard 1993; Schluter 2000; Dayan and Simberloff 2005). Three classes of such observational evidence are

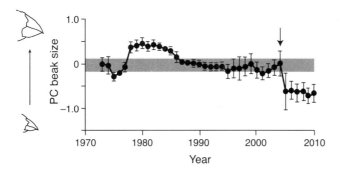

FIGURE 1.6.
Direct evidence of character displacement in a population of the medium ground finch, *Geospiza fortis*, seen when it encountered competition from the large ground finch, *G. magnirostris*. In 1982, *G. magnirostris* arrived on an undisturbed Galápagos island and began to compete with the resident population of *G. fortis* for food (seeds). By the time a severe drought struck in 2003, the population size of *G. magnirostris* had increased substantially. Character displacement occurred in *G. fortis* during 2004–05 (arrow), when selection acting against large-beaked *G. fortis* led to the evolution of greatly reduced beak size. Black dots denote the mean beak size (upper and lower 95% confidence intervals) during a particular year; the gray bar marks the upper and lower 95% confidence limits on the estimate of the mean beak size in 1973 to illustrate subsequent changes in the mean. Reproduced, with permission, from Grant and Grant (2010).

widely used to infer character displacement (Schluter 2000). First, the most frequently cited class of observational evidence is finding that two species differ more in traits associated with resource use or reproduction where they are sympatric than where they are allopatric (for example, Figures 1.4, 1.5). A second (related) class of observational evidence is finding that such traits are overdispersed in multispecies assemblages (Figure 1.8; see also Dayan et al. 1990; Chek et al. 2003). Finally, a third class of observational evidence comes from instances of parallel evolution, in which an unusual degree of similarity in guild structure exists between sets of species that have evolved independently (that is, species-for-species matching). For example, different pairs of fish species found in many postglacial lakes have repeatedly—and independently—diverged into benthivore and planktivore ecomorphs (see Table 3.2 in Schluter 2000), a pattern that is widely interpreted as evidence for character displacement (Robinson and Wilson 1994).

In addition to these three classes of observational evidence, support for character displacement is strengthened when observational data also allow one to test more specific predictions. For example, finding that the degree of divergence between two species is positively associated with the frequency of interspecific interactions in any given popu-

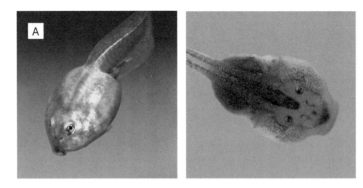

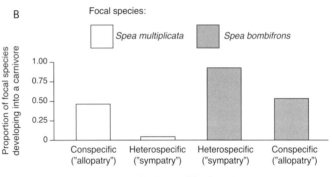

FIGURE 1.7.

Species that respond to competition through phenotypic plasticity can be used to demonstrate a causal link between the presence of a heterospecific and character change. For example, two species of spadefoot toads—Plains spadefoot toads, *Spea bombifrons*, and Mexican spadefoot toads, *S. multiplicata*—co-occur in the southwestern United States, where they have undergone character displacement with each other. Specifically, each species shifts from being a dietary generalist in allopatry to becoming more restrictive in resource use in sympatry, with (A) *S. multiplicata* producing a distinctive omnivore ecomorph that feeds mostly on detritus (left), and *S. bombifrons* producing a distinctive carnivore ecomorph, whose large jaw musculature equips it to prey on anostracan shrimp and other tadpoles (right). (B) Experiments reveal that allopatric populations of each species harbor plasticity that enables them to produce similar, intermediate frequencies of both ecomorphs. Indeed, when allopatric individuals of each species are reared alone with both shrimp and detritus, they produce similar proportions of both morphs. By contrast, when the two species are experimentally combined (simulating "sympatry"), *S. bombifrons* outcompete *S. multiplicata* for shrimp. As a consequence of consuming less shrimp (a cue that induces the production of the carnivore ecomorph), *S. multiplicata* develop mostly into omnivores in the presence of *S. bombifrons*. However, as a consequence of consuming *more* shrimp, *S. bombifrons* develop almost entirely into carnivores in the presence of *S. multiplicata*. Thus these experimentally demonstrated niche shifts mirror niche shifts observed in natural populations. Photos by D. Pfennig.

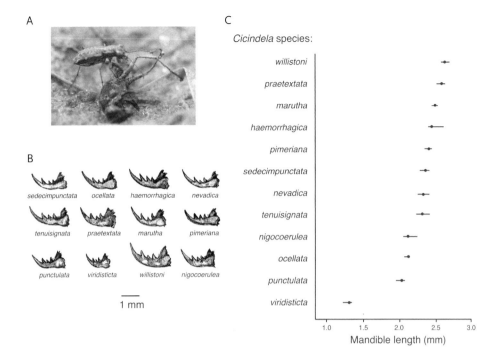

FIGURE 1.8.
Evidence of ecological character displacement, as revealed by trait overdispersion in tiger beetles. (A) A tiger beetle (*Cicindela* sp.) scavenging a dead cricket. (B) Different species of *Cicindela* that occur together in the Sulphur Springs Valley of southeastern Arizona appear to reduce competition by consuming different prey. The mandible length of adult tiger beetles is highly correlated with the size of the prey that they capture and eat, and these lengths vary among species. (C) Co-occurring species differ more in mandible length than expected by chance (dots: median, bars: 95% confidence interval). Photo in panel A by D. Pfennig. Data in panel C from Pearson and Mury (1979). Beetle mandible drawings in panel B reproduced, with permission, from Pearson and Mury (1979).

lation provides compelling evidence for character displacement (for example, see Pfennig and Murphy 2002; Tynkkynen et al. 2004; Pfennig and Pfennig 2005; Goldberg and Lande 2006; Anderson and Grether 2010).

Finally, a complementary approach to those outlined above is to look for a molecular signature of character displacement in natural populations. For example, once one has identified candidate genes involved in resource use or reproduction (see chapter 4), one could analyze patterns of genetic variation in these genes (relative to neutral genetic loci) to determine if selection has acted on any of them (for example, see Charlesworth et al. 1997; Charlesworth et al. 2003; Durrett and Schweinsberg 2004; Nielsen 2005).

As an illustration of this approach, Hopkins et al. (2012) investigated reproductive character displacement between two Texas wildflowers, *Phlox drummondii* and *P. cus-*

pidata. Previous experiments had demonstrated that competitively mediated selection favors floral color divergence as a means of minimizing hybridization between these species (Levin 1985). Hopkins and Rausher (2011) identified two genes involved in the flower-color change in sympatric populations of *P. drummondii*. An analysis of sequence variation in these genes revealed evidence of a selective sweep in one of the two genes, implicating selection as the cause of floral color divergence. Augmenting experimental or observational data with molecular data in this way is a particularly powerful approach for demonstrating character displacement. This is because molecular data integrate the effects of both neutral and adaptive divergence over much longer time periods and greater spatial scales than is possible with classical direct measurements of divergent selection (Hopkins et al. 2012).

In sum, various approaches have been used to document character displacement. The result of these endeavors is the establishment of character displacement as a phenomenon that occurs and that is taxonomically widespread.

PHENOMENA MISTAKEN FOR CHARACTER DISPLACEMENT

Above, we described how much of the support for character displacement comes from observations of patterns in nature. We further noted that the most frequently cited such observation is a geographical pattern of trait variation in which species show more exaggerated divergence in resource-use or reproductive traits when they are sympatric with each other than when they occur in allopatry (Figure 1.3).

However, comparing sympatric and allopatric populations to deduce that competition has promoted trait evolution has limitations. Specifically, although a pattern of exaggerated divergence between sympatric species relative to allopatry is consistent with the notion that selection has driven trait evolution and thereby lessened resource competition or reproductive interactions, such a pattern is also consistent with the action of various other selective—and nonselective—processes. Here we discuss some selective and nonselective processes that could produce this pattern and that could therefore erroneously be taken as evidence that character displacement had occurred.

Consider first that species and populations may differ for evolutionary reasons other than selection. For example, a divergent phenotype may come to predominate in a single ancestral sympatric population because of genetic drift. Following the evolution of this divergent trait in a single population, gene flow from this population into other sympatric populations may result in multiple sympatric populations exhibiting this trait. Yet the presence of divergent traits in multiple sympatric populations would be the consequence of nonselective processes (that is, genetic drift and gene flow) and would therefore not constitute character displacement (for a possible empirical example, see Marko 2005).

One way around this problem is to test for "parallel" character displacement. Parallel character displacement is a special case of parallel evolution, which occurs when a similar trait evolves repeatedly in independently evolving lineages (see above). Such data constitute compelling evidence of selection's role in trait evolution (Clarke 1975; Endler 1986; Schluter and Nagel 1995). Thus, finding that the same divergent trait has evolved independently in replicate sympatric populations—and that each such sympatric population is most closely related to a nearby allopatric (pre-displacement) population rather than to another sympatric (post-displacement) population—implicates selection as having caused the divergence (for example, see Hansen et al. 2000; Matocq and Murphy 2007; Rice et al. 2009; Adams 2010).

Species and populations may also differ more in sympatry than in allopatry for selective reasons other than that of lessening resource competition or reproductive interactions with another species. Indeed, several ecological interactions can generate patterns that may be mistaken for character displacement. One such interaction is "intraguild predation," which occurs when species that share common resources also prey on each other (Polis et al. 1989; Holt and Polis 1997). Intraguild predation is common in nature (Polis et al. 1989), and it may generate a pattern of exaggerated differences between sympatric species that could be mistaken for character displacement. Specifically, if members of one species are most likely to prey on heterospecifics that are the most phenotypically similar to them (as might occur in many species; see Pfennig 2000a), then the two species might come to differ more in sympatry than in allopatry (Holt and Polis 1997). However, in such a situation, competitively mediated selection would have played no role in promoting divergence between these species.

Another ecological interaction that can produce a pattern of divergent traits in sympatry relative to allopatry that could be mistakenly attributed to character displacement is "apparent competition." Apparent competition occurs when divergence between species arises because of shared predators or parasites, not because of shared resources (Holt 1977). With apparent competition, as one species increases in abundance, its predators and parasites also increase in abundance. Consequently, predation and parasitism on the other species increases as well, thereby potentially favoring divergence between species in resource use as a means of escaping predators or parasites (Brown and Vincent 1992; Abrams 2000). For example, enhanced predation and parasitism may drive each species into different habitats, which could then favor different resource-acquisition traits. A similar process could also generate divergence in reproductive characters. Indeed, because predators and parasites frequently locate their victims by keying in on their sexual signals (reviewed in Andersson 1994), selection to avoid such costly interactions may favor the production of divergent mating signals in sympatric species. This divergence might be mistakenly interpreted as reflecting reproductive character displacement.

Above, we focused on the problem of comparing traits in sympatry and allopatry to infer character displacement. A pattern of overdispersion of traits in sympatry is also problematic for inferring character displacement, however. In particular, "species sorting" can produce patterns that could be mistakenly construed as evidence that character displacement had occurred. Species sorting was first anticipated by Rensch (1933; cited in Mayr 1947) and occurs when species that just happen to differ in resource use are more likely to coexist than those that do not differ. Species sorting can arise either through the differential invasion into a habitat of species that happen to differ or through the differential extinction of species that happen to be too similar to coexist. For example, each of the small islands of the Lesser Antilles harbors either one or two species of *Anolis* lizards (Losos 1992). On two-species islands, the species differ in body size. However, large and small species come from two distinct evolutionary clades, suggesting that they already differed in body size before they came into contact with each other (Losos 1992). Thus, size differences between species appear to have arisen through species sorting rather than through character displacement, in which each species evolved a different body size owing to the action of competitively mediated selection. Note, however, that on the large islands of the Greater Antilles, size differences among sympatric species appear to have arisen through character displacement (reviewed in Losos 2009), and, as we describe in chapter 2, character displacement and species sorting are not mutually exclusive.

As the above paragraphs highlight, when relying on observational data to infer character displacement, a major challenge is to rule out other hypotheses that could produce the same patterns as character displacement. One way to meet this challenge is to establish rigorous criteria that, when satisfied, would strongly suggest that character displacement had occurred. Schluter and McPhail (1992), following the suggestions of Grant (1972), Connell (1980), and Arthur (1982), proposed six criteria for demonstrating ecological character displacement (see also Taper and Case 1992). By contrast, no formal criteria have been adopted for reproductive character displacement (but see Howard 1993 and Andersson 1994 for a discussion of how to demonstrate reproductive character displacement). In Table 1.2, we have extended the six criteria for ecological character displacement to include reproductive character displacement. Such criteria are valuable for ruling out alternative causes of trait divergence.

Before leaving this topic, however, we stress that an overly rigorous application of any such criteria can be as problematic as not applying the criteria in the first place. Indeed, by applying these criteria too strictly, one runs the opposite risk of falsely rejecting some actual instances of character displacement. We return to this issue in chapters 2 and 4, when we discuss the possible role of environmentally induced shifts in mediating character displacement. Additionally, as will be discussed in chapter 3, different forms of character displacement—both ecological and reproductive—can potentially produce similar patterns, and it is important to be aware of this possibility.

In sum, using a pattern of trait divergence as a proxy for the process of competitively mediated divergence—that is, conflating a predicted pattern with the process hypothesized to produce it—is problematic without the addition of corroborating evidence to indicate that the hypothesized process has actually caused the pattern. This conflation of pattern and process runs even deeper than the issue of how to demonstrate character

TABLE 1.2. Criteria for demonstrating character displacement from observational data

Criterion	Justification for criterion	Commonly used approaches for testing criterion
(1) There should be independent evidence that the displaced trait is involved in resource use (for ECD) or reproduction (for RCD).	By definition, competitively mediated selection acts on traits associated with resource use or reproduction.	Behavioral observations.
(2) Chance should be ruled out as an explanation for the pattern.	Geographical patterns of trait variation that are predicted to arise from character displacement (see main text) can also arise by chance.	Tests for replicated (i.e., "parallel") character displacement (e.g., see Schluter and Nagel 1995; Hansen et al. 2000; Marko 2005; Matocq and Murphy 2007; Rice et al. 2009; Adams 2010) or measures of selection on the focal trait (Schluter 1994; Pfennig et al. 2007; Smith and Rausher 2008; Martin and Pfennig 2011).
(3) Phenotypic differences between populations and species must be demonstrated to reflect an evolutionary adjustment by one species to the other.	Species sorting, resulting from either the biased extinction of species that are too similar to coexist or the biased colonization of species that are different, could produce a pattern similar to that predicted for character displacement. Species sorting is not a mechanism for character displacement, however, because it does not entail trait evolution per se.	Phylogenetic data are used to test for nonrandom extinction or colonization of species (e.g., see Losos 1990).

(continued)

TABLE 1.2. *(continued)*

(4)	Shifts in resource use (for ECD) or reproduction (for RCD) should mirror changes in phenotypic traits.	To demonstrate that the focal traits have undergone evolutionary shifts because of selection stemming from resource competition or reproductive interactions between species, a link must be established between trait distributions and overlap between species in either resource use (for ECD) or the incidence of reproductive encounters (for RCD).	Behavioral observations, often coupled with morphometric measurements.
(5)	Any trait differences between allopatric and sympatric populations should not be ascribed to differences between sites in biotic factors (other than the presence of a competitor) or in abiotic factors.	Sympatric and allopatric sites may differ ecologically, and these differences (not the presence or absence of a competitor) may account for observed trait differences expressed in the two types of sites.	Detailed ecological measurements of various biotic and abiotic factors between allopatric and sympatric sites (e.g., see Pfennig and Murphy 2003; Gray et al. 2005).
(6)	Individuals that are the most similar phenotypically compete most for resources (for ECD) or interfere most in each other's reproduction (for RCD).	Evolutionary shifts may result from interspecific interactions other than resource competition or reproductive interactions.	Experiments are used to test whether selection disfavors individuals that express resource-use phenotypes (ECD) or reproductive phenotypes (RCD) that most closely resemble those of their heterospecific competitor (e.g., see Pacala and Roughgarden 1985; Pritchard and Schluter 2001; Gray and Robinson 2002; Schluter 2003; Pfennig et al. 2007).

NOTE: Modified from Schluter and McPhail (1992); Andersson (1994).

Abbreviations: ECD = ecological character displacement. RCD = reproductive character displacement.

displacement. This conflation also reflects longstanding confusion over how to *define* character displacement—the topic to which we turn next.

WHAT CONSTITUTES CHARACTER DISPLACEMENT? CONFLATION OF PROCESS AND PATTERN

As described above, Darwin (1859 [2009]) hypothesized that competition drives trait divergence. To test this hypothesis, Lack (1947), and subsequently Brown and Wilson (1956), relied on geographic *patterns* of trait variation to identify situations in which competitively mediated trait divergence might have unfolded. The logic of their approach was that selection arising from resource competition or reproductive interactions between species should lead to: (1) a *process* of divergence of resource-exploiting or reproductive traits between sympatric species; and, consequently, (2) a resulting *pattern* of geographical variation in which species show exaggerated divergence in these traits in sympatry (Figure 1.3).

As we have seen, however, using a pattern to infer a process is problematic, because patterns of exaggerated divergence can be explained by the process of competitively mediated divergence, but they can also be explained by various processes other than competitively mediated selection. Yet, such a conflation of process and pattern has long plagued the study of character displacement. Indeed, in the case of ecological character displacement, this conflation of process and pattern provoked a spirited debate over what constituted "true" character displacement (Grant 1972; Strong et al. 1979; Simberloff and Boecklen 1981; Arthur 1982; Diamond et al. 1989; reviewed in Schluter 2000). Central to this debate was the realization that numerous selective and nonselective factors can cause species to exhibit exaggerated divergence in sympatry (see above). In fact, serious reservations arose about whether competition played *any* role in generating trait divergence (Strong et al. 1979). What emerged from this debate was the realization that defining character displacement in terms of a pattern of geographical variation breeds confusion over which (if any) cases of exaggerated divergence in sympatry actually reflect selection stemming from resource competition or reproductive interactions between species.

Grant (1972) advocated distinguishing between pattern and process when defining character displacement. He recommended restricting use of the term "character displacement" to the selective *process* that actually causes divergence between competing species. In particular, Grant defined character displacement as:

> . . . the process by which a morphological character state of a species changes under natural selection arising from the presence, in the same environment, of one or more species similar to it ecologically and/or reproductively. (Grant 1972, p. 44)

Most researchers in the realm of ecological character displacement now follow Grant's suggestion and define ecological character displacement in terms of process, thereby reducing confusion over what does, and what does not, constitute character displacement (for example, see Schluter 2001, 2002).

An added benefit of defining ecological character displacement in terms of process is that it has become apparent that competitively mediated selection can generate patterns of trait evolution other than exaggerated divergence in sympatry. Indeed, as described in Box 1.1 and illustrated in Box 1.1 Figure 1, character displacement may give rise to additional, seemingly unrelated phenomena ranging from the evolution of ecological specialization and the evolution of increased competitive ability and associated traits (such as increased body size) to character convergence.

Although the field of ecological character displacement has largely agreed on what constitutes character displacement, the field of reproductive character displacement has achieved no such resolution (despite the fact that Grant's [1972] definition also applied to reproductive characters; see above). Consequently, conflation of pattern and process is widespread when dealing with reproductive character displacement. Indeed, some define reproductive character displacement in terms of pattern rather than process.

Yet, as we stressed above, *patterns of trait divergence can be generated via processes other than selection to avoid interactions with competitors.* As we noted above, apparent competition, for example, could cause reproductive trait divergence between sympatry and allopatry. Defining reproductive character displacement in terms of process, rather than pattern, therefore has the same benefit that a process-oriented definition has for ecological character displacement: researchers can focus specifically on the ecological and evolutionary implications of competitive interactions between species driving reproductive trait divergence. Moreover, defining ecological and reproductive character displacement in the same way is critical for understanding how the two processes interact. If the two fields define character displacement differently, then the ability to address questions of how and why ecological and reproductive character displacement interact will be hampered.

REPRODUCTIVE CHARACTER DISPLACEMENT VERSUS REINFORCEMENT

A problem that has uniquely faced the field of reproductive character displacement must necessarily be addressed before we conclude with the formal definition of character displacement that we will use throughout the book. Namely, what is the relationship between reproductive character displacement and reinforcement? "Reinforcement" is generally defined as trait evolution stemming from selection that minimizes hybridization between species (Howard 1993; Servedio and Noor 2003; Coyne and Orr 2004).

Our usage of the phrase "character displacement" so far would suggest that we do not consider reproductive character displacement and reinforcement to be separate phenomena. Indeed, we hold with many of the founders of the field who did not see these as distinct processes (Brown and Wilson 1956; Blair 1974). For example, W. F. Blair—who is credited with coining the term "reinforcement" (Brown and Wilson 1956; Coyne and Orr 2004)—considered reinforcement "a rather restricted form of character displacement" (Blair 1974, p. 1119).

How then did reinforcement come to be distinguished as a separate process from reproductive character displacement? One of the most influential definitions of reproductive character displacement was that of Butlin (1987) and subsequently Butlin and Ritchie, who stated that reproductive character displacement is "the process of divergence in mating signal systems between reproductively isolated species" (Butlin and Ritchie 1994, p. 62). By defining reproductive character displacement in this way, they distinguished it from reinforcement. Thus, whereas reinforcement was defined as arising from interactions where species could actually exchange genes during mating, reproductive character displacement was deemed to arise from mating interactions between species that did not exchange genes (Butlin 1987; Butlin and Ritchie 1994). At the time, theoretical models suggested that reinforcement was unlikely to occur. In forging this distinction, Butlin (1987) and Butlin and Ritchie (1994) established a critically important standard for demonstrating reinforcement.

Recent theoretical and empirical advances have confirmed, however, that reinforcement is possible and more common than once assumed (reviewed in Servedio and Noor 2003), thus reducing the motivation for treating reproductive character displacement and reinforcement as distinct processes. Indeed, as we highlight in the next chapter, whether via hybridization or by other reproductive interactions, both the underlying agent of selection (reproductive interactions with other species) and the underlying target of selection (traits associated with reproduction) are the same, regardless of whether the process is called "reinforcement" or "reproductive character displacement." Nevertheless, many still maintain that reinforcement should be considered a distinct process because of its importance in speciation. As we shall see in chapter 8, however, character displacement (whether by reinforcement or not) plays a critical role in both finalizing and initiating speciation.

We therefore hold with the view that reinforcement is a case where reproductive character displacement unfolds owing to selection that lessens a specific type of reproductive interaction with heterospecifics: hybridization. That is, reinforcement is a "restricted form of reproductive character displacement" (Blair 1974).

Nevertheless, because this book cannot provide a final resolution to this debate, we will use the term "reinforcement" as appropriate so that readers with alternative definitions can still access the concepts and ideas we develop without confusion.

TERMINOLOGY

Above, we emphasized the need to distinguish between character displacement and the patterns that potentially arise from it. Because the process underlying a pattern is what matters in terms of understanding how and why traits evolve, we hold with

BOX 1.1. Alternative Manifestations of Character Displacement

Competitively mediated selection can generate several forms of trait evolution other than divergence between lineages (Box 1.1 Figure 1A). Here we discuss three alternative manifestations of character displacement.

First, competitively mediated selection may promote the evolution of ecological (or reproductive) specialization (Box 1.1 Figure 1B). For instance, species that compete for shared resources may evolve different resource-acquisition traits such that each utilizes a more restrictive range of resources in sympatry. Similarly, females may become more discriminating in their choice of mate. In other words, there might be a shift in trait *variance* but not a shift in trait *mean* (as would be the case when there is divergence between species). We discuss this manifestation of character displacement in greater detail in chapters 3 and 6.

Second, competition may favor the evolution of increased competitive ability (Moore 1952). For instance, rather than shift to an alternative resource, a focal species might evolve such that it becomes increasingly better at acquiring its current resource in the face of competition. In such a situation, competitively mediated selection may lead to the evolution of enhanced competitive ability in one or more interacting species (see Arthur 1982 and references therein; for a classic experimental demonstration of this process, see Pimentel et al. 1965).

The evolution of increased competitive ability has additional and important evolutionary ramifications, however (Box 1.1 Figure 1C). Although the evolution of enhanced competitive ability in one species may result in the extinction of the other species (the weaker competitor; for example, see Pimentel et al. 1965), this need not always be the case. Indeed, competitors may coexist and become locked in a co-evolutionary arms race of mutually increasing competitive ability (Pimentel et al. 1965; Arthur 1982; Taper and Case 1992). Essentially, once a lineage evolves an increased competitive ability, it exerts selection on any lineages with which it is interacting to also become better competitors, which imposes additional selection on the original lineage, and so on (Taper and Case 1992). We explore such competitively mediated "escalation" (*sensu* Vermeij 1987) in greater detail in chapter 9. For now, we note that such competitively mediated escalation may explain a diversity of phenomena, such as the evolution of virulence in pathogen populations (Ewald 1994; Kamada et al. 2012), the evolution of increased aggressiveness between competitors (Adams 2004; Deitloff et al. 2009; Grether et al. 2009), the evolution of increased body size (Bonner 1988), and, possibly, the evolution of complexity (see chapter 9).

(continued)

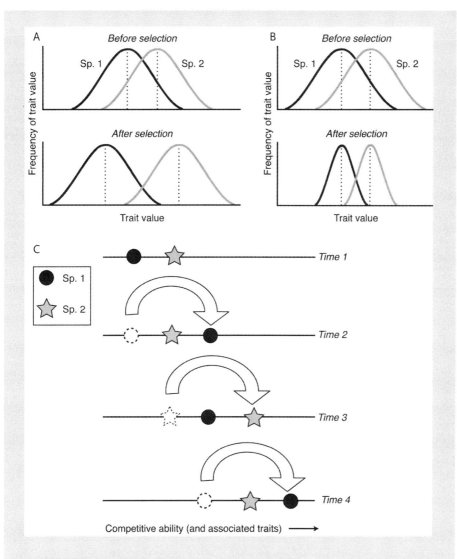

BOX 1.1. FIGURE 1
Different forms of trait evolution wrought by competitively mediated selection—that is, different forms of character displacement. Selection favoring traits that reduce competitive interactions between species may lead to: (A) divergence between species, where species differ more in mean trait value (dotted line) after selection than before selection; (B) increased specialization, where trait distributions of interacting species are narrower, and therefore overlap less, after selection than before selection; and (C) increased competitive ability, where mutual selection (and where only one direction of change is adaptive) fuels a co-evolutionary arms race that promotes greater competitive ability and larger values of associated traits, such as body size, aggression, or developmental or behavioral complexity.

BOX 1.1. *(continued)*

A third, alternative manifestation of character displacement (related to the second manifestation above) is that competing species may become *more* similar to one another (Grant 1972; Cody 1973). Such character convergence may occur when selection favors heterospecifics to converge on similar traits used in agonistic encounters (for example, see Cody 1969; Deitloff et al. 2009; Grether et al. 2009). For example, if two or more species fight directly over resources, competitively mediated selection may favor in each such species the evolution of similar large body size, toxins, or weaponry such as horns or antlers (Emlen 2008). Moreover, such selection may favor heterospecifics to copy each other's acoustic or behavioral signals that are used to maintain territoriality (Secondi et al. 2003; Haavie et al. 2004; Gorissen et al. 2006). In short, contest competition may promote character convergence as an adaptive response to such competition.

---

previous authors (Grant 1972; Howard 1993; Schluter 2001, 2002; Grether et al. 2009) that character displacement—whether arising from competition for resources or access to successful reproduction—should be defined as a process and not a pattern.

We therefore define "character displacement" as *trait evolution that arises as an adaptive response to resource competition or deleterious reproductive interactions between species.* Later, in chapter 5, we broaden this definition to include selection that acts *within* species, a process termed "intraspecific character displacement." Whether selection operates within or between species, however, we define character displacement in terms of *process*.

We do not coin a special term to describe the patterns that are predicted to arise from the process of character displacement; instead, we describe the actual pattern that is generated by character displacement (for example, when discussing exaggerated divergence between sympatric species, we use the phrase "exaggerated divergence between sympatric species"). Our justification for doing so is twofold: first, the process of character displacement can produce patterns other than exaggerated divergence (see Box 1.1); and, second, we hope to minimize the potential for conflating pattern and process as described above.

Moreover, as highlighted above, two distinct forms of character displacement are widely recognized. The first of these, "ecological character displacement" (*sensu* Slatkin 1980; Schluter 2001), refers to trait evolution stemming from selection arising from resource competition between species. This process acts on traits associated with resource acquisition (for example, morphological structures such as beaks and jaws or even foraging behavior). A second, widely recognized form, "reproductive character displacement" (*sensu* Crozier 1974), refers to trait evolution stemming from selection that minimizes deleterious reproductive interactions between species. This process acts on traits associated with reproduction (for example, sexual traits or female mate preferences; see Figure 1.5). In addition to these two forms, Grether et al. (2009) recently proposed a third form of character displacement: "agonistic character displacement." Although we will not use this term in our book (in our view, all cases of character displacement can be categorized as either ecological or reproductive character displacement), the basic idea behind agonistic character displacement is useful and one that we return to in the next chapter. This idea is that agonistic interactions stemming from competition—whether for access to resources (Adams 2004; Peiman and Robinson 2007, 2010) or successful reproduction (Tynkkynen et al. 2004; Anderson and Grether 2010)—can promote trait evolution.

Throughout the book, we will use the term "character displacement" when the principles that we are discussing apply to both ecological and reproductive character displacement. However, when the principles apply to only one form of character displacement, we specify which form we are discussing.

More generally, a major theme of this book is that the basic proximate and evolutionary mechanisms by which ecological and reproductive character displacement unfold are similar. Moreover, as we explain in chapter 3, these two forms of character displacement can interact in consequential ways. Indeed, the interaction between these two processes can leave an important imprint on the history of life.

A UNIFYING PRINCIPLE

In this book, we maintain that the study of character displacement can be used to answer fundamental questions in evolutionary biology and ecology regarding the origins of biodiversity. In particular, we address seven key questions throughout the book: Why do closely related species typically express different resource-acquisition or reproductive traits? How do novel, complex traits associated with resource-acquisition or reproduction originate? Why do individual species harbor so much trait variation in the first place? How and why do new species arise? Once they do arise, how do species coexist with each other? Why is evolution generally divergent? Finally, why is evolution frequently escalatory?

In addressing these questions, we will highlight how the study of character displacement can help unify alternative perspectives on the origins and maintenance of diversity in two key ways. First, the study of character displacement can illuminate how evolutionary processes acting within species translate into major evolutionary change above the level of species. That is, character displacement can link microevolution and macroevolution. The relationship between microevolution and macroevolution has long been the focus of debate in evolutionary biology (for recent discussions, see Gould [2002] and Carroll [2005]). Throughout the book, we will suggest how character displacement—a

> **BOX 1.2.** Suggestions for Future Research
>
> - Identify the various agents of selection (other than competition) that can generate patterns that could be mistaken for character displacement. Devise approaches for discriminating between these agents and competition as causing similar forms of trait evolution (for example, see MacColl 2011).
> - Conduct field experiments to determine whether naturally occurring populations undergo character displacement in direct response to the experimental introduction of a heterospecific competitor (that is, attempt to observe character displacement unfold *directly* in field experiments).
> - Look for the signature of character displacement at the DNA coding sequence level. Specifically, identify candidate genes involved in resource use or reproduction and analyze patterns of genetic variation to determine whether selection has acted on these genes (for example, see Hopkins et al. 2012).
> - Identify additional cases of "parallel" character displacement.
> - Determine the conditions under which competitively mediated selection promotes divergence rather than other patterns of trait evolution (for example, the evolution of specialization, increased competitive ability, and even convergence).
> - Using phylogenetic approaches, re-evaluate instances of character displacement to determine whether the allopatric or the sympatric condition is ancestral. Finding that divergence is derived would suggest that character displacement had occurred between species. Finding that divergence is ancestral, however, would point to "character release," in which intraspecific character displacement occurs *within* allopatric populations of each species (see Figure 1.4 and also chapter 5 for further discussion).

microevolutionary process—can potentially explain certain macroevolutionary patterns, such as the divergent nature of evolution (Figure 1.2B), the widespread tendency for evolutionary lineages to increase in size over time (Alroy 1998; Benton 2002; Payne et al. 2009), and the tendency for certain clades to diversify more widely and more rapidly than others (for example, Sepkoski 1984; Mitter et al. 1988; Foote 2000; Sahney et al. 2010).

Second, character displacement inexorably links the fields of ecology and evolution. We underscore throughout the book how character displacement can bridge our understanding of ecological and evolutionary processes. Moreover, although the study of resource competition in relationship to ecological patterns (particularly community dynamics) has a long history (see MacArthur 1972 and references therein), relatively little work has examined how *reproductive* interactions between species affect ecological processes. This is particularly surprising given that, compared to resource competition, reproductive interactions between species may be no less important—and, in some cases, may even be more important—for explaining ecological patterns and processes (for example, see Gonzalez-Voyer and Kolm 2011).

SUMMARY

As an explanation for biodiversity, Darwin first proposed that all organisms face recurring competition for scarce resources, and that this competition favors individuals that are least like their competitors. Thus, according to Darwin, groups of organisms that compete will become increasingly dissimilar over time. This evolutionary process is now known as "character displacement." Character displacement is expected to produce a pattern of exaggerated divergence in sympatry, which is how it has often been detected. However, as a consequence, the process of character displacement has been confounded with the pattern that it is expected to produce. To avoid such confusion, we therefore define character displacement as trait evolution that arises as an adaptive response to resource competition or deleterious reproductive interactions between species. Character displacement takes two forms: ecological character displacement, which is trait evolution stemming from selection that reduces resource competition, and reproductive character displacement, which is trait evolution stemming from selection that reduces costly reproductive interactions. Because character displacement is central to an explanation of the origins of biodiversity, the study of character displacement can help unify disparate areas of evolutionary biology as well as the fields of evolution and ecology. In Box 1.2, we list some suggestions for future research.

FURTHER READING

Brown, W. L., and E. O. Wilson. 1956. Character displacement. *Systematic Zoology* 5:49–64. This is the paper that first brought the concept of character displacement to the attention of many scientists.

Darwin, C. 1859 (2009). *The annotated Origin: A facsimile of the first edition of On the origin of species.* The Belknap Press of Harvard University Press: Cambridge, MA. This is the book in which Darwin first laid out his principle of divergence of character (that is, character displacement). The annotated version by J. T. Costa is highly recommended.

Grant, P. R., and B. R. Grant. 2008. *How and why species multiply: The radiation of Darwin's finches.* Princeton University Press: Princeton, NJ. An easily read overview of the ecology and evolution of Darwin's finches.

Ridley, M. 2005. *How to read Darwin.* Norton: New York. This short, highly readable book describes Darwin's principle of divergence of character. The author also focuses on how Darwin viewed this principle as crucial in explaining the divergent nature of evolution.

Schluter, D. 2000. *The ecology of adaptive radiation.* Oxford University Press: Oxford, UK. This scholarly monograph explores in detail the ecological causes of adaptive radiation, including the role of character displacement. The author also describes how to detect character displacement in natural populations.

Wilson, E. O. 1992. *The diversity of life.* Harvard University Press: Cambridge, MA. This book presents an excellent overview of the causes and consequences of biodiversity. It also includes a clear description of how character displacement arises.

2

WHY CHARACTER DISPLACEMENT OCCURS

In chapter 1, we defined "character displacement" as trait evolution that arises as an adaptive response to resource competition or deleterious reproductive interactions between species. Left unaddressed in that chapter, however, were the *causes* of resource competition or deleterious reproductive interactions between species. Specifically, how do such interactions come about, why are they costly, and—in the absence of character displacement—what is the ultimate fate of populations that experience such interactions? In other words, *why* does character displacement occur?

In this chapter, we examine the selective bases of character displacement. As we describe, regardless of whether competitive interactions occur over resources or successful reproduction, these interactions can have severe fitness consequences; in some cases, competition can even cause a species to become extinct in a particular location. We also discuss how these costs abate—and the chances of species coexistence thereby increase—when species evolve distinctive resource-use or reproductive traits that enable them to obtain resources or reproduce successfully in the face of competition with a heterospecific. Further, we highlight how such resource or reproductive "partitioning" can arise either through character displacement or, alternatively, through species sorting.

In our discussion below, we treat ecological character displacement and reproductive character displacement separately. We do so because, historically, each process has been studied independently of the other, resulting in two distinct literatures. Nevertheless, we highlight the many similarities between ecological character displacement and reproductive character displacement (see Figure 2.1).

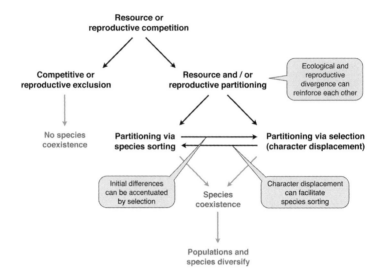

FIGURE 2.1.

Flowchart showing some immediate outcomes of competition (black arrows and black text) and their potential downstream effects (gray arrows and gray text). The general outcomes of both resource and reproductive competition are potentially the same (recall from chapter 1 that "competition" refers to interactions that are deleterious to both parties; thus, deleterious reproductive interactions between species are broadly classified as "reproductive competition"). Other factors may enable species coexistence even in the absence of resource or reproductive partitioning (see Box 2.1). The possibility that populations and species can diversify once they coexist is the focus of the second half of the book.

WHY ECOLOGICAL CHARACTER DISPLACEMENT OCCURS

Because the underlying cause of ecological character displacement is competition for resources, we begin our discussion by examining the causes and consequences of resource competition between species.

RESOURCE COMPETITION

Organisms experience resource competition when they share a resource that is in some way limited (Grover 1997; Keddy 2001). By "resources" we mean any feature of the environment—such as food, water, light, or space—that organisms need to survive and reproduce and that can be consumed or otherwise used to the point of depletion (exclusive of gametes or mates, which we discuss later in this chapter). By "limited" we mean that the resource's supply is short relative to demand.

Because resource competition ultimately depends on this ratio of demand to supply, the process of competing for resources is not an all-or-nothing process. Instead, the intensity of resource competition will vary continuously with the ratio of demand to supply, and will be especially strong when resources are scarce relative to demand. Several meta-analyses have examined the incidence and intensity of resource competition in natural communities and found it to be common and frequently strong (Connell 1983; Goldberg and Barton 1992; Gurevitch et al. 1992; Denno et al. 1995; Kaplan and Denno 2007; Dybzinski and Tilman 2009). Moreover, even in systems where the intensity of competition may fluctuate, short bouts of intense competition can be important in influencing population persistence and promoting trait evolution (for example, see Figure 1.6; reviewed in Dhondt 2012).

Resource competition can assume two general forms: interference competition or exploitative competition. Which of these forms resource competition assumes depends on whether organisms compete via direct (interference competition) or indirect (exploitative competition) interactions. Interference competition typically involves antagonistic encounters in which organisms directly impede the foraging of others, such as when animals aggressively prevent members of other species from entering their feeding territory (for example, see Hairston 1987; Adams 2004; Peiman and Robinson 2007, 2010; Grether et al. 2009). Interference competition also occurs when organisms directly deter others from becoming established in a portion of their habitat, such as when certain plants release toxins that hinder the establishment of competitors (Harper 1977). For example, garlic mustard (*Alliaria petiolata*), an invasive plant found in the understory of North American forests, secretes a chemical into the soil that stifles growth of mycorrhizal fungi (Stinson et al. 2006; Wolfe et al. 2008). By doing so, it suppresses the growth of other competing species (such as tree seedlings) that depend on these fungi for the uptake of nutrients. Similarly, microorganisms engage in interference competition by releasing toxins that reduce the growth of competitors (antibiotics are familiar examples of such toxins; see Waksman 1947).

In contrast to interference competition, exploitative competition occurs indirectly through the consumption of a shared, limited resource. In particular, when the members of one species consume a resource, they concomitantly reduce this resource for the members of other species. For instance, when neighboring plants compete for food or water, the acquisition of these resources by the members of one species limits their availability to heterospecifics. Organisms may also engage in exploitative competition by making space unavailable to other species, as occurs when plants or the members of encrusting organisms (for example, lichens, bryozoans, barnacles) take up all of the available room to grow, thereby precluding heterospecifics from becoming established in the same space (Rubin 1985). Although organisms engaging in exploitative competition may not actually ever encounter one another, exploitative competition can have wide-ranging effects: with exploitative competition, *all* members of a population can be impacted if a competitor depletes a shared, limited resource.

Both interference competition and exploitative competition are common in nature.

Indeed, in an analysis of 164 field experiments in which competition had been demonstrated (Schoener 1983), interference competition was documented in just over half of the cases. Note, however, that it is often difficult to categorize competitive interactions as being of one form or the other (Schoener 1983). For instance, space—like any other resource—can be consumed. In this sense, space competition is indirect and exploitative. However, space competition often also involves direct interference interactions, such as fighting (in animals) and the release of toxins (in plants and microorganisms). Thus, interference competition and exploitative competition are not mutually exclusive.

Regardless of its precise form, resource competition, by definition (see Table 1.1), is always detrimental to both parties. In particular, by using up a scarce resource, competitors reduce the amount available to others for maintenance and reproduction. Additionally, by allocating time, energy, or material toward acquiring contested resources, competitors have less time, energy, or material remaining to invest in their own maintenance and reproduction (Pianka 2000). Moreover, when fighting is involved, combatants risk injury or even death. Even when there is extreme asymmetry between competitors in fighting ability—where a superior competitor seemingly pays few costs—the superior competitor must pay the costs of producing and maintaining traits (for example, weaponry, large body size) associated with being a stronger competitor in the first place (Emlen 2008). Competition therefore entails fitness costs at two levels: the costs of losing access to resources, and the costs invested in the acquisition of those resources.

Because of these costs, selection should favor the evolution of traits that reduce competition or its effects (that is, character displacement should occur). Generally, competitively mediated selection (regardless of whether it arises from either exploitative competition or interference competition) should favor traits that maximize access to vital resources while simultaneously minimizing the investments associated with obtaining these resources. One solution to this optimization problem is for selection to promote enhanced competitive ability (Pimentel et al. 1965). As noted in Box 1.1, however, this would not represent a long-term resolution if competing species become locked in a coevolutionary arms race of escalating competitive ability. Moreover, enhanced competitive ability may not be "worth it" if alternative resources are available. Thus, an alternative evolutionary solution is for a species to either utilize different resources than those used by heterospecifics or utilize the contested resources in a slightly different manner.

Below, we examine how these differences in resource use come about. But first we explain why, in the absence of such differences, the cost of resource competition can become so severe that it may cause a species to go locally extinct. We do so for two reasons. First, the possibility that competition can result in extinction underscores the severe negative consequences that can arise from competition. Second, the fact that extinction is a potential alternative to character displacement emphasizes character displacement's role in maintaining and even promoting diversity (a point to which we will return throughout this book).

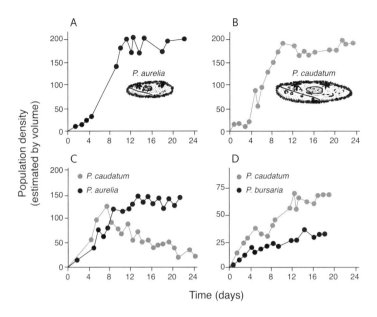

FIGURE 2.2.
Gause's (1934) classic experiments with *Paramecium* demonstrated competitive exclusion and coexistence. (A) When reared alone, *P. aurelia* and (B) *P. caudatum* each grew until each species reached its stable carrying capacity. (C) By contrast, when these same two species were reared together, only one species, *P. aurelia*, achieved its stable carrying capacity; the other species, *P. caudatum*, was driven to extinction. (D) However, when *P. caudatum* (the species driven to extinction in panel C) was reared with another species, *P. bursaria*, the two species were able to coexist by dividing up the food resources. Based on data in Gause (1934).

COMPETITIVE EXCLUSION

If the costs of competition are sufficiently severe, population birth rates will decrease and death rates will increase, such that population size will decline to a point where the population is no longer viable and becomes extinct. If this happens, "competitive exclusion" is said to occur.

The classic demonstration of competitive exclusion comes from Gause's (1934) pioneering laboratory experiments with *Paramecium*. In these experiments, Gause (1934) established that two species, *P. aurelia* and *P. caudatum*, each survived well on its own (Figure 2.2A, B). However, when he reared these two species together, he found that only one species, *P. aurelia*, grew well; the population size of the other species, *P. caudatum*, collapsed (Figure 2.2C). Gause (1934) concluded that *P. caudatum* was outcompeted for food by *P. aurelia*, and thereby driven extinct.

Competitive exclusion has also been demonstrated in natural populations. A prime example comes from Connell's (1961a, b) work. Connell sought to identify the factors that govern the local distributions of two species of barnacles, *Chthamalus stellatus* and *Semibalanus balanoides*, along the rocky intertidal coast of Scotland. Connell observed that the larvae of these species co-occur in both the upper and middle intertidal zones. However, adult *Chthamalus* are restricted to the upper intertidal zone, whereas adult *Semibalanus* are restricted to the adjacent middle intertidal zone. By removing adult *Semibalanus* from the middle intertidal zone, Connell showed that adult *Chthamalus* could survive there, even though they are normally not present in the middle intertidal zone (Connell 1961a, b). He concluded that *Semibalanus* competitively excluded *Chthamalus* from this portion of its habitat. In addition to these controlled experiments, competitive exclusion has been inferred in natural populations of many species, when species are absent from areas where they should occur had heterospecific competitors not been present (for classic examples, see Lack 1945; MacArthur 1958; Hairston 1987).

SPECIES COEXISTENCE

Clearly, competitive exclusion is not the only outcome of resource competition. Most natural communities consist of numerous species, some of which use the same limited resource (note that the mere co-occurrence of two or more species does not necessarily imply that these species are coexisting *stably*; some such species may be declining, but at too slow a rate to be detected easily; see Dybzinski and Tilman 2009). Indeed, in some cases, seemingly competing species have coexisted for millions of years (for example, see Grant and Grant 2008), suggesting that long-term coexistence is possible.

As we describe in more detail below, coexistence can arise when species subdivide or partition resources in a way that reduces competition or its effects. Alternatively, coexistence can arise owing to the differential suppression of the superior competitor's population size—for example, due to disturbance (Box 2.1).

Darwin (1859 [2009]) anticipated both alternatives and was the first to recognize that disturbance could alter the outcome of competition and thereby promote species coexistence. However, it is Darwin's theory regarding *partitioning* on which we focus here. Specifically, in *The Origin of Species*, Darwin (1859 [2009], p. 127) maintained that "... more living beings can be supported on the same area the more they diverge in structure, habits, and constitution." Essentially, Darwin proposed that competitors could coexist when they use resources differently. But why do differences in resource use promote species coexistence?

As illustrated in Figure 2.3, when species differ in resource use, each species' population size will be regulated more by *intra*specific competition than by interspecific competition. As first demonstrated mathematically by Lotka (1932) and Volterra (1926), when populations are regulated more by intraspecific competition than by interspe-

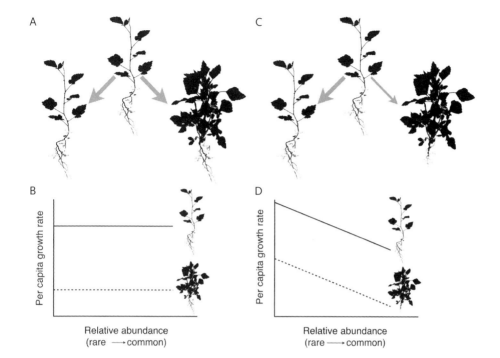

FIGURE 2.3.
How differences between species in resource use promote species coexistence. (A) When species do *not* differ in resource use (in this case, by having roots that are of the same length), individuals impact the growth rate of conspecifics and heterospecifics equally (the width of the arrow reflects the degree to which a focal individual [middle] limits the growth rate of conspecifics [left] versus heterospecifics [right]). (B) Consequently, differences between species in per capita growth rate reflect inherent differences in each species' competitive ability. Under these circumstances, the more-rapidly growing species may drive the other species to extinction through competitive exclusion. (C) By contrast, when species *do differ* in resource use (in this case, by having roots that differ in length), individuals limit the growth rate of conspecifics more than that of heterospecifics. (D) Each species therefore has greater per capita growth rates when rare. Potential competitors coexist, because the abundance of each species never becomes so high that it threatens other species with competitive exclusion. Based on Levine and HilleRisLambers (2009).

cific competition, intraspecific competition prevents any one species from becoming so numerous that it can exclude heterospecifics. In such situations, potential competitors coexist, because even rare species can increase in abundance rather than be eliminated in the face of competition.

This idea forms the basis for one of the central ideas in ecology: the competitive exclusion principle (Hardin 1960). This principle is often stated as the proposition that two species cannot coexist stably if they occupy the same ecological niche. Although there are multiple definitions of the term "niche" (for example, see Grinnell 1917, 1924,

BOX 2.1. Alternative Models of Species Coexistence

Various models have been proposed to explain species coexistence (reviewed in Tilman and Pacala 1993; Chesson 2000). In the main text, we describe one such theory, which holds that competitors may coexist when they differ in how they utilize resources.

An alternative set of models suggests that species coexistence is possible when factors such as environmental variability, disturbance, and exploitation differentially suppress the superior competitor's population size and thereby prevent it from driving inferior competitors extinct (Hutchinson 1961). For instance, species may coexist if different species are superior competitors at different times and under different conditions (assuming that the environment fluctuates sufficiently rapidly to reverse the competitive advantage before one species drives the other extinct; Hutchinson 1961). Likewise, a species that is a relatively poor competitor could coexist with a superior competitor if it tolerates periodic disturbance (such as fires, disease outbreaks, or storms) better than does the superior competitor (Paine 1979; for an example, see Byers 2002). Essentially, each disturbance decreases the abundance of a superior competitor before it drives the inferior competitor extinct. Alternatively, predation, herbivory, or parasitism may promote species coexistence if the exploiter differentially suppresses the superior competitor, thereby allowing inferior competitors to persist (Paine 1966; Lubchenco 1978). Thus, in these "non-equilibrium" models, coexistence is achieved only because of the suppression of a superior competitor before it reaches its equilibrium population size (that is, its carrying capacity).

1928; Elton 1927 [1943]; Hutchinson 1957; Odum 1959; Pianka 1976; Schoener 2009), for our purposes, a "species' niche" can be thought of as the physical and biological conditions that the species needs in order to grow, survive, and reproduce (*sensu* Hutchinson 1957). Thus, in the absence of other environmental factors (such as the factors described in Box 2.1), two species that use a limiting resource in the same way cannot coexist indefinitely.

But this raises an important question: how much ecological overlap can two species tolerate and still coexist? MacArthur and Levins (1967) answered this question by proposing the notion of "limiting similarity"—the degree of similarity between two species in resource use at which the two species can just coexist. Essentially, their theory showed that, even if two species are not identical in resource use, they still might not be able to coexist if they are too similar; that is, if there is too much niche overlap (Figure

Finally, another theory for species coexistence (Gross 2008) rests on the supposition that "positive" species interactions, such as mutualism (see Table 1.1), are common in many ecological communities (Stachowicz 2001; Bruno et al. 2003). According to this theory, potential competitors may coexist if the fitness benefits of positive interactions compensate for the costs of competition.

A recent empirical test of this facilitation hypothesis, however, found a pattern opposite from that predicted by the theory. In particular, Alexandrou and colleagues (2011) evaluated the relative importance of competition versus Müllerian mimicry in determining the local community structure in a species-rich group of Neotropical catfishes that have converged on both an effective predator deterrent—retractable venomous spines—and a similar warning signal—bold color patterns (Müllerian mimics are mutualists that share the mortality costs of predator education concerning their unprofitability as prey; see Ruxton et al. 2004). The researchers identified 52 different species belonging to 24 different mimicry rings, each composed of two or three sympatric species that shared the same warning coloration. They found that the vast majority of co-mimics differed in resource use, indicating that the benefits of mutualism do not outweigh the need for potential competitors to partition resources (Alexandrou et al. 2011). Interestingly, co-mimics were also less closely related than expected by chance (Alexandrou et al. 2011), suggesting that species-specific dietary differences arose through species sorting (generally, more closely related species should be more ecologically similar and therefore compete more intensely; reviewed in Losos 2008; Wiens et al. 2010; Burns and Strauss 2011). Thus the degree to which long-term coexistence can be explained by positive ecological interactions remains unclear (see also Elias et al. 2008).

2.4). Thus, species can stably coexist only when they differ in resource use by some minimum amount (MacArthur and Levins 1967).

In support of these ideas, Gause (1934) demonstrated stable coexistence between two species of *Paramecium* when they differed in how they acquired resources (Figure 2.2D). More recent experiments with different species of *Paramecium* have revealed that the frequency of competitive exclusion increases with increasing similarity between species in mouth size, a trait related to resource use (Figure 2.5). Furthermore, field experiments—using more complex communities of multiple, competing species—have confirmed that stable coexistence occurs when species differ in resource use (for a recent example, see Levine and HilleRisLambers 2009). Additional experiments using different species of vertebrates have demonstrated that competitively mediated selection favors those individuals that are the most dissimilar to their competitor in resource use

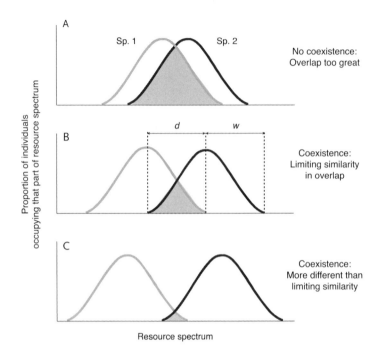

FIGURE 2.4.
In the absence of other environmental factors, theory predicts that species can coexist stably only when they differ in resource use by some minimum amount. For example, the resource utilization functions of two species that share a common resource spectrum can be depicted as two bell-shaped curves. (A) If niche overlap is too great, coexistence may not be possible. (B) However, coexistence may be possible if niches overlap—but not by too much. (C) When overlap is small, coexistence becomes even more likely. The minimum overlap at which coexistence is possible (any closer and one species is eliminated)—the "limiting similarity"—can be measured as the ratio of the distance between the two species' niches (d, indicated here as the distance between the peaks of the two species' distributions) to niche width (w). MacArthur and Levins (1967) predicted that values of $d/w > 1$ would result in coexistence, a prediction generally supported by empirical work. Redrawn from Schoener (2009).

(Figure 2.6; see also Pacala and Roughgarden 1985; Pritchard and Schluter 2001; Gray and Robinson 2002; Schluter 2003). Such data are important, because it is on individual organisms that competitively mediated selection acts during the evolution of character displacement. Thus, ample evidence suggests that the more dissimilar any two species, populations, and individuals are in resource use, the less intense the resource competition between them and, therefore, the more likely they will be to coexist (all else being equal).

In the next section, we examine how species-specific differences in resource use come about. In particular, we describe how species differences can arise through ecological character displacement or, alternatively, through species sorting.

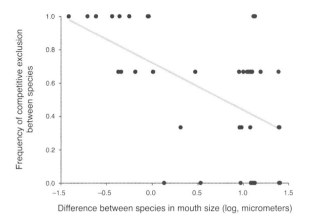

FIGURE 2.5.
Experimental evidence that the likelihood of competitive exclusion occurring between any two species depends on their similarity in traits associated with resource use. Here, the focal species are bacterivorous *Paramecium*, and the resource-use trait is mouth size. In this experiment, researchers introduced into a microcosm about 100 individuals each of two different species. Each species was matched against nine other species, for a total of 45 two-species combinations. After eight weeks, either competitive exclusion or stable coexistence was observed (the frequency of competitive exclusion was estimated as the fraction of times each species pair exhibited competitive exclusion out of the three replicates for that pairwise species interaction treatment). Redrawn from Violle et al. (2011).

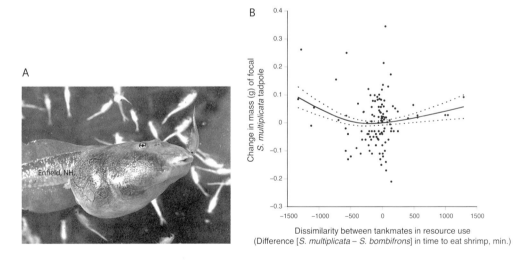

FIGURE 2.6.
Experimental evidence that selection favors those individuals that are the most dissimilar from their competitor in resource use. (A) Mexican spadefoot toad tadpoles, *Spea multiplicata*, compete with Plains spadefoot toad tadpoles, *S. bombifrons*, for anostracan fairy shrimp, and both species exhibit heritable variation in how quickly they can eat these shrimp. (B) In an experiment when individual *S. multiplicata* tadpoles were housed with individual *S. bombifrons* tadpoles, the more similar the two individuals were, the lower the growth of the focal *S. multiplicata* tadpole (regardless of which individual ate shrimp faster; these shrimp-eating times were estimated before the tadpoles were placed in competition with each other). The curved line is a cubic spline estimate of growth (a proxy for fitness) as a function of how similar the focal tadpole was to its tankmate in resource use (dashed lines denote upper and lower 95% confidence intervals). Panel B redrawn from Pfennig et al. (2007). Photo by David Saunders.

RESOURCE PARTITIONING VIA CHARACTER DISPLACEMENT VERSUS SPECIES SORTING

Co-occurring species often subdivide shared resources through a process known as "resource partitioning" (*sensu* Schoener 1974, 1986). Resource partitioning refers to the situation in which potentially interacting species differ in some way in resource use, regardless of how these differences come about (we describe the proximate and evolutionary bases of resource partitioning below).

Resource partitioning can be manifested in three non–mutually exclusive ways. First, it may be expressed as *temporal* partitioning, in which potentially interacting species experience reduced competition by utilizing the same resource at different times. For example, bumblebees (*Bombus*) and solitary bees (andrenid bees) often forage on the same plant at different times of the day, thereby minimizing direct interactions with each other (Heinrich 1976).

Second, resource partitioning may occur through *spatial* partitioning, in which potentially competing species subdivide resources by occupying different microhabitats, maintaining territories, or otherwise spatially segregating resource acquisition. For example, in a classic study, MacArthur (1958) found that multiple species of warblers coexisted within the same forest by feeding and nesting in different microhabitats (for example, different heights and/or interior versus exterior portions of the same trees). Moreover, spatial resource partitioning may even occur in environments where seemingly little opportunity exists for habitat segregation, such as in well-mixed coastal waters. For instance, sympatric strains of coastal bacterioplankton of the family Vibrionaceae occupy distinct microhabitats—for example, some strains associate with zooplankton; others associate with large, suspended particles; and still others are free-living (Hunt et al. 2008).

Finally, resource partitioning may be manifest as *phenotypic* partitioning, in which organisms produce different morphological, physiological, or behavioral traits that enable them to utilize alternate resources (see Figure 1.8, and Barluenga et al. 2006), monopolize access to resources (see Grether et al. 2009), or acquire slightly different proportions of the same limiting resource (Tilman 1977). In some cases, phenotypic partitioning may result in the evolution of "ecomorphs" (*sensu* Williams 1972), which are distinctive phenotypes that are adapted to different resources. For instance, on the Greater Antilles, different species of *Anolis* occur as different ecomorphs that utilize different microhabitats (Figure 2.7). Consequently, these species minimize direct competition with each other for two critical resources: insect prey and basking sites. That these *Anolis* show both phenotypic and spatial partitioning emphasizes how species may use different forms of partitioning to subdivide shared resources.

Resource partitioning can arise through two different proximate mechanisms. First, any given individual may express a particular resource-use phenotype (including behavioral traits) constitutively. That is, the trait may be canalized, such that the particular way

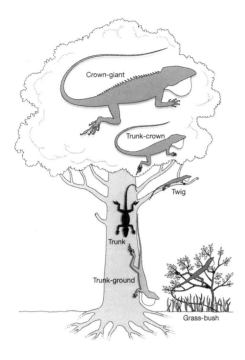

FIGURE 2.7.
Different species of *Anolis* lizards from the Greater Antilles subdivide resources through both phenotypic and spatial resource partitioning. On islands where they are sympatric, different species occur as different ecomorphs, which partition their habitat by residing in tree crowns (crown-giant ecomorph), on trunk/crowns (trunk-crown ecomorph), on the trunk (trunk ecomorph), on the trunk/ground (trunk-ground ecomorph), on twigs (twig ecomorph), and in grass or on bushes (grass-bush ecomorph). By expressing different morphologies and occurring in different microhabitats, different species minimize direct competition with each other for insect prey and basking sites. Reproduced from Losos (2009), with the kind permission of the author and publisher.

in which any given individual utilizes a resource is constant or "fixed" for that individual, regardless of whether or not heterospecifics are actually present. Alternatively, individuals may facultatively adjust their resource-use phenotype depending upon whether a potential heterospecific competitor is present or absent. Specifically, phenotypic plasticity may enable organisms to produce—through environmentally sensitive development—morphological, behavioral, and/or physiological traits that enhance resource acquisition in response to a heterospecific competitor (for examples, see Table 2.1).

In chapter 4, we explore in greater detail the potential contributions of phenotypic plasticity versus canalized differences in promoting adaptive responses to competition. For now, we stress that adaptive responses to competition (that is, character displacement) can be expressed either constitutively or facultatively; that both constitutively expressed and facultatively expressed responses can evolve (West-Eberhard 2003); and that both proximate mechanisms can therefore mediate character displacement. Indeed, in Box 2.2, we describe the conditions under which competitively mediated plasticity constitutes character displacement.

Regardless of the proximate basis of resource partitioning, it can arise evolutionarily via two, non–mutually exclusive routes (Figure 2.8). First, resource partitioning may evolve by competitively mediated selection in which one or both of the interacting species undergoes an evolutionary adjustment to the other species' presence. That is, ecological character displacement transpires.

More precisely, ecological character displacement occurs if: (1) individuals in a

TABLE 2.1. Reduction of resource competition with heterospecifics: Examples in which individuals potentially reduce resource competition with heterospecifics through phenotypic plasticity

Organism	Phenotypic shift	Source
Many species of plants	In response to the presence of neighboring plants (including heterospecifics), many plant species facultatively alter the spatial position of their roots; examples include growing roots away from potential competitors; decreasing or increasing root growth; and altering depth of roots.	Mahall and Callaway (1991); Nobel (1997); Gersani et al. (2001); Cahill et al. (2010)
Many species of plants	In response to low light conditions (such as those induced by the presence of a heterospecific competitor), many plant species facultatively elongate their stems, suppress branching, alter the width of their leaves, and/or accelerate flowering ("shade avoidance syndrome").	Smith and Whitelam (1997); Schmitt et al. (1999); Sultan (2007)
Many species of echinoids (sea urchins)	In response to low levels of available food (such as levels induced by the presence of a heterospecific competitor), larvae facultatively allocate more tissue to their ciliated "arms," which are their feeding structures.	Hart and Strathmann (1994)
Many species of freshwater fishes	In the presence of heterospecific competitors, many species facultatively shift to utilizing either the benthic (bottom-dwelling) or limnetic (open-water) niches, often by producing distinct ecomorphs.	Robinson and Wilson (1994)
Spadefoot toad tadpoles (genus *Spea*)	In response to the presence of a heterospecific competitor, *Spea* tadpoles facultatively shift from being dietary generalists to becoming dietary specialists that produce either an omnivore ecomorph, which feeds on organic detritus on the pond bottom, or a distinctive carnivore ecomorph, which specializes on shrimp.	Pfennig et al. (2007)
Anole lizards (*Anolis carolinensis, A. sagrei*)	Individuals facultatively alter hind limb length in response to being reared on different surfaces, as may occur when heterospecifics compete for perch sites.	Losos et al. (2000); Kolbe and Losos (2005)

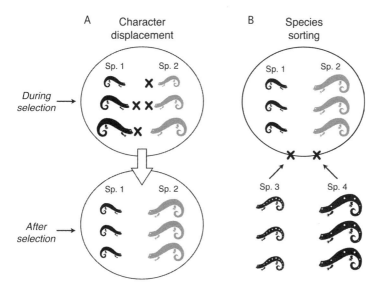

FIGURE 2.8.
Resource partitioning and reproductive partitioning can arise via either character displacement or species sorting. In these hypothetical examples, similar-sized salamanders compete for resources, reproduction, or both, such that similar-sized species cannot coexist (individuals that are selectively disfavored are denoted by an "×"). (A) Character displacement occurs when individuals within each species exhibit heritable variation in body size, and competitively mediated selection favors those individuals that differ the most from the other species in body size, thereby promoting the evolution of species differences in body size. (B) Species sorting arises when individuals within each species do not vary in body size (or when body size is not heritable) and when those species that happen to differ in body size are able to coexist (for example, neither species 3, which is similar to species 1 in body size, nor species 4, which is similar to species 2 in body size, can invade the habitat of species 1 and 2).

population vary in how they acquire resources in the face of competition; (2) this variation is inherited; and (3) some individuals have higher fitness than others because of their distinctive traits used to obtain resources in the face of competition (Figure 2.8A). For instance, individuals that overlap the least with a heterospecific in traits associated with resource use should suffer the least interspecific competition. Moreover, certain members of one species might possess traits that enable them to outcompete the heterospecific for a shared resource. Species may even *converge* in trait expression if doing so mediates resource partitioning, such as when two species converge on similar signaling to demarcate feeding territories (Grether et al. 2009). Regardless of the traits' form, ecological character displacement unfolds as these traits accumulate within a population over time (for a review of the theory of ecological character displacement, see Slatkin

BOX 2.2. Is Competitively Induced Plasticity Character Displacement?

Resource partitioning is often expressed facultatively. In particular, the individuals of many species respond adaptively to interspecific competition by facultatively modifying their resource-use traits through phenotypic plasticity (for examples, see Table 2.1). Here, we discuss whether such environmentally contingent phenotypic shifts can constitute character displacement.

Consider the following situation involving a social bee and a wasp, which often compete aggressively over access to flowers (reviewed in Rashed and Sherratt 2007). Suppose a bee that is initially foraging alone on a flower (Box 2.2 Figure 1A) is joined on the same flower by a competitively superior wasp (Box 2.2 Figure 1B), whereupon the bee moves to a different type of flower (Box 2.2 Figure 1C). By making this shift to a new floral resource, the bee and wasp engage in resource partitioning. But does this resource partitioning constitute ecological character displacement?

Traditionally, competitively induced plasticity has not been regarded as character displacement because of an oft-held view that phenotypic plasticity is a "nongenetic" response incapable of mediating adaptive evolution (Arthur 1982; Schluter and McPhail 1992; Taper and Case 1992). Yet individuals within the same population often harbor genetic variation in the degree to which they respond to environmental cues—that is, different genotypes typically express different environmentally contingent phenotypic responses ("reaction norms"; see, for example, Gupta and Lewontin 1982; Sultan and Bazzaz 1993; Kingsolver et al. 2004). Thus the tendency to respond to competitors in the first place, and the manner in which individuals express these responses, can serve as alternative axes of heritable variation on which selection can act to promote competitively mediated trait *evolution* (that is, character displacement). In short, any reaction norm that evolves in response to competitively mediated selection would constitute character displacement.

For example, if bees with different genotypes differ in their propensity to respond to wasps—and if those that leave a flower in response to wasps and find alternative floral resources are therefore best at avoiding competition and come to predominate in the population—then such induced shifts in foraging behavior would constitute ecological character displacement. Eventually this facultative character displacement may even promote fixed (that is, constitutively expressed) differences between species in foraging behavior (for examples in which different species have evolved such fixed differences in foraging behavior, see Nagamitsu and Inoue 1997; Biesmeijer et al. 1999; for a general discussion of how such fixation may come about, see chapter 4). Although differences among species in when and where they forage may arise through the alternative process of species sorting (in which species that happen to differ, before contact, are the ones that are able to coexist), such differences could also come about through ecological character displacement.

Finally, as an intriguing twist to this story, several species of hoverflies (harmless syrphid flies) have evolved remarkable resemblances to stinging Hymenoptera (that is,

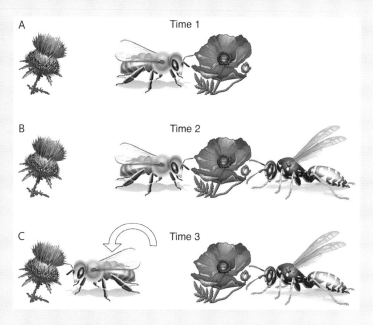

BOX 2.2 FIGURE 1
Resource partitioning can be mediated by facultative changes in resource-use phenotypes, such as when (A) a bee feeding on a flower is (B) joined by a superior competitor, a wasp, which (C) causes the bee to switch to an alternative type of flower (alternatively, the bee may forage on the original flower at a time when the wasp is not likely to be present). Such a sequence of events may constitute character displacement if certain conditions are met (as described in Box 2.2).

bees and wasps; reviewed in Gilbert 2005). Hoverflies have likely evolved such mimicry because it gives them protection from predation (see references in Rashed and Sherratt 2007). It is also possible, however, that these resemblances confer a competitive advantage to the mimetic hoverflies: by resembling a stinging insect, they may intimidate other species that seek to forage on the same flower (Rashed and Sherratt 2007). Although empirical tests have thus far failed to support this "competitive mimicry" hypothesis (Rashed and Sherratt 2007), it remains a compelling explanation for why hoverflies might have converged on the same phenotype as their potential bee and wasp competitors. Indeed, the shift from a non-mimetic phenotype to a mimetic phenotype may represent an example of character displacement promoting *convergent* trait evolution, in which a focal species converges on the phenotype of a superior competitor in order to outcompete other species (for further discussion of the interplay between competition and mimicry, see Rainey and Grether 2007; Pfennig and Kikuchi 2012).

Returning to the main point of this box, competitively induced phenotypic plasticity can constitute character displacement under certain circumstances. Indeed, as we describe in greater detail in chapter 4, environmentally induced shifts in resource-use or reproductive traits might play a key role in mediating character displacement.

FIGURE 2.9. Species sorting and character displacement are not mutually exclusive mechanisms of resource partitioning. Indeed, both processes may operate together. For example, in (A), species sorting may occur first and enable species to coexist, whereupon character displacement may occur and accentuate pre-existing differences between species. Alternatively, in (B), character displacement may occur first, whereupon each species may subsequently invade new habitats where already-present species differences enable them to coexist with other species via species sorting.

A Species sorting then character displacement

Time 1. A single species (Sp. 1) exists alone in a habitat.

Time 2. A larger species (Sp. 2) invades; both species initially coexist via species sorting.

Time 3. Selection accentuates existing size differences and reduces competition further.

B Character displacement then species sorting

Time 1. A single species (Sp. 1) exists alone in a habitat.

Time 2. A similar-sized species (Sp. 2) invades and competes with Sp. 1.

Time 3. Selection promotes size differences and reduces competition.

Time 4. Either species can subsequently invade new habitats and coexist with alternatively sized resident species via species sorting.

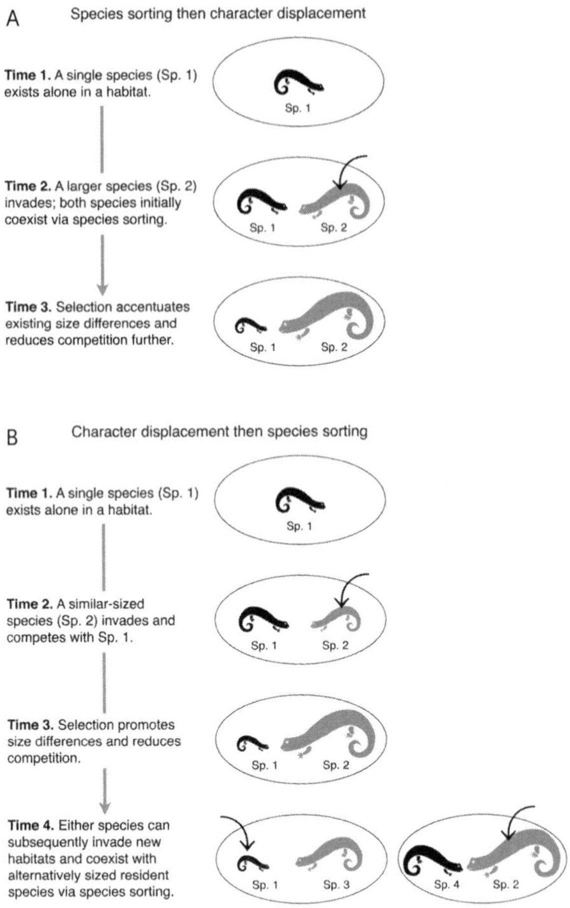

1980; Taper and Case 1985, 1992; Abrams 1987; Doebeli 1996; Schluter 2000; Dayan and Simberloff 2005; note that in contrast to many models, which assume that exploitative competition is the sole agent of ecological character displacement, we view both exploitative competition and *interference* competition as potential drivers of ecological character displacement).

Second, and as an alternative to character displacement, resource partitioning may arise via species sorting (Figure 2.8B). Recall from chapter 1 that species that just happen to differ in resource use will be more likely to coexist than those that do not differ. Thus, resource partitioning may arise in nature through the differential invasion into a habitat of species that happen to differ, or through the differential extinction of species that happen to be too similar to coexist.

Although resource partitioning arising in response to competitively mediated selection constitutes character displacement, resource partitioning arising through species sorting does not. Species sorting is not a mechanism of *adaptive* evolution (Strong et al. 1979; Case and Sidell 1983; Waser 1983; Schluter 2000) because selection does not favor

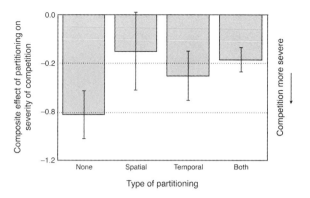

FIGURE 2.10.
The effect of spatial and/or temporal resource partitioning on the strength of interspecific competition among plant-feeding (phytophagous) insects. Shown are the results of a meta-analysis of published studies, in which researchers estimated the effects of spatial and/or temporal separation between different species of phytophagous insects on their combined growth rate, body size, survival, fecundity, and abundance (the "composite effect"). Note that both spatial and temporal partitioning—together or separately—reduce the severity of interspecific competition. Mean effect sizes are presented with upper and lower 95% bootstrap confidence intervals ($P < 0.01$ among groups). Based on data in Kaplan and Denno (2007).

certain traits relative to others (as it does during character displacement; see above). Instead, species that just happen to differ are more likely to coexist (Figure 2.8B).

Note, however, that species sorting and ecological character displacement are not mutually exclusive alternatives. Pre-existing differences between species can be accentuated by competitively mediated selection during character displacement; that is, both mechanisms may occur, with species sorting occurring first (Figure 2.9A). Conversely, character displacement may promote differences between species in one region, and these differences may enable these two species to coexist when they encounter each other (or other competing species) in another region; that is, both mechanisms may occur, with character displacement occurring first (Figure 2.9B).

Regardless of how resource partitioning unfolds developmentally, or how it arises evolutionarily, resource partitioning promotes species coexistence by enabling species to optimize resource acquisition when confronted with interspecific competition. Indeed, although resource partitioning does not prevent interspecific competition from occurring, it can greatly reduce competition's intensity (Figure 2.10). Generally, the more any two species differ in resource use, the less intense the competition between them (Pacala and Roughgarden 1982).

We return to resource partitioning's role in species coexistence in chapter 6, when we describe how it fosters species richness within communities. For now, we emphasize that: (1) resource partitioning reduces competition or its effects and thereby promotes species coexistence; (2) resource partitioning can evolve via species sorting or competitively mediated selection (that is, ecological character displacement); and (3) species sorting and ecological character displacement are not mutually exclusive mechanisms of resource partitioning.

WHY REPRODUCTIVE CHARACTER DISPLACEMENT OCCURS

We now focus on why *reproductive* character displacement occurs. Given that reproductive character displacement is a type of trait evolution stemming from selection that lessens deleterious reproductive interactions between species, we begin by examining the causes and consequences of such reproductive competition.

REPRODUCTIVE COMPETITION

Species often interact in ways that impede the ability of individuals of each species to reproduce successfully (Andersson 1994). Such interactions can arise in two key ways. First, one species can inhibit another species from reproducing successfully through hybridization; that is, when individuals from genetically differentiated populations or species mate with each other (Harrison 1993). If hybrid pairings fail to produce any progeny (as frequently occurs—see Barton and Hewitt 1989; Arnold 1997), then individuals waste their entire reproductive effort. Even when hybrid matings produce progeny, hybridization can be disfavored because hybrid offspring typically have lower survivorship or fecundity than pure-species offspring (Barton and Hewitt 1989; Arnold 1997).

Second, sympatric species can interfere with each other's ability to identify high-quality conspecific mates (that is, mates that can provide the choosy individual, its offspring, or both with fitness benefits; see chapter 7). Such interference arises when heterospecifics compete for access to: (1) the physical space in which they attract or locate mates; (2) the "reproductive-trait space" (for example, "signaling space") that they use to attract or locate mates; (3) the times and locations that they release their gametes; or (4) the intermediaries (for example, pollinators) that transfer gametes to mates (reviewed in Gerhardt and Huber 2002; for example, Waser 1983; Levin 1985; Armbruster et al. 1994; Caruso 2000; Smith and Rausher 2008).

Although reproductive interference does not necessarily generate the same severe costs as hybridization, reproductive interference can be costly in terms of increased signaling effort, search time/costs, and agonistic interactions over territories (reviewed in Andersson 1994; Grether et al. 2009). Moreover, because signal "noise" from heterospecifics may interfere with an individual's ability to identify high-quality mates and thereby cause an individual to mistakenly select a low-quality conspecific mate (Pfennig 1998, 2000b), such individuals may suffer reduced numbers or quality of offspring (Pfennig 2000b, 2008).

For example, in many plants, pollination is crucial for reproduction, and competition for pollen can impose strong selection on coexisting species (Waser 1983). Indeed, plants in regions with the greatest species richness face the highest levels of pollen limitation (Vamosi et al. 2006), which can reduce a plant's reproductive success. Such costs can accrue to both females (which suffer when heterospecific pollen is deposited on stigmas, thereby resulting in blockage to conspecific pollen; Caruso and Alfaro 2000) and males (which waste gametes when pollen is lost to heterospecific flowers; Muchhala et al. 2010). Thus, even if individuals never actually risk hybridization, heterospecific interactions can impose significant fitness costs.

In the next section, we describe how reproductive interactions between species can become so costly that they may drive one of the species locally extinct.

REPRODUCTIVE EXCLUSION

Reproductive interactions between species can lead to "reproductive exclusion" (also called "sexual exclusion"; Gröning and Hochkirch 2008). In particular, when species hybridize or interfere with each other's ability to identify high-quality mates, each species' reproductive output may decline (Waser 1978; Sved 1981; Spencer et al. 1986; Kuno 1992; Liou and Price 1994; Takafuji et al. 1997; Hochkirch et al. 2007; Kishi et al. 2009). The resulting negative population growth can ultimately cause a species to become locally extinct through the reproductive analog of competitive exclusion (see above).

For example, in a laboratory experiment that manipulated the relative abundance of two weevil species, Kishi et al. (2009) found that one species, *Callosobruchus maculatus*, was more likely to be driven to extinction than another species, *C. chinensis*, with which the former interacted reproductively. In particular, males of both species were indiscriminate in their mating attempts toward females of either species. However, *C. maculatus* females were less tolerant of mating attempts by heterospecifics than were *C. chinensis* females (Kishi et al. 2009 and references therein). Indeed, because *C. maculatus* females spent more time escaping heterospecifics than did *C. chinensis* females, they invested less in total reproduction than did *C. chinensis* females. As a result of this reproductive interference, *C. maculatus* experienced population declines that ultimately led to this species' local extinction.

Interestingly, in this weevil example, the authors could rule out *resource* competition as driving the exclusion of *C. maculatus*. Of the two species, *C. maculatus* is the superior competitor for resources, so exclusion stemming from resource competition should have produced the opposite effect (Kishi et al. 2009). This example therefore illustrates that reproductive interactions can potentially be more important than resource competition in promoting extinction (Kishi et al. 2009). Thus, although reproductive exclusion remains relatively poorly studied, it can play a key role in explaining why species sometimes fail to coexist.

In the same way that differences between species in resource use lessens the chances of competitive exclusion, differences between species in reproductive traits renders reproductive exclusion less likely to occur (see, for example, Liou and Price 1994). Moreover, just as with resource competition, there should be a limit to how similar two species can be in reproductive traits and still manage to coexist (recall our discussion above of the theory of limiting similarity; see also Figure 2.4). Thus, as with resource partitioning, species that engage in reproductive partitioning should be more likely to coexist (all else being equal). In the next section, we examine how reproductive partitioning comes about.

REPRODUCTIVE PARTITIONING VIA CHARACTER DISPLACEMENT VERSUS SPECIES SORTING

Reproductive partitioning is the reproductive equivalent of resource partitioning. It refers to the situation where potentially interacting species differ in mating behavior or other traits associated with reproduction, regardless of how these differences arise (we describe the proximate and evolutionary bases of reproductive partitioning below).

Reproductive partitioning, like resource partitioning, can be manifested in three non–mutually exclusive ways. First, it may be expressed as temporal partitioning, in which species reduce reproductive interactions by seeking mates and/or reproducing at different times. For example, in habitats that contain multiple species of frogs (which call to attract mates), different species typically call at slightly different times of the year or even different times of the same day (see, for example, Drewry and Rand 1983). Similarly, plants can reduce competition for pollination by flowering at different times of the day or the year (Pleasants 1980; Ashton et al. 1988; Stone et al. 1998).

Second, reproductive partitioning may occur through spatial partitioning, in which species reproduce in different microhabitats, defend territories against each other, or otherwise spatially segregate in reproduction. For example, Gray treefrogs (*Hyla versicolor*) and Cope's Gray treefrogs (*Hyla chrysoscelis*) co-occur in certain areas of central North America, where they risk hybridizing with each other. In sympatry, *H. versicolor* males tend to call from sites that are more than 50 cm above the ground, whereas *H. chrysoscelis* males tend to call from sites that are less than 50 cm above the ground. However, in allopatry, both species call from *both* types of sites (Ptacek 1992).

A third manifestation of reproductive partitioning is phenotypic partitioning, where individuals express behavioral, morphological, or physiological traits that minimize reproductive interactions. Both males and females of interacting species may express phenotypic partitioning in diverse ways. For example, females may evolve (1) preferential use of conspecific gametes (Howard 1999); (2) mate preferences for conspecific male traits that are most dissimilar from heterospecific males (see, for example, Pfen-

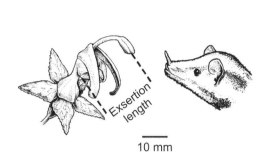

FIGURE 2.11.
Phenotypic (morphological) reproductive partitioning in plants. Different species of *Burmeistera* plants differ in the degree to which the reproductive parts are exserted outside the corolla, which determines where the pollen is deposited on a bat's head (bats are the sole pollinators of these plants). Because different species of *Burmeistera* differ more in degree of exsertion than expected by chance, they likely experience reduced pollen interference from heterospecific plants. Shown here are a typical *Burmeistera* flower (*B. borjensis*) and the head of one of its pollinators, a nectar-feeding bat, *Anoura geoffroyi*. Reproduced from Muchhala and Potts (2007), with the kind permission of the authors and publisher.

nig 2000b), which may also promote species divergence in male traits (reviewed in Andersson 1994); or (3) enhanced abilities to discriminate between conspecific and heterospecific males (Kelly and Noor 1996). As with resource partitioning, reproductive partitioning can even involve the paradoxical evolution of convergent traits that mediate the establishment and maintenance of territories between species (Grether et al. 2009).

In cases in which reproductive partitioning arises when females evolve enhanced abilities to discriminate between conspecific and heterospecific males, such enhanced discrimination need not be mediated by higher order cognitive processing. Instead, auditory, visual, or chemosensory organs, such as antennae or ear anatomy, may evolve to become better tuned to conspecific stimuli and thereby filter out signal noise from heterospecifics. For instance, many insects and anurans show "matched filtering" whereby the peripheral auditory system is most sensitive to the frequencies found in conspecific—but not heterospecific—calls (reviewed in Gerhardt and Huber 2002).

Finally, a remarkable example of phenotypic partitioning occurs among certain plant species that share pollinators and overlap in flowering time: these species diverge in floral morphology. For instance, *Burmeistera*, a diverse genus of Neotropical plants, share bats as their primary pollinators. Coexisting species exhibit substantial interspecific variation in the degree to which their reproductive parts are exserted outside the corolla; exsertion length determines where the pollen is deposited on a bat's head (Figure 2.11). Sympatric species of *Burmeistera* differ more in exsertion length than expected by chance, suggesting that competition for pollinators may have promoted differences between species in floral morphology (Muchhala and Potts 2007).

As with resource partitioning, reproductive partitioning can arise developmentally through either fixed changes in phenotype or through phenotypic plasticity. As an example of the latter process, the fitness consequences of hybridization between two species of spadefoot toads—Plains spadefoot toads (*Spea bombifrons*) and Mexican spadefoot toads (*S. multiplicata*)—vary across different habitats (Pfennig and Simovich 2002). Controlled mate-choice tests have revealed that females of *S. bombifrons* facultatively change their preferences for the male calls of conspecifics versus heterospecifics (that is, *S. multiplicata*) when exposed to conditions that favor hybridization (Pfennig 2007), demonstrating that reproductive partitioning can be mediated by phenotypic plasticity.

Regardless of how reproductive partitioning arises developmentally, it can come about evolutionarily through the same two non–mutually exclusive routes as resource partitioning: competitively mediated selection and species sorting (Figure 2.8). Indeed, competitively mediated selection should promote the evolution of reproductive traits that minimize costly reproductive interactions between species (reproductive character displacement) in the same way that it should promote the evolution of ecological traits as an adaptive response to resource competition.

In particular, reproductive character displacement occurs if: (1) individuals in a population vary in how they engage in reproduction; (2) this variation is inherited; and (3) some individuals have higher fitness than others because the distinctive way that they engage in reproduction minimizes deleterious reproductive interactions with heterospecifics. However, because reproductive character displacement often occurs between species that exchange genes (Pfennig 2003; Grant and Grant 2008; Behm et al. 2010), hybridization can impede or even prevent species-specific traits from evolving (Howard 1993; Servedio and Noor 2003; Coyne and Orr 2004), a point to which we will return in the next chapter. Nevertheless, the evolution of traits as an adaptive response to costly reproductive interactions between species constitutes reproductive character displacement (Figure 2.8A).

Alternatively, reproductive partitioning can also arise via species sorting (Figure 2.8B). In particular, species that come into contact with each other may coexist because they each already possess divergent reproductive traits that arose for reasons other than to reduce reproductive interactions with heterospecifics. For instance, females that encounter novel heterospecific males may not mate with them simply because these heterospecifics possess signals that are unattractive to females relative to those possessed by conspecifics (Paterson 1978, 1982; West-Eberhard 1979, 1983; Rodriguez et al. 2004; Pfennig and Ryan 2007). Such a case may arise if, for example, females possess preferences for high values of a given trait in conspecifics (for example, longer calls, longer tails) and the novel heterospecific males have trait values that are much lower than any traits possessed by conspecific males (Rodriguez et al 2004; Pfennig and Ryan 2007).

As with resource partitioning, selection and species sorting are not mutually exclusive routes to reproductive partitioning. Pre-existing differences between species in reproductive characters can be accentuated by competitively mediated selection during character displacement (Figure 2.9A; Noor 1995; Pfennig and Ryan 2006). Alternatively, character displacement may promote differences between species in reproductive

> **BOX 2.3.** Suggestions for Future Research
>
> - Using theoretical and empirical approaches, determine the relative importance of: (1) exploitative versus interference competition in promoting ecological character displacement; and (2) hybridization versus reproductive interference in promoting reproductive character displacement (for example, see Grether et al. 2009).
>
> - Using theoretical and empirical approaches, determine the conditions under which partitioning (of resources or reproduction) arises through selection rather than species sorting. Assess empirically which route to partitioning occurs more frequently.
>
> - Incorporate facultative niche shifts into models of ecological character displacement to determine whether the existence of flexible niche use increases (or decreases) the conditions under which character displacement is expected to evolve.
>
> - Use experimental studies to determine whether the intensity of competition is greater when two species are more closely related (for example, see Violle et al. 2011).
>
> - Determine whether reproductive exclusion occurs in natural populations.
>
> - Identify the conditions under which exclusion, as opposed to character displacement, is more likely to occur.
>
> - Use experimental approaches to determine whether selection stemming from deleterious reproductive interactions between species is as strong, and possibly even stronger, than that stemming from resource competition (see Kishi et al. 2009).

characters in one region, and these differences may enable these two species to coexist when they encounter each other or different species in another region (Figure 2.9B).

Finally, although we have discussed resource and reproductive partitioning separately in this chapter, they are not mutually exclusive. Indeed, they may impact each other. For example, reproductive partitioning may arise as a by-product of resource partitioning between species. As noted above, selection stemming from resource competition may cause species to diverge in the times or locations in which they seek resources. Once this happens, these species may also be less likely to encounter each other during reproduction. Of course, the converse is also possible: resource partitioning can arise as a by-product of reproductive partitioning (Dame and Petren 2006; Okuzaki et al. 2010). We explore these issues in greater detail in the next chapter.

SUMMARY

Resource competition is common in nature and often strong. Additionally, heterospecifics frequently interact in ways that impede each species' ability to reproduce successfully. Both types of competitive interactions can have severe fitness costs. Indeed, in some cases, competition can even cause a species to become locally extinct. The fact that competition can result in extinction emphasizes competition's role in affecting species diversity. As a consequence of the deleterious effects of competition, selection should generally favor resource-use or reproductive traits that enable individuals to obtain resources or successful reproduction in the face of interspecific competition. In particular, character displacement occurs if: (1) individuals in a population vary in how they acquire resources or engage in reproduction in the face of interspecific competition; (2) this variation is inherited; and (3) some individuals have higher fitness than others because of their distinctive traits that are used to obtain resources or reproduction in the face of competition. Character displacement thereby leads to resource and reproductive partitioning that fosters species coexistence. Although character displacement is not the only means by which partitioning arises or by which species coexist, character displacement can contribute to diversity in important ways, as we will see in subsequent chapters. We list some key challenges for future research in Box 2.3.

FURTHER READING

Dhondt, A. A. 2012. *Interspecific competition in birds*. Oxford University Press: Oxford, UK. Although this book focuses on birds, it does an excellent job of highlighting general principles that are applicable to *all* organisms.

Grether, G. F., N. Losin, C. N. Anderson, and K. Okamoto. 2009. The role of interspecific interference competition in character displacement and the evolution of competitor recognition. *Biological Reviews* 84:617–635. This paper considers the role of interference competition in both ecological and reproductive interactions. The authors advocate for a third form of character displacement: agonistic character displacement.

Gröning, J., and A. Hochkirch. 2008. Reproductive interference between animal species. *Quarterly Review of Biology* 83:257–282. This paper discusses how reproductive interactions between species can lead to exclusion through a process that is analogous to competitive exclusion.

Kaplan, I., and R. F. Denno. 2007. Interspecific interactions in phytophagous insects revisited: A quantitative assessment of competition theory. *Ecology Letters* 10:977–994. This paper evaluates the evidence for interspecific competition in an important and diverse group of organisms—phytophagous insects—using a meta-analysis of published studies. The paper specifically describes tests of several key assumptions underlying traditional competition theory.

Levine, J. M., and J. HilleRisLambers. 2009. The importance of niches for the maintenance of species diversity. *Nature* 461:254–257. This paper describes an elegant field experimental demonstration of why niche differences are important for species coexistence.

Schoener, T. W. 1974. Resource partitioning in ecological communities. *Science* 185:27–39. This paper is the definitive treatment of resource partitioning.

Servedio, M. R., and M. A. F. Noor. 2003. The role of reinforcement in speciation: Theory and data. *Annual Review of Ecology, Evolution, and Systematics* 34:339–364. This paper provides a comprehensive review of the theory of, and empirical support for, reinforcement.

3

WHEN CHARACTER DISPLACEMENT OCCURS

In the previous two chapters, we examined how character displacement arises as an adaptive response to resource competition or deleterious reproductive interactions between species. A key unresolved issue, however, is why some populations and species are more likely to undergo character displacement as opposed to the alternative outcomes of competitive exclusion or reproductive exclusion (Schluter 2000; Rice and Pfennig 2007). In this chapter, we address this issue by considering *when* character displacement occurs.

We begin by discussing six general factors that facilitate character displacement. Like other forms of local adaptation, character displacement is more likely to occur when various proximate and evolutionary factors are in place; here, we examine these factors in detail. Furthermore, we describe how variation in the occurrence of these factors can generate variation in the incidence and extent of character displacement. Indeed, variation in these facilitative factors can explain why different interacting species, populations, and even conspecific individuals within the same population may differ in the expression of traits that minimize competitive interactions with other species.

We then consider how one form of character displacement (be it ecological or reproductive) potentially facilitates the alternative form. Because species that are similar enough to compete for resources will also likely impede each other's ability to reproduce successfully (and vice versa), populations that experience one form of character displacement would generally be expected to experience the other form. Once one form of character displacement has occurred, it can then facilitate the other form by generating variation on which selection can act. Indeed, either form of character displacement may be decisive in determining whether or not the alternative form also transpires.

Understanding when character displacement is more likely to proceed is important, because differences in its occurrence could explain ecological and evolutionary patterns of diversity. In particular, although character displacement is taxonomically widespread, it appears to be more prevalent in some taxa, communities, and populations than in others (Schluter 2000). Those taxa or communities that are more prone to undergo character displacement should also be more diverse, for at least two reasons. First, species that undergo character displacement are less likely to become extinct through competitive or reproductive exclusion. Second, as we describe in chapter 8, character displacement may promote speciation. Therefore identifying the facilitators of character displacement should ultimately provide key insights into the origins of diversity.

FACILITATORS OF CHARACTER DISPLACEMENT

Six non–mutually exclusive factors appear to be important in facilitating character displacement: (1) standing variation; (2) strong selection favoring the avoidance of interactions with heterospecifics; (3) ecological opportunity; (4) initial trait differences between species; (5) gene flow; and (6) a lack of antagonistic genetic correlations. Below, we briefly discuss each factor's impacts on character displacement.

STANDING VARIATION

A key facilitator of character displacement is standing variation (Roughgarden 1976; Slatkin 1979, 1980; Taper and Case 1985, 1992; Doebeli 1996). Here, we use the term "standing variation" to mean pre-existing phenotypic variants within species that ultimately minimize competitive interactions with other species when they come into contact. We focus on standing *phenotypic* variation, because it is phenotypic variation on which selection acts. We assume, however, that this phenotypic variation is underlain by *genetic* variation, which drives an evolutionary response to this selection (in the next chapter, we discuss the proximate causes of phenotypic variation). Although character displacement can eventually occur without standing variation, when such variation is present, character displacement should transpire more rapidly and be more likely to come about before the alternative outcome of competitive or reproductive exclusion takes place (Rice and Pfennig 2007; Pfennig and Pfennig 2009).

To understand why standing variation facilitates character displacement, consider that character displacement can evolve through two non–mutually exclusive routes, which differ in whether or not standing variation is present (Figure 3.1; Rice and Pfennig 2007). In the first route—which we refer to as "in situ evolution of novel phenotypes"—novel traits that happen to arise in sympatry (for example, through a new mutation) *after* competitors encounter each other are selectively favored and increase in frequency if they minimize competitive interactions (Rice and Pfennig 2007). Thus, this route occurs when competitively mediated selection favors phenotypes in sympatry that were not initially present in allopatry but that arise after competitors encounter each other. Cases in which character displacement appears to have evolved primarily through in situ evolution of novel phenotypes include the evolution of divergent tadpole trophic morphology in the spadefoot toads, *Spea bombifrons* and *S. multiplicata* (Pfennig and Murphy 2003) and the evolution of divergent beak size in the nightingales, *Luscinia megarhynchos* and *L. luscinia* (Reifová et al. 2011).

In the second route, phenotypic variants that exist in allopatry *before* competitors encounter each other can subsequently be selectively favored in sympatry if they minimize competitive interactions with heterospecifics. This second route—which we refer to as the "sorting of pre-existing variation"—occurs when selection filters phenotypes in sympatry that were already present in allopatry; that is, competitively mediated selection acts on *standing variation* (Rice and Pfennig 2007). With the latter route, phenotypic variants already exist before competitors encounter each other, as these variants have previously evolved for reasons other than selection arising from competitive interactions between species. For example, as we describe in chapter 5, phenotypic variants may evolve in allopatry as an adaptive response to *intra*specific competition for resources or mates. Cases in which character displacement appears to have evolved primarily through the sorting of pre-existing variation include the evolution of divergent shell length in the mud snails *Hydrobia ulvae* and *H. ventrosa* (Fenchel 1975) and the evolution of divergent body size in the giant rhinoceros beetles, *Chalcosoma caucasus* and *C. atlas* (Kawano 2002).

This sorting process, described in the previous paragraph, should not be confused with *species* sorting (Figure 2.8B). Although species sorting is not a mechanism of adaptive evolution (Schluter 2000), the sorting of pre-existing phenotypic variation *is* adaptive evolution in that selection favors some variants over others and those variants thereby spread in the population (Barrett and Schluter 2008). Thus, the sorting of pre-existing variants constitutes character displacement provided that an evolutionary change in the frequencies of phenotypes occurs as a direct consequence of competitively mediated selection acting in sympatry (Rice and Pfennig 2007).

Of course, because selection can act only on existing variation, it might be contended that all instances of character displacement involve the sorting of pre-existing variation, at least to some extent. For example, in Figure 3.1C, individuals of the focal species whose trait values are in the left tail of the distribution will be selectively favored through such a sorting process. However, in situ evolution of novel phenotypes goes beyond this initial sorting process and favors novel and increasingly divergent phenotypes that happen to arise in sympatry subsequent to heterospecific competitors coming into contact with each other (Figure 3.1E). Nevertheless, when abundant standing variation is present before contact with a heterospecific, the sorting of pre-existing phenotypes *alone* can be sufficient to drive character displacement (see Figure 3.1F).

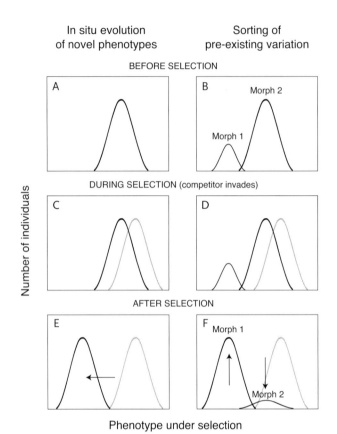

FIGURE 3.1.
Character displacement can proceed through two non–mutually exclusive evolutionary routes: in situ evolution of novel phenotypes and sorting of pre-existing (standing) variation. Initially, a focal species (species 1) occurs alone in allopatry, either (A) as a monomorphic species, or (B) as a polymorphic species consisting of alternative resource use or mating tactic morphs (morphs 1, 2), one of which is initially rarer than the other. (C, D) A superior heterospecific competitor (species 2, the trait distribution of which is indicated by the gray line) then comes into sympatry with species 1. (E, F) Because of selection imposed by species 2, species 1 undergoes an evolutionary shift in resource use and associated phenotypic features (ecological character displacement) or in mating signals/preferences (reproductive character displacement). With in situ evolution of novel phenotypes (A, C, E), character displacement occurs when novel phenotypes that are more dissimilar to the competitor arise and spread in sympatry following the invasion of species 2. As a result, the entire distribution of species 1 shifts to the left—that is, away from the competitor. By contrast, with sorting of pre-existing variation (B, D, F), character displacement occurs when the morph that is more dissimilar to the competitor (here, morph 1) is selectively favored and thereby increases in frequency at the expense of the alternative morph. As a result, the entire distribution of species 1 again shifts to the left. Although we have illustrated sorting of pre-existing phenotypes as involving discrete morphs, it could also occur in populations expressing continuously distributed phenotypes. In both cases (E, F), the outcome of character displacement is identical, even though the two populations undertook two different routes. Redrawn from Rice and Pfennig (2007).

Generally, character displacement would be expected to transpire more slowly when it occurs through in situ evolution of novel phenotypes (Rice and Pfennig 2007). This is because new genetically based variants must arise for in situ evolution of novel phenotypes to proceed. These variants arise via three non–mutually exclusive processes (Grant and Grant 2008). First, new variants may arise following one or more favorable new mutations. Second, new variants may appear following sexual reproduction and genetic recombination. Finally, new variants may be brought into a population from another conspecific population (through gene flow) or another species (through hybridization and introgression; see Rieseberg et al. 1999; Seehausen 2004). All of these processes are likely to transpire relatively slowly. For instance, the waiting time for favorable, new mutations to arise and spread in a population can take many generations (Phillips 1996). Although recombination and gene flow/introgression can potentially introduce new variants into a population more quickly than can new mutations, these processes will typically impact only a few individuals in any given generation. Perhaps more critically, theory has shown that high gene flow and recombination actually reduce a population's ability to respond to competitors, because maladaptive gene combinations swamp adaptive combinations over time (Case and Taper 2000; Goldberg and Lande 2006; reviewed in Servedio and Noor 2003; Coyne and Orr 2004; see also the discussion on gene flow below). Thus, even though variants may be introduced that are favored by competitively mediated selection, the population may continue to receive an influx of gene combinations that ultimately prevents local adaptation to competition. However, if new divergent traits spread through a population too slowly (or not at all), the population may undergo competitive or reproductive exclusion before character displacement can proceed.

This is where sorting of pre-existing variation becomes crucial. Because this process operates on *standing* variation, character displacement occurs rapidly as frequencies of pre-existing phenotypes change in sympatry. Indeed, this process could, in theory, unfold in only two generations. Thus, because the presence of standing variation can be decisive in enabling character displacement to unfold rapidly, standing variation should ultimately render character displacement—as opposed to the alternative outcome of competitive or reproductive exclusion—more likely to occur (Rice and Pfennig 2007; Pfennig and Pfennig 2009).

STRONG SELECTION

Character displacement is more likely to occur when selection that favors avoidance of heterospecifics is strong (see, for example, Liou and Price 1994; Doebeli 1996). Whether or not such selection is strong depends, in turn, on the degree of overlap in phenotypes between competitors (Roughgarden 1972, 1976; Taper and Case 1985; Schluter 2003), the costs of interactions between them (Pfennig 2007), and the frequency with which

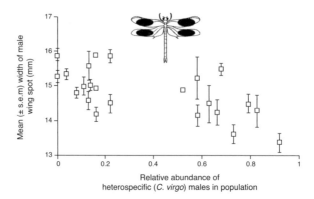

FIGURE 3.2.

An example of frequency-dependent reproductive character displacement, in which the degree of divergence in reproductive characters reflects the likelihood of encountering a heterospecific individual. In the damselfly, the banded demoiselle, *Calopteryx splendens*, males possess a secondary sexual character that females prefer: large pigmented spots on both wings. However, the largest-spotted males resemble the males of another, sympatric species, the beautiful demoiselle, *C. virgo*, interfering with species recognition. Possibly as an adaptive response to the presence of *C. virgo*, the wing-spot size of *C. splendens* in populations decreases with increasing relative abundance of *C. virgo*. Redrawn from Tynkkynen et al. (2004).

encounters occur (Pfennig and Murphy 2002; Tynkkynen et al. 2004; Pfennig and Pfennig 2005; Goldberg and Lande 2006; Anderson and Grether 2010).

Regarding overlap in phenotypes, the intensity of competition is generally greater, and character displacement therefore is more likely to occur, when competitors are more closely matched in phenotypes. Generally, as highlighted in chapter 2, the more similar any two competitors are in phenotype, the more severely they are expected to compete with each other (for example, see Figures 2.5, 2.6; Pacala and Roughgarden 1985; Pritchard and Schluter 2001; Gray and Robinson 2002; Schluter 2003; see also Burns and Strauss 2011).

However, the costs of competitive interactions often vary. For example, in times and places where resources are relatively abundant, competitive interactions are not likely to be costly, even when competitors are phenotypically similar. Additionally, species that are closely related may pay lower costs of hybridizing than species that are more genetically divergent (Harrison 1993; Arnold 1997). When costs of heterospecific interactions are low, character displacement will be less likely to occur.

Finally, even when interactions with heterospecifics are extremely costly, selection that favors avoidance of such interactions will be weak if heterospecifics are rarely encountered (Pfennig and Murphy 2002; Tynkkynen et al. 2004; Pfennig and Pfennig 2005; Goldberg and Lande 2006; Anderson and Grether 2010). Thus, the strength of selection favoring traits that reduce heterospecific interactions should be greater, and character displacement more likely to occur, when competitors are relatively abundant (for an example, see Figure 3.2).

ECOLOGICAL OPPORTUNITY

Ecological opportunity facilitates character displacement (Schluter 2000; Losos and Mahler 2010). Traditionally, ecological opportunity has been defined as "a wealth of evolutionarily accessible resources little used by competing taxa" (Schluter 2000, p. 69). However, a population may also experience ecological opportunity through enhanced access to "reproductive-trait space" (for example, signaling space for attracting mates that is not already used by other species; Pfennig and Pfennig 2009; Leal and Losos 2010; Arnegard et al. 2010). For character displacement to occur, exploitable resources or reproductive-trait space must be available. In their absence, competitive or reproductive exclusion can result (Pfennig et al. 2006; Hochkirch et al. 2007).

To illustrate the importance of ecological opportunity for character displacement, we turn to spadefoot toads. In the southwestern United States, two species, *Spea multiplicata* and *S. bombifrons*, have broadly overlapping ranges but, on a local scale, they co-occur only where ecological opportunity allows ecological character displacement to transpire among their tadpoles (Pfennig et al. 2006). This character displacement occurs when both species switch from being dietary generalists in allopatry to specialists in sympatry, with *S. multiplicata* specializing on detritus, and *S. bombifrons* specializing on fairy shrimp (see Figure 1.7). Not surprisingly, in ponds where both the detritus and shrimp resources are present (that is, when ecological opportunity is high), the two species undergo character displacement (Pfennig et al. 2006). However, in nearby ponds where only one of these two resources is available (that is, when ecological opportunity is low), character displacement is precluded and only one species (the species that is the specialist for the available resource) is present (Pfennig et al. 2006).

Ecological opportunity is especially important when character displacement proceeds through in situ evolution of novel phenotypes as opposed to the sorting of pre-existing variation (see above). When ecological opportunity is low, novel phenotypes arising through in situ evolution might not succeed, owing to a dearth of available resources or reproductive-trait space for these novel phenotypes to utilize. By contrast, when character displacement proceeds by sorting pre-existing phenotypes, selection filters phenotypes that are already present (Figure 3.1F). In such a case, individuals expressing these pre-existing phenotypes would likely already have resources or reproductive-trait space to utilize—that is, they have already taken advantage of what ecological opportunity there is.

INITIAL DIFFERENCES BETWEEN SPECIES

Character displacement is facilitated by initial trait differences between species (Slatkin 1980; Milligan 1985; Liou and Price 1994; Doebeli 1996; Schluter 2000). In the absence of initial differences—that is, in the unlikely event that species overlap in phenotype *completely*—all individuals would experience equally strong competition. In such circumstances, one species would drive the other locally extinct through competitive or reproductive exclusion (Milligan 1985). Thus, interacting species are more prone to experience character displacement when they already differ from one another in traits associated with resource use or reproduction.

That species possessing initial differences are more prone to undergo character displacement poses an apparent paradox, given that selection favoring divergence is strongest when interacting species are most similar (see above). The resolution to this paradox comes from realizing that, for character displacement to occur, some variants must exist in a sympatric population that experience less competition, or less costly effects, than others (see chapter 2). These variants are thereby favored in the early stages of character displacement, because the initial sorting of this pre-existing variation buys time while each species accumulates new variants that fuel further differentiation. Thus character displacement is most likely to occur when species possess both similarities that generate strong competition *and* differences on which selection can act to foster traits that mitigate such competition or its effects.

GENE FLOW

Gene flow can either facilitate or impede character displacement (Case and Taper 2000). Gene flow facilitates character displacement if: (1) a trait that enables individuals to obtain resources or successful reproduction in the face of interspecific competition evolves in a single sympatric population; (2) the alleles underlying such a trait subsequently spread into additional sympatric populations via gene flow; whereupon (3) they increase in frequency owing to competetively mediated selection.

Alternatively, gene flow may impede character displacement through two different routes. First, gene flow potentially inhibits character displacement when it occurs between conspecific populations that are in sympatry with a heterospecific competitor versus those in allopatry. In such cases, maladaptive alleles from allopatry can swamp alleles for divergent traits in sympatry, thereby hindering character displacement (Case and Taper 2000; Goldberg and Lande 2006).

Second, gene flow can potentially inhibit character displacement when interacting species hybridize. Hybridization can impede or even preclude character displacement for three reasons (Howard 1993; Servedio and Noor 2003; Coyne and Orr 2004).

First, hybridization generally causes interacting species to become more (not less) similar (Wilson 1965; for full discussion see Kelly and Noor 1996; Coyne and Orr 2004). In extreme cases, the genomes of interacting species may fuse entirely, and each species may lose its distinctive characteristics (Seehausen et al. 1997; Behm et al. 2010). Such fusion is most likely to occur when species experience weak selection to avoid hybridization (for example, because hybrids are relatively fit; Arnold 1997) and/or when relatively little divergence in reproductive traits exists between species (resulting in a high incidence of hybridization; Wilson 1965; Liou and Price 1994; Kelly and Noor 1996).

Additionally, hybridization hinders character displacement by breaking up adaptive gene complexes that underlie divergent resource-use or reproductive traits in each species. For example, when species exchange genes, alleles encoding traits that contribute to assortative mating would be continuously dissociated from alleles that generate low hybrid fitness (Felsenstein 1981). Consequently, the spread of alternative alleles for traits that promote assortative mating among pure species types is precluded.

Finally, antagonistic selection can preclude character displacement between hybridizing species when alleles encoding for traits that contribute to assortative mating or differential resource use in one species (for example, preferences for traits possessed only by that species) are disfavored when expressed in the other species. Because potentially favorable alleles in one species would be purged in the alternative species, species are less likely to become differentiated.

In the next chapter, we discuss how some types of genetic architecture can overcome or lessen these inhibitory effects of hybridization on character displacement.

LACK OF ANTAGONISTIC GENETIC CORRELATIONS

Finally, the occurrence of character displacement may also depend on whether or not antagonistic genetic correlations exist among the targets of selection. By "antagonistic" genetic correlations, we mean correlations among loci that preclude evolutionary responses to competitively mediated selection. Generally, character displacement involves the evolution of a complex suite of traits that must be functionally coordinated for individuals to succeed either in acquiring resources or in successful reproduction (see, for example, Schluter 1993). Genetic correlations among the genes that encode these traits could inhibit character displacement (Smith and Rausher 2008). Note, however, that genetic correlations may actually facilitate character displacement if these correlations are concordant with the direction of selection acting on each trait. In such a situation, direct selection acting on one trait may actually facilitate character displacement if it promotes indirect selection on the other traits (for example, see Martin and Pfennig 2011).

VARIATION IN THE EXPRESSION OF CHARACTER DISPLACEMENT

Above, we highlighted six factors that facilitate character displacement. Variation in any of these factors can cause character displacement to unfold to a greater degree—and produce more pronounced trait evolution—in some species, populations, and groups of individuals than in others. To illustrate this point, we focus here on the impacts of one facilitator: strong competitively mediated selection. As noted above, the strength of competitively mediated selection depends on the degree of trait overlap between interacting species; the frequency with which heterospecifics are encountered; and the costs of heterospecific interactions. All of these features can vary between species, populations, or even individuals within the same population.

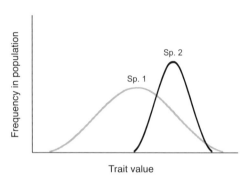

FIGURE 3.3.
Two competing species may differ in the degree to which they overlap with each other in resource-use or reproductive traits. In this case, species 1 is the more phenotypically variable; the trait distribution of species 2, the less variable species, may be completely included within that of species 1. Consequently, one species (species 2) may experience stronger competitively mediated selection.

For instance, one species may possess a narrow trait distribution that is overlapped to a greater extent than that of a competing species with a wider trait distribution (Figure 3.3). Likewise, species may differ in abundance, whereupon the rarer species will experience costly interactions more frequently (for example, see Pfennig and Murphy 2002). Finally, the costs of the interactions may differ for each species.

To illustrate how the costs of competitive interactions may differ for each of two interacting species, recall from chapter 2 that two species of spadefoot toads—Plains spadefoot toads (*Spea bombifrons*) and Mexican spadefoot toads (*S. multiplicata*)—potentially risk hybridizing with each other in arid regions of southwestern North America. The fitness consequences of hybridization vary for these two species, however. Although hybrid tadpoles develop more slowly than *S. multiplicata* tadpoles, they develop more rapidly than pure *S. bombifrons* tadpoles (Pfennig and Simovich 2002). Consequently, hybrid tadpoles may be able to achieve metamorphosis, whereas pure *S. bombifrons* tadpoles may not—*Spea* tadpoles often occur in highly ephemeral ponds that dry before tadpoles can achieve metamorphosis (Bragg 1965; Pfennig and Simovich 2002). By hybridizing with *S. multiplicata*, *S. bombifrons* may therefore have higher fitness (although hybrid sons may be sterile, hybrid daughters are fertile and capable of interbreeding with either species). By contrast, *S. multiplicata* achieve no such benefit by hybridizing with *S. bombifrons* (Pfennig and Simovich 2002; Pfennig 2007).

Because the fitness consequences of hybridization differ for each of these two interacting species, not unexpectedly they differ in their likelihood of engaging in hybridization, with *S. bombifrons* being more likely to hybridize than *S. multiplicata* (Pfennig and Simovich 2002; Pfennig 2007). Moreover, the incidence of hybridization is greatest in the smallest ponds, which tend to be the most ephemeral (Pfennig and Simovich 2002). Thus, because the fitness consequences of hybridization can differ for different interacting species, the strength of selection favoring the avoidance of such interactions—and the response to such selection—will also vary.

Generally, variation in any of the above three factors that influence the strength of competitively mediated selection can produce asymmetries in the incidence and magni-

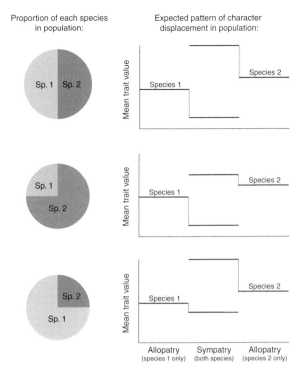

FIGURE 3.4.
All else being equal, variation in the relative frequencies of two competing species in different populations may lead to variation among these populations in the magnitude of character displacement that each species undergoes. Such variation may even occur among neighboring populations.

tude of character displacement (Cooley 2007). Indeed, the species that is experiencing stronger competitively mediated selection is expected to show more pronounced divergence than the species that is experiencing weaker selection (Cooley 2007; Price and Kirkpatrick 2009). For example, among the two species of spadefoots described above, controlled mate-choice tests have shown that *S. multiplicata*, the species that experiences stronger selection to avoid heterospecific interactions, discriminates more strongly against the male calls of the other species than does *S. bombifrons*, the species that sometimes benefits by hybridizing (Pfennig 2007). Indeed, *S. bombifrons* sometimes *prefers* the calls of the other species over those produced by its own species (Pfennig 2007).

Selection favoring avoidance of heterospecifics can differ not only between the two species, but also among conspecific populations. For example, conspecific populations may occur in different habitats that vary in resource abundance or quality and that thereby vary in the extent of character displacement (Goldberg and Lande 2006). Moreover, different conspecific populations may differ in the likelihood with which individuals encounter heterospecifics (Figure 3.4; for an empirical example, see Figure 3.2). In such cases, the strength of competitively mediated selection may differ—even among conspecific populations—resulting in greater divergence in those populations where heterospecific encounters are more frequent (see, for example, Pfennig and Murphy 2002; Tynkkynen et al. 2004; Pfennig and Pfennig 2005; Goldberg and Lande 2006; Anderson and Grether 2010).

Selection favoring avoidance of heterospecifics may even vary for different *individuals*, including (potentially) those within the *same* population. Males and females, for example, will not necessarily experience the same selective pressures to avoid reproductive interactions with heterospecifics (Parker and Partridge 1998; Wirtz 1999). Indeed, in many species, males may experience much weaker selection to avoid heterospecifics, if mating with a heterospecific is better than not mating at all (reviewed in Wirtz 1999). Thus, female mate-choice preferences may undergo reproductive character displacement without a concomitant shift in male advertisement traits (Gerhardt 1991, 1994; Gerhardt and Huber 2002).

Even same-sex individuals may differ in the strength of selection favoring avoidance of heterospecifics. For example, individual *S. bombifrons* females (the spadefoot toad species that potentially benefits from hybridization) facultatively switch their mate preferences depending on pond depth. In deeper, long-lasting pools (an environment that does not favor hybridization), *S. bombifrons* females prefer conspecific males whereas in shallow, highly ephemeral pools (an environment that favors hybridization), *S. bombifrons* females are more likely to prefer *S. multiplicata* males (Pfennig 2007). However, whether or not an individual *S. bombifrons* female actually switches her preference to favor a *S. multiplicata* male depends on her own body condition. In particular, the poorest-condition females are most likely to switch preferences (Figure 3.5A; Pfennig 2007). This individual variation in avoidance of heterospecific signals reflects variation in selection favoring such avoidance: *S. bombifrons* females in the poorest condition benefit the most from hybridizing with *S. multiplicata*, because these poorest-condition females produce the most slowly developing tadpoles (Figure 3.5B), which may fail to escape an ephemeral pond before it goes dry. By preferring a *S. multiplicata* male and producing hybrid tadpoles, these females increase the chances that their offspring will survive to metamorphosis (see above).

In sum, the strength of selection favoring the avoidance of heterospecific interactions can vary between species, between conspecific populations, between sexes, and—perhaps most surprisingly—even among individuals of the same sex. Such variable selection can promote variation in the occurrence of character displacement among different populations; it can also promote variation in the expression of traits that minimize interactions with heterospecifics among individuals within a population. Presumably, variation in the other five facilitators of character displacement would similarly produce variation in the expression of character displacement and its targeted traits. Additional studies are needed to evaluate this prediction, however. Such studies are important, because (as noted in the introduction to this chapter) variation in the occurrence of character displacement could ultimately explain ecological and evolutionary patterns of diversity.

Many of the facilitators of character displacement discussed until this point in the chapter promote local adaptation generally and are therefore not specific to character displacement. A factor that is specific to character displacement, however, is whether

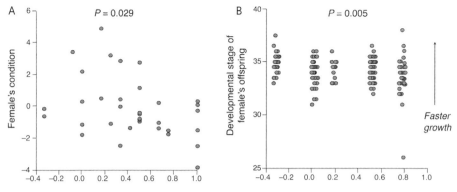

FIGURE 3.5.
Individuals within a single population may differ in the strength of selection favoring avoidance of heterospecific interactions. For example, in spadefoot toads, (A) *Spea bombifrons* females that are in the poorest condition are the most likely to switch their mate preferences, from selecting a conspecific male to selecting a heterospecific (specifically, a *S. multiplicata*) male. (B) Presumably females that are the most likely to engage in hybridization do so because they produce the most slowly developing tadpoles with conspecifics. Thus, the poorest-condition females can benefit the most from hybridizing and producing more rapidly developing hybrid tadpoles. Reproduced from Pfennig (2007), with the kind permission of the publisher.

one form of character displacement—either ecological character displacement or reproductive character displacement—facilitates the alternative form. We consider this issue next.

HOW ECOLOGICAL AND REPRODUCTIVE CHARACTER DISPLACEMENT FACILITATE EACH OTHER

Generally, species that are similar enough to compete for resources will also likely impede each other's ability to reproduce successfully (and vice versa). Therefore, populations that experience selection that fosters a particular form of character displacement (either ecological or reproductive) should also experience selection that fosters the alternative form. Here we describe how, once one form of character displacement has occurred, it can promote changes in either phenotypes or habitat that render the alternative form of character displacement more likely to unfold. Although these two forms may impede each other (see below), each can facilitate the other by generating variation on which selection can act.

Before we examine how the different forms of character displacement impact each other, we must clear up two potential sources of confusion. First, although the discussion below is couched in terms of how one form of character displacement initiates the other, the two processes are likely to occur *simultaneously*.

Second, the correlated evolution of both resource use and reproductive traits does not necessarily imply that both ecological and reproductive character displacement have occurred. Character displacement is trait evolution resulting from *competitively mediated selection* (see chapter 1). Without evidence that competitively mediated selection has promoted the observed trait evolution, one cannot conclude that character displacement has transpired. For example, imagine that selection favors traits that minimize resource competition, and that such selection leads to ecological character displacement. However, imagine further that this adaptive evolution of resource-use traits leads to the *correlated* evolution of *reproductive* traits. In such a situation, one cannot conclude that reproductive character displacement has occurred unless there is evidence that selection has *also* favored reproductive traits that lessen costly reproductive interactions between species (Rundle and Schluter 1998, 2004). In other words, species and populations may diverge in reproductive traits purely as a by-product of niche divergence caused by ecological character displacement, but such incidental reproductive divergence would not constitute reproductive character displacement. Thus, in the discussion below, when we state that one form of character displacement *initiates* the other form, we are specifically referring to those circumstances in which selection minimizes *both* resource competition and reproductive interactions (we consider the effects of character displacement on sexual selection separately in chapter 7).

With these caveats in mind, we begin by describing ways that ecological character displacement can promote reproductive character displacement.

ECOLOGICAL DIVERGENCE AS A FACILITATOR OF REPRODUCTIVE DIVERGENCE

Whenever the expression of reproductive traits is associated with traits involved in resource acquisition (such as when the same trait is used for both resource acquisition and mate acquisition), ecological character displacement has the potential to initiate reproductive character displacement (Figure 3.6A, B). In particular, evolutionary shifts in resource-acquisition traits can bring about evolutionary changes in male sexual traits, female preferences, or the habitats that individuals use for reproduction. If such changes in reproductive traits or habitat also reduce deleterious reproductive interactions between species, then ecological selection could jump-start reproductive character displacement.

Consider, for example, that shifts in bird beak size resulting from ecological character displacement (see Figure 1.6) can, in turn, cause concomitant shifts in a reproductive trait that is crucial in mate recognition—bird song (reviewed in Podos and Nowicki 2004; Grant and Grant 2008; Price 2008). Indeed, in the medium ground finch, *Geospiza fortis*, the population at El Garrapatero (Santa Cruz Island) in the Galápagos

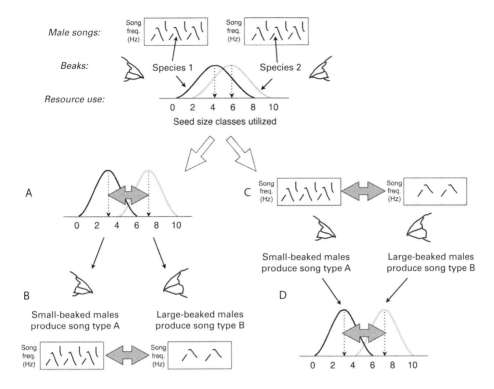

FIGURE 3.6.
How ecological and reproductive character displacement can facilitate each other. Imagine two species of seed-eating birds that initially share similar resources and songs. Imagine further that these two species are initially similar in their chief resource-acquisition trait, their beak. Finally, suppose that beak morphology influences the trait that males use to attract females: his song. Ecological character displacement can facilitate reproductive character displacement if (A) selection for traits that lessen resource competition promotes divergence between species in traits associated with resource use—that is, beak morphology. (B) As a consequence of shifts in beak morphology, the two species would likely diverge in male song. If females use song to recognize their own species, a shift in song type would promote assortative mating. If this shift in song type is maintained or elaborated by selection for traits that minimize reproductive interactions, then ecological character displacement would have jump-started reproductive character displacement. Reproductive character displacement can facilitate ecological character displacement if (C) selection for traits that lessen reproductive interactions (for example, hybridization, signal interference) favors divergence between species in features associated with song production—that is, beak morphology. (D) As a consequence of these shifts in beak morphology, the two species would likely also diverge in types of seeds for which each species is best adapted. If this shift in resource use is maintained or elaborated by selection for traits that minimize resource competitive, then reproductive character displacement would have jump-started ecological character displacement.

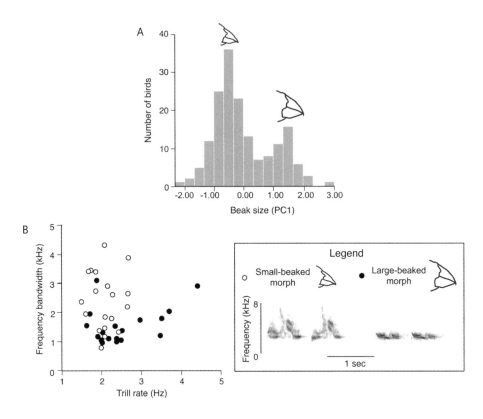

FIGURE 3.7.
Beak morphology (a trait involved in resource acquisition) and song features (a trait involved in reproduction) covary in a population of Darwin's finches (*Geospiza fortis*). (A) The adult population at El Garrapatero (Santa Cruz Island) consists of a small-beaked morph and a large-beaked morph (beak size PC1 is the first principle component of an analysis that included beak length, beak depth, and beak width as variables). (B) Songs are more similar within morphs than between morphs, as revealed by the relationship between song trill rate and frequency bandwidth (representative sonograms from a large-beaked morph and a small-beaked morph are shown in the legend). Based on data in Huber and Podos (2006).

archipelago consists of a large-beaked morph and a small-beaked morph (Figure 3.7A), which preferentially feed on large and small seeds, respectively (J. Podos, personal communication). Because song structure changes with beak morphology (Podos 2001; Podos and Nowicki 2004), these two morphs also produce distinct song types (Figure 3.7B). Females apparently use these different song types during mate choice, and they mate assortatively with males of their own beak type (Huber et al. 2007). Thus, selection acting on resource-acquisition traits can also promote changes in reproductive traits, which in turn reduce reproductive interactions between groups (Podos 2001; Podos and Nowicki 2004; Podos et al. 2004; Huber et al. 2007; Grant and Grant 2008).

Although this example focuses on *within* species interactions—and is therefore not an instance of character displacement *between* species—similar dynamics could play out between species. Indeed, any shifts in resource use may foster changes in reproductive traits that could be further exaggerated via reproductive character displacement. Shifts in resource use could affect not only auditory traits used for reproduction (as above), but such shifts could also affect visual or olfactory traits, especially when dietary components (for example, carotenoids) are incorporated into sexual displays (see chapter 7). In many fish species, for example, male coloration is both diet-dependent and important in species recognition (Seehausen and van Alphen 1998; Boughman 2001). If ecological character displacement causes a shift in diet, male signaling may be affected if the dietary components used to generate a given trait are no longer available or are too costly to obtain (see, for example, Boughman 2007).

Our discussion so far has concentrated on how resource shifts can alter male traits. Yet phenotypic changes wrought by ecological character displacement can also affect female perception of male sexual traits, and, consequently, female mate preferences. Indeed, changes caused by ecological character displacement—such as shifts in jaw morphology used to capture prey, in olfactory or visual sensitivity used to localize prey, and in overall body size for specializing on different resources—may often simultaneously alter female perception and discrimination of male traits (*sensu* Bradbury and Vehrencamp 1998; Endler and Basolo 1998; Ryan 1998). Because a female's ability to perceive and discriminate among male traits dramatically affects the expression of female mate preferences, any changes in female sensory systems induced by ecological character displacement will necessarily alter female mate choice (Ryan 1998). Ecological character displacement can therefore reveal new female-choice variants on which selection can act to minimize *reproductive* interactions between species.

Thus, selection that promotes divergent resource-acquisition traits between species potentially also promotes divergence between species in reproductive traits. If such divergent traits reduce reproductive interactions between species, then they may be maintained and elaborated via reproductive character displacement (Figure 3.6B).

Ecological character displacement can also promote reproductive character displacement by changing the *habitat* in which mating or reproduction takes place. In chapter 2, we described how ecological character displacement could instigate evolutionary shifts in the habitat in which individuals of one or both species or populations seek resources. Because an organism's habitat critically affects the attenuation and perception of its sexual traits (reviewed in Wiley 1994; Bradbury and Vehrencamp 1998; Boughman 2002; Gerhardt and Huber 2002), shifts in habitat use (such as those that may be brought about by ecological character displacement) can favor the expression and evolution of novel reproductive traits that can become further elaborated via reproductive character displacement.

Anole lizards from Puerto Rico offer a possible example of how shifts in habitat can alter reproductive traits. Two sympatric species, *Anolis cooki* and *A. cristatellus*, both of which occur as the trunk-ground ecomorph (see Figure 2.7), display divergent ultraviolet (UV) light sensitivity that appears to enable them to occupy slightly different light microenvironments (Leal and Fleishman 2002). Such divergent microhabitat use may facilitate co-occurrence (Leal and Fleishman 2002). These two species have also diverged, however, in the UV reflectance of male dewlaps. Specifically, each species' dewlap provides the strongest contrast in the light microhabitat in which that species resides, thereby facilitating species recognition (Leal and Fleishman 2002). Presumably, divergent habitat use simultaneously selects for male sexual traits that optimize communication in the novel habitat while also minimizing reproductive interactions between species.

Changes in habitat use brought about by ecological character displacement may also generate changes in female mate preferences. Such changes in female preferences could promote reproductive character displacement in two different ways. First, novel habitats may exert selection on females to evolve preferences for male traits that are most efficiently detected in those new habitats (Endler and Basolo 1998; Boughman 2002). Second, novel habitats or resources may also impose natural selection on female sensory systems to better identify prey. These shifts in sensory sensitivity could indirectly alter patterns of female mate choice that thereby minimize deleterious reproductive interactions with heterospecifics (Endler and Basolo 1998; Boughman 2002, 2007).

For example, three-spined sticklebacks (*Gasterosteus aculeatus* complex) are small fishes that occur in both marine environments and freshwater lakes and streams throughout temperate regions of the Northern Hemisphere. In certain small coastal lakes in southwestern Canada, two species co-occur: one species expresses a distinctive benthic phenotype that forages and mates in the lake's littoral zone, whereas the other expresses a distinctive limnetic phenotype that forages and mates in open water (see Rundle and Schluter 2004 and references therein). These two ecomorphs are thought to have arisen following the invasion from the ocean of an ancestral limnetic ecomorph into lakes already containing an intermediate ecomorph (which is thought to have evolved from a previous marine invasion of an ancestral limnetic ecomorph). Ecological character displacement then promoted the evolution of a new benthic ecomorph within each such lake, replacing the ancestral, intermediate ecomorph (Rundle and Schluter 2004).

Not only have these habitat shifts (wrought by ecological character displacement) caused the two ecomorphs to minimize dietary overlap with each other, but they have also forced these two ecomorphs into different light environments, which has affected the evolution of their reproductive traits. In the littoral zone, where the benthic ecomorph forages and mates, red coloration is more difficult to detect (Boughman 2001). In contrast, in open water, where the limnetic ecomorph forages and mates, red coloration is more discernible (Boughman 2001). Benthic females are less sensitive to variation in red than are limnetic females and, unlike limnetic females, benthic females do not tend to prefer redder males (Boughman 2001, 2007; reviewed in Boughman 2002). Male red coloration, in turn, is "tuned" to female perception of red color: males are redder in populations where females are actually sensitive to, and therefore prefer, redder males (Boughman 2001, 2007). Thus, shifts in mate preference that are tied to different habitats dictate the degree to which reproductive divergence has occurred (Boughman 2001). Moreover, controlled experiments have shown that such reproductive divergence reflects selection to minimize reproductive interactions per se (Rundle and Schluter 1998, 2004). Generally, shifts in habitat use that are brought about by ecological character displacement may play a critical role in initiating and promoting reproductive character displacement.

REPRODUCTIVE DIVERGENCE AS A FACILITATOR OF ECOLOGICAL DIVERGENCE

Above, we emphasized how ecological character displacement could jump-start reproductive character displacement. Of the relatively few empirical studies that have examined how these alternative forms of character displacement interact, most have concentrated on ecological character displacement's potential impacts on the evolution of reproductive traits. We therefore focused initially on ecological divergence as a facilitator of reproductive divergence (as opposed to the reverse direction of causation). Yet reproductive character displacement may also facilitate ecological character displacement (Konuma and Chiba 2007). As is the case for ecological divergence, such facilitation may occur through shifts in either phenotypes or habitat.

We begin by considering how reproductive character displacement can promote ecological character displacement through phenotypic shifts (Figure 3.6C, D). In particular, if, as a result of reproductive character displacement, a population undergoes a shift in a resource-use trait that reduces resource competition with a heterospecific competitor, then reproductive character displacement would have jump-started ecological character displacement. For instance, mate preferences to avoid interactions with heterospecifics may promote the evolution of reproductive traits—such as a change in body size or beak morphology—that could, in turn, cause a shift in resource use (Konuma and Chiba 2007; for example, see Figure 3.6D).

Reproductive character displacement can also facilitate ecological character displacement by promoting shifts in *habitat*. Consider that, if species segregate in space or time to avoid reproductive interactions (see chapter 2), then they may encounter novel, underutilized resources. If utilization of these resources is selectively favored to minimize resource competition between species, then reproductive character displacement will have facilitated ecological character displacement.

A possible example comes from the anole lizards described in the previous section. Although ecological character displacement could have promoted reproductive character displacement (as described above), the reverse could also have occurred. That is, reproductive interactions may have generated divergence in female perception and male traits (recall from above that the two species display divergent UV light sensitivity as well as divergent UV reflectance of male dewlaps). Such divergence in female perception and male traits could have, in turn, fostered habitat partitioning, in which each species occupies a slightly different light environment and resource competition is thereby minimized (Leal and Fleishman 2002). Thus reproductive character displacement may have facilitated ecological character displacement by promoting shifts in habitat use. This example also illustrates another point: it is often hard to determine whether ecological character displacement occurred first and subsequently promoted reproductive character displacement, or the opposite sequence of events transpired. Disentangling the two would best be achieved by combining measures of selection in the field with experimental work that explicitly controls for the effects of ecological character displacement versus reproductive character displacement (see, for example, Rundle and Schluter 1998).

In sum, although few empirical studies have examined reproductive character displacement's impacts on ecological character displacement, reproductive character displacement has the potential to facilitate ecological character displacement. As with ecological character displacement's possible effects on reproductive character displacement, reproductive character displacement may advance ecological character displacement by promoting shifts in either phenotypes or habitat use.

WHY ONE FORM IS NECESSARY TO FACILITATE THE OTHER

Our discussion above implies that the absence of one form of character displacement sometimes inhibits the alternative form from occurring. Yet, in situations such as those described above, why is one form of character displacement required to jump-start the other? Why does the latter process not unfold on its own?

An obvious answer to this question is that at least one of the facilitators of character displacement outlined earlier in this chapter may not be in place for one form of character displacement to transpire. For instance, there may be abundant standing variation in either resource-use traits or reproductive traits—but not in both—and such a lack of variation in one type of trait may prevent that form of character displacement from occurring on its own. However, once one form of character displacement occurs, it may facilitate the alternative form by revealing formerly cryptic genetic variation on which the alternative form of selection could act (we discuss this scenario in greater detail in chapter 4).

Moreover, the unfolding of one process may actually generate the selective pressure that fuels the alternative process. For example, as resource use diverges between species, matings between individuals of both species could result in offspring that are inferior competitors relative to pure species types (for example, see Hatfield and Schluter 1999; Rundle 2002; Pfennig and Rice 2007; Fuller 2008). The production of such competetively poor offspring may then constitute the *only* selective basis for reproductive character displacement.

In sum, either form of character displacement may be decisive in determining whether or not the alternative form also transpires. A major impetus for future work in character displacement is the need to understand how the two forms of character displacement interact, if at all, and the resulting effects such interaction has on trait evolution across different populations, communities, and species.

HOW ECOLOGICAL AND REPRODUCTIVE CHARACTER DISPLACEMENT CAN IMPEDE EACH OTHER

Finally, rather than one form of character displacement facilitating the other, ecological or reproductive character displacement may preclude the occurrence of the alternative form of character displacement. Such inhibition may come about in one of two ways. First, one form of character displacement may cause a population to shift to a new habitat in order to seek resources or to reproduce. If this shift results in the alternative form of competitive interactions (that is, resource or reproductive competition) being attenuated (such as when the heterospecific is no longer encountered), then the operation of one process essentially removes the selective pressure for the other process to occur.

This point seemingly contradicts the notion, described above, that habitat partitioning could allow one form of character displacement to facilitate the other form. However, whether such facilitation occurs depends on the degree to which species become segregated in space or time. For example, ecological character displacement can cause two species to shift into separate habitats or seek resources at different times. However, the two species may nevertheless risk hybridizing or interfering with each other's reproduction, albeit at a lower frequency. Consequently, selection should favor divergence in reproductive traits; that is, reproductive character displacement. If, however, as a result of such habitat shifts, the two species do not actually encounter or interfere with one another, then selection would not favor divergence in reproductive traits. In such a case, reproductive character displacement would not be expected to occur for the simple reason that there would be no selective basis for it.

A second route by which ecological and reproductive character displacement could impede each other is when selection on a given trait under one form of character displacement opposes selection on that same trait exerted under the alternative form. For example, ecological character displacement could generate habitat or phenotypic shifts that cause convergence of reproductive traits (such as when traits used in male-male aggression over resources generate convergent traits that confound females' ability to identify conspecific mates; Grether et al. 2009). Such convergence may be opposed by selection that minimizes reproductive interactions between species. Thus, changes brought about by one form of character displacement can disrupt the alternative form if the patterns of selection exerted by ecological or reproductive interactions are in opposition. Ultimately, however, selection would be expected to favor novel traits that minimize *both* resource competition and reproductive interactions between species. Yet, as

> **BOX 3.1.** Suggestions for Future Research
>
> - Use meta-analyses of character displacement to determine, first, whether character displacement is more prevalent in some taxa or communities than in others; and second, what factors explain this variation, if indeed there is any variation.
> - Gather data from natural populations to identify the proximate and evolutionary factors that are the most important in facilitating character displacement.
> - Ascertain whether the sorting of pre-existing variation is crucial in mediating character displacement in situations where ecological opportunity is low.
> - Establish whether (and, more importantly, *why*) variation often exists among interacting species and populations in the expression of character displacement, and even among individuals in the expression of traits that minimize interactions with heterospecifics.
> - Conduct theoretical and empirical studies to determine whether and how ecological and reproductive character displacement affect each other.
> - Determine whether ecological character displacement jump-starts reproductive character displacement or vice versa (for example, see Konuma and Chiba 2007).
> - In situations where one form of character displacement is required to initiate the other, identify the reasons that the latter process could not unfold on its own.

we emphasized earlier in this chapter, such in situ evolution of novel phenotypes—as opposed to competitive or reproductive exclusion—would be slow and thereby reduce the likelihood that character displacement would transpire.

SUMMARY

Because character displacement is central to the generation, maintenance, and distribution of biodiversity, clarifying the factors that facilitate character displacement is crucial for illuminating the origins of diversity. Six non–mutually exclusive factors facilitate character displacement: (1) standing variation; (2) strong selection favoring the avoidance of interactions with heterospecifics; (3) ecological opportunity; (4) initial trait differences between species; (5) gene flow; and (6) a lack of antagonistic genetic correlations. Yet character displacement is not an all-or-nothing process; variation in the occurrence of these factors can, in turn, generate variation in the incidence of character displacement. Indeed, variation in the above facilitative factors can explain why different species, populations, and even conspecific individuals within the same population often differ in the expression of traits that minimize interactions with heterospecifics. Additionally, whether or not ecological or reproductive character displacement transpires may also depend upon whether or not the alternative form of character displacement has also occurred. In particular, once one form of character displacement occurs initially, it may then facilitate the other form by generating variation on which selection can act. However, either form of character displacement may also preclude the alternative form. More research is needed to identify the facilitators of character displacement and to clarify how one form impacts the other. We present some key challenges for future research in Box 3.1.

FURTHER READING

Boughman, J. W. 2002. How sensory drive can promote speciation. *Trends in Ecology and Evolution* 17:571–577. This paper illustrates how resource shifts can be accompanied by shifts in sexual traits, which can be maintained and further elaborated via reproductive character displacement.

Leal, M., and L. J. Fleishman. 2002. Evidence for habitat partitioning based on adaptation to environmental light in a pair of sympatric lizard species. *Proceedings of the Royal Society B: Biological Sciences* 269:351–259. This paper presents a case in which character displacement has led to habitat partitioning, which reduces both resource competition and reproductive interactions between species. However, it is unclear which form of character displacement occurred first.

Podos, J. 2001. Correlated evolution of morphology and vocal signal structure in Darwin's finches. *Nature* 409:185–188. This paper provides a compelling example of how changes in a trait involved in resource acquisition—the beak of the finch—affect a reproductive trait—the song of the finch.

Rice, A. M., and D. W. Pfennig. 2007. Character displacement: In situ evolution of novel phenotypes or sorting of pre-existing variation? *Journal of Evolutionary Biology* 20:448–459. This paper describes how character displacement can proceed through two nonexclusive routes that differ in the source of phenotypic variation, and, hence, in the ease with which character displacement might unfold.

Schluter, D. 2000. *The ecology of adaptive radiation*. Oxford University Press: Oxford, UK. This book reviews the extensive theory on "when" ecological character displacement occurs.

Slatkin, M. 1980. Ecological character displacement. *Ecology* 61:163–177. This paper serves as the starting point for recent theoretical discussions of ecological character displacement. Slatkin presents a two-species model that has been dubbed a "null" model for character displacement because it is very general and has few assumptions.

4

HOW CHARACTER DISPLACEMENT UNFOLDS

Recall from chapter 3 that a key facilitator of character displacement is the presence of standing phenotypic variation on which competitively mediated selection can act. Yet, relatively little is known about the source(s) of such variation or how different sources affect character displacement's tempo and mode. In this chapter, we address these issues by examining the underlying proximate (that is, genetic and developmental) mechanisms of character displacement.

We begin by considering whether character displacement comes about strictly through genetically canalized change (that is, change that reflects allelic or genotype frequency changes and that is relatively insensitive to the environment) or whether, alternatively, it can arise through phenotypic plasticity. Although genetically canalized change is generally presumed to be the sole mediator of character displacement (Grant 1972; Connell 1980; Arthur 1982; Schluter and McPhail 1992; Taper and Case 1992; Losos 2000; Schluter 2000; Coyne and Orr 2004), environmentally initiated niche shifts can promote character displacement under certain circumstances, such as when reaction norms become a target of competitively mediated selection (Box 2.2).

After outlining the various ways by which both genetically canalized change and phenotypic plasticity can mediate adaptive divergence between competitors, we then discuss how each mechanism can impact character displacement's *speed*. The speed at

---

A modified version of this chapter appeared in Pfennig, D. W., and K. S. Pfennig. 2012. Development and evolution of character displacement. *Annals of the New York Academy of Sciences: The Year in Evolutionary Biology* 1256: 89–107.

which divergence evolves is crucial because character displacement is a time-sensitive process: if it occurs too slowly, a population risks becoming extinct through competitive or reproductive exclusion. As we describe, phenotypic plasticity not only can potentially mediate character displacement, but it may also promote *rapid* character displacement, and thereby play a key role in buffering populations from competitive or reproductive exclusion until genetically canalized change can evolve.

Finally, we consider how these alternative mechanisms interact and affect character displacement's mode. In particular, because environmentally contingent phenotypes can lose their environmental sensitivity over evolutionary time, divergent traits that initially arise via phenotypic plasticity may eventually become genetically canalized (West-Eberhard 2003, 2005; Lande 2009; Moczek et al. 2011). Character displacement may therefore proceed through an initial phase in which trait differences are environmentally induced to a later phase in which such differences are expressed constitutively (Pfennig and Martin 2010).

Identifying the mechanisms underlying character displacement is important for at least two reasons. First, because the principles described in this chapter may apply to other forms of adaptive trait evolution (for example, see Scoville and Pfrender 2010), understanding the mechanisms of character displacement may help illuminate the mechanisms underlying phenotypic evolution generally. Second, in contrast to some other forms of trait evolution, where proximate mechanisms can be treated as a "black box" (as where the particular mechanism generating a trait does not necessarily affect whether or not adaptive evolution occurs), different proximate mechanisms can impact the speed of character displacement and, therefore, affect whether or not it unfolds to completion. Thus a full understanding of character displacement's role in evolutionary diversification may ultimately rest on identifying its underlying mechanisms.

MECHANISMS OF DIVERGENCE

Most evolutionary biologists and ecologists recognize two distinct proximate mechanisms for explaining why populations or species may differ phenotypically (see, for example, Futuyma 2009, chapter 9; Cain et al. 2008, p. 85). In the first mechanism, the individual members of different populations or species may possess different alleles or genotypes at loci that regulate the expression of the focal trait. In other words, divergence may reflect genetically canalized change. In the second mechanism, trait differences may arise through phenotypic plasticity, in which a single genotype expresses different phenotypes in response to different environmental cues (Sultan 2011). Phenotypic plasticity is pervasive (West-Eberhard 2003), and traits associated with resource use or reproduction are no exception. For example, in many species, divergence in resource-use traits can be induced by the presence of a heterospecific competitor (for examples, see Figure 1.7 and Table 2.1).

These two mechanisms are not as distinct from one another as is sometimes por-

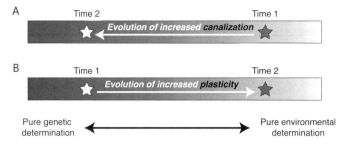

FIGURE 4.1.
A schematic that emphasizes the continuum—and interchangeability—between genetic and environmental influences on phenotype production. Individuals may vary phenotypically either because they possess different alleles or because they experienced contrasting environments and trait differences arose through phenotypic plasticity. These two proximate mechanisms are best thought of as occupying different positions along a continuum in which strict genetic determination of trait production resides at one end and pure environmental induction resides at the opposite end. Most (perhaps all) traits, however, lie between these two extremes. Moreover, a trait's position along this continuum can change over evolutionary time. For example, in (A) a trait evolves to become more canalized (that is, it undergoes genetic assimilation), whereas in (B) a trait evolves to be more plastic. Such shifts may play an important role in character displacement. Reproduced from Pfennig et al. (2010), with the kind permission of the publisher.

trayed (as discussed in Futuyma 2009; Pfennig et al. 2010; Pigliucci 2010). On the one hand, even with genetically canalized differences, the environment is a normal and necessary agent in trait production (Gilbert and Epel 2009). On the other hand, the ability to detect cues in the environment, the threshold level of cue that elicits a phenotypic response, and the properties of the induced phenotypes are all features of plasticity that have a genetic basis and that typically exhibit heritable variation within populations (reviewed in Schlichting and Pigliucci 1998; Windig et al. 2004; Schlichting 2008). Therefore plasticity, like any other trait that shows heritable variation, can respond to selection by undergoing evolutionary change (West-Eberhard 2003; Pigliucci 2010).

The above two proximate mechanisms are best thought of as occupying opposite ends along a continuum of environmental influences on trait production (Figure 4.1), along which most (if not all) traits lie in the middle (Gilbert and Epel 2009). Moreover, a trait's position along this continuum can evolve. For instance, as first demonstrated by Waddington (1953), a trait that is initially highly dependent on environmental input may evolve reduced environmental sensitivity. As we explain below, such evolutionary shifts in the degree to which a trait's expression is sensitive to the environment may be crucial in the evolution of character displacement.

Before developing this idea, however, we first describe how either genetically canalized shifts or phenotypic plasticity, *acting alone*, can promote character displacement.

GENETICALLY CANALIZED DIVERGENCE

As noted in the Introduction, genetically canalized changes are the oft-assumed mechanism of character displacement. Consequently, substantial effort has gone into identifying genes that influence the expression of traits that promote species-specific differences in resource use or reproduction (reviewed in Nosil and Schluter 2011). Recent studies have implicated four different genetic mechanisms as playing a role in character displacement.

The simplest such mechanism is a single-allele mechanism (Felsenstein 1981; reviewed in Servedio and Noor 2003; Coyne and Orr 2004). With such a mechanism, divergence between interacting species arises when they share the *same* allele that enhances existing differences between species. By heightening existing differences between species, such an allele could reduce species interactions and thereby promote character displacement. For example, if an allele enhances sensory sensitivity to male sexual signals, females may become better at identifying conspecific males and thereby become less likely to mate with heterospecifics. The selectively favored spread of such an allele that reduces heterospecific matings constitutes reproductive character displacement.

Single-allele mechanisms may also mediate ecological character displacement. Indeed, in the same way that heightened sensory sensitivity may hone the tendency to mate with conspecifics, enhanced sensitivity may also refine pre-existing food preferences within each species and thereby reduce overlap between species in resource use. Moreover, for both resource competition and reproductive interactions, a single allele may reduce dispersal tendencies; this reduction could exaggerate differences between species in habitat use and thereby preclude interactions between them.

Single-allele mechanisms have been primarily posited in theory (Servedio and Noor 2003; Coyne and Orr 2004), but they have also received some empirical support. In the fruit flies *Drosophila pseudoobscura* and *D. persimilis*, a single allele potentially affects mate discrimination and thereby minimizes hybridization between these two species (Ortíz-Barrientos and Noor 2005). In particular, females of both species from sympatric populations show enhanced discrimination against heterospecifics compared with females from populations in allopatry (Noor 1995). This ability to discriminate has been traced, at least in part, to a single allele at the *Coy-2* chromosomal region that enhances discrimination against heterospecifics (likely via olfactory cues) in *both* species (Ortíz-Barrientos and Noor 2005; see also Ortíz-Barrientos et al. 2004). Although discrimination in natural populations probably stems from additional mechanisms (Ortíz-Barrientos et al. 2004; Barnwell and Noor 2008), these flies demonstrate that interactions between species can be reduced—and assortative mating within species enhanced—by the action of a single allele shared by both species.

A second, slightly more complex mechanism—but one that still involves a single locus—entails the evolution of *alternative* alleles in different species at a single locus that specifies divergent resource-use or reproductive traits. Such allelic differentiation at one locus may suffice to reduce competitive or reproductive interactions between species. A possible example comes from passion-vine butterflies (genus *Heliconius*), which inhabit the Neotropics and show great diversity in wing color patterns. Members of this genus have undergone rapid speciation (Brower 1996), but they also show considerable convergence in wing pattern owing to Müllerian mimicry (Turner 1981; Mallet and Gilbert 1995). Specifically, these butterflies are distasteful to predators, and they advertise this noxiousness with bright wing coloration (aposematism). Where sympatric, different species often converge on the same wing color pattern, which can be adaptive if this convergence reinforces the aposematic signal among potential predators. However, by converging on the same wing color pattern, species that use wing coloration in mate choice increase the risk of mistakenly mating with the wrong species (Estrada and Jiggins 2008). Thus, as a means of minimizing such mistakes, reproductive character displacement is expected to occur.

Male *H. cydno* and *H. pachinus* use wing color to discriminate conspecific from heterospecific mates (Kronforst et al. 2006). Interestingly, the genetic loci that influence both mate preference and wing coloration have been mapped to the same chromosomal region. Although this may reflect two separate loci that are tightly linked, it could also indicate that both mate preference and wing coloration are influenced by the same gene (Kronforst et al. 2006). In these butterflies, pigments involved in wing coloration are also present in the eye and thereby affect perception of wing coloration (Kronforst et al. 2006 and references therein). Thus a single gene could have pleiotropic effects by influencing both the perception of, and the preference for, that coloration (Chamberlain et al. 2009). In this way, species that possess alternative alleles at a single locus can diverge in traits such as sexual signaling and mate choice that mediate character displacement. If that locus has pleiotropic effects, as in *Heliconius* butterflies, then divergence in multiple traits can arise simultaneously.

In contrast to the two mechanisms outlined above that depend either on a single allele or on alternative alleles at a single locus (with or without pleiotropy), a third mechanism of character displacement involves multiple, divergent loci. With such a mechanism, the traits that serve as the targets of competitively mediated selection are under the direct control of two or more loci, and character displacement occurs when species diverge in these loci. For example, in areas where they co-occur, pied flycatcher birds (*Ficedula hypoleuca*) and collared flycatchers (*F. albicollis*) have undergone reproductive character displacement (specifically, reinforcement) in female preferences and male coloration (Sætre et al. 1997). This divergence in reproductive traits appears to be mediated by multiple, divergent loci (Backstrom et al. 2010). Another example comes from stickleback fishes (*Gasterosteus aculeatus* complex), in which morphological and ecological divergence between sympatric benthic and limnetic species has evolved through character displacement (Schluter and McPhail 1992). That F_1 hybrids between these two species are intermediate in morphology suggests that this divergence involves numerous loci (McPhail 1984, 1992; Hatfield 1997; Hatfield and Schluter 1999).

In the three mechanisms of character displacement described thus far, character displacement would likely involve changes in structural genes (such as regions of the genome encoding enzymes or cellular components). However, differences in structural genes are not the only mechanism that could cause species to become differentiated. Instead, species may diverge through changes in regulatory genes (such as regions of the genome encoding transcription factors or signaling molecules that affect expression of other genes; King and Wilson 1975; Raff and Kaufman 1983; Carroll et al. 2001; Wilkins 2002; Wittkopp et al. 2004; Wray 2007, 2010).

A possible example comes from Darwin's finches. The Galápagos archipelago hosts 14 endemic species of these finches (Grant and Grant 2008), all of which are closely related to each other, and all of which are similar to one another morphologically, except for the size and shape of their beaks (Grant 1986; Grant and Grant 2008). The finch's beak is its chief resource-acquisition trait (Grant 1986; Grant and Grant 2008). Moreover, the size and shape of the beak can also affect the production of male song, which is used in territorial defense and mate attraction (reviewed in Podos and Nowicki 2004). Where they occur together, different species of finches often show exaggerated differences in beak morphology (see, for example, Figure 1.6), and, in those cases, both ecological and reproductive character displacement are thought to contribute to species-specific differences in beak morphology (Lack 1947; Grant and Grant 2008).

Recent research has begun to reveal how beaks form developmentally, and these studies point to the possible genetic targets of divergent selection in this system (Figure 4.2). Beak development is influenced by several genes that encode a series of signaling molecules. Two such gene products, fibroblast growth factor 8 (*Fgf8*) and sonic hedgehog (*Shh*), induce the expression of another signaling molecule, bone morphogenetic protein 4 (*Bmp4*). *Bmp4* is important in the origin of differences in beak shape between species (Abzhanov et al. 2004, 2006). For example, *Bmp4* gene expression occurs earlier in the large ground finch, *G. magnirostris*, than in other species, resulting in a beak that is deep and broad (Abzhanov et al. 2006). The above two gene products—*Fgf8* and *Shh*—also affect the expression of the gene *calmodulin* (*CaM*), which encodes for a calcium-binding protein involved in apoptosis (programmed cell death) that thereby influences beak length (Abzhanov et al. 2006).

Any one of these genes (Figure 4.2) could potentially serve as a target of selection during character displacement. However, because these same genes also mediate several crucial metabolic processes (in addition to influencing the shape and size of the beak), it is unlikely that selection would favor different alleles of these genes in different

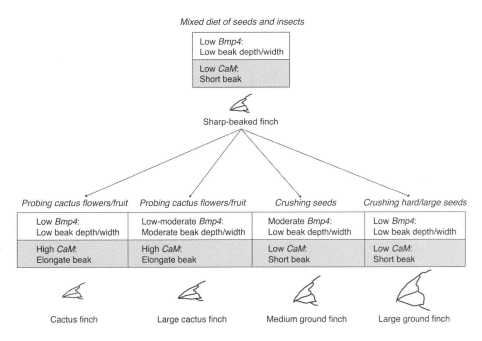

FIGURE 4.2.

Hypothesized effects of two genes, *Bmp4* and *CaM*, on beak development in Darwin's finches. Species-specific differences in beak morphology may have arisen when competitively mediated selection promoted changes in the regulation of these two genes. Based on Abzhanov et al. (2006).

species and populations (Grant and Grant 2008). Instead, character displacement has likely occurred via selection fostering changes in the *regulation* of these genes (Grant and Grant 2008).

Regulatory mutations have also been implicated in mediating divergence of pheromones (cuticular hydrocarbons) used in mate choice in fruit fly species of the *Drosophila serrata* complex (McGraw et al. 2011). They have likewise been implicated in divergence of floral color in the Texas wildflower *Phlox drummondii* (Hopkins and Rausher 2011), which constitutes one of the best-documented instances of reproductive character displacement (specifically, reinforcement) in plants (Levin 1985). Indeed, the fixation of alternative alleles in upstream regulatory genes could represent a common means by which character displacement occurs. Although the contribution of regulatory mutations to adaptive evolution has been controversial (Hoekstra and Coyne 2007; but see Razeto-Barry and Maldonado 2011; Stern 2011), increasing evidence suggests that they may play a key role in mediating adaptive population divergence (Wittkopp et al. 2004; Pavey et al. 2010; Wittkopp and Kalay 2012), including character displacement.

Before leaving this discussion of genetically canalized divergence, it is important to note that the four mechanisms described above may differ in the ease with which they can promote divergence between interacting species. Indeed, each mechanism has its likely advantages and disadvantages in mediating character displacement. Consider first that, relative to the other three mechanisms, single-allele mechanisms may have two key advantages. First, the emergence and spread of a single allele at a key gene may be achieved more readily than the simultaneous production of multiple alleles, especially when these alleles reside at multiple loci. Second, in systems in which hybridization occurs, gene flow between species is thought to be a key inhibitor to character displacement because species-specific traits that are favored in one species would be disfavored in the other (reviewed in Howard 1993; Servedio and Noor 2003; Coyne and Orr 2004). Single-allele mechanisms are not faced with this problem, however, because they are expressed and favored in the same way in both species (Servedio and Noor 2003; Coyne and Orr 2004).

Both single-allele mechanisms and single-locus, multiple-allele mechanisms likely have two additional advantages over multiple-locus mechanisms. First, a key challenge confronting multiple-locus mechanisms is to explain how such mechanisms could persist in the face of recombination, which should tend to break up co-adapted gene complexes that are responsible for species differences (Butlin 2005; Hoffmann and Rieseberg 2008; Nosil and Schluter 2011). This is particularly problematic in species that exchange genes, because (as noted in the previous chapter) hybridization could "scramble" allelic combinations at loci that isolate species. One possible solution to this apparent problem is that these loci may reside in areas that are protected from recombination (Noor et al. 2001; Butlin 2005; Hoffmann and Rieseberg 2008; Noor and Bennett 2009), including inside chromosomal inversions (Noor et al. 2001), near the chromosome's centromere (Noor and Bennett 2009), or on sex chromosomes (Sæther et al. 2007). However, even if this first problem is overcome, a second potential disadvantage for a multiple-locus mechanism is that genetic correlations among loci may impede character displacement (see chapter 3).

At the same time, both single-allele and single-locus, multiple-allele mechanisms themselves face a challenge not faced by either multiple-locus or regulatory-mutation mechanisms. Because character displacement often involves changes in complex suites of traits (see, for example, Smith and Rausher 2008; Martin and Pfennig 2011), it is difficult to imagine how a single locus could, by itself, mediate such shifts. In such cases, mechanisms involving either multiple loci or regulatory mutations (which can promote changes in diverse traits; Wray 2010) may be expected to mediate character displacement. Yet, at present, it is unclear how commonly character displacement arises through a single locus with large effect versus many loci with small effects (Wolf et al. 2010).

To summarize this section, investigations into the genetic targets of competitively mediated selection have uncovered a diversity of genetic mechanisms. Moreover, theory suggests that these mechanisms likely differ in the ease with which they promote character displacement (Servedio and Noor 2003). Although putative examples exist for each mechanism, additional work across a wide variety of species is needed to ascertain whether some genetic mechanisms are more likely than others to underpin character displacement, as predicted by theory. In the next section, we consider an alternative mechanism that may be particularly effective at mediating character displacement.

ENVIRONMENTALLY INDUCED DIVERGENCE

In contrast to reflecting genetically canalized differences—that is, rather than being expressed constitutively—divergent traits may be environmentally induced, such that they arise developmentally through phenotypic plasticity. Specifically, divergent traits may be produced in any given individual *only when they are needed*; that is, only when that individual experiences competition from a heterospecific (for examples in which individuals reduce resource competition with heterospecifics through phenotypic plasticity, see Table 2.1; for an example in which individuals reduce costly *reproductive* interactions with heterospecifics through phenotypic plasticity, see Pfennig 2007). Although environmentally induced changes have not generally been considered in the theory of character displacement (except in models involving learning and cultural transmission; see, for example, Servedio et al. 2009), environmentally contingent niche shifts may play an under-appreciated role in mediating character displacement. Below we describe two ways that plasticity may promote character displacement. First, however, we discuss an important caveat.

Recall from Box 2.2 that competitively mediated plasticity may or may not constitute character displacement. "Character displacement" is defined as trait evolution that arises as an adaptive response to resource competition or deleterious reproductive interactions between species (chapter 1). Therefore, competitively mediated plasticity constitutes character displacement only when it has actually evolved in direct response to competitively mediated selection.

Even when environmentally induced traits have not themselves undergone character displacement, they can still play a key role in ultimately facilitating character displacement. In particular, species (or populations) that respond to heterospecific competitors through phenotypic plasticity (whether or not it is an evolved response to heterospecifics per se) may be less likely to become extinct through competitive or reproductive exclusion. This is because plasticity may enable individuals to produce resource-use or reproductive phenotypes that are less like the phenotypes expressed by their competitors, thereby reducing the frequency and intensity of competitive interactions. Consequently, plasticity may increase population persistence. Indeed, plasticity has long been viewed as critical in shielding populations from extinction when confronted with rapidly changing environmental circumstances (Baldwin 1896; Morgan 1896; Robinson and Dukas 1999; Price et al. 2003; Richards et al. 2006; Ghalambor et al. 2007; Pfennig and McGee 2010; Chevin et al. 2010; Davidson et al. 2011), including exposure to a new heterospecific competitor (Agrawal 2001; Fordyce 2006). Thus, even when environmentally induced niche shifts do not actually constitute character displacement, they may still promote resource partitioning or reproductive partitioning, and thereby foster species coexistence. Essentially, such environmentally initiated niche shifts may act as a mechanism of species sorting.

A number of studies have demonstrated, however, that environmentally induced shifts do indeed constitute character displacement (Robinson and Wilson 1994; Pfennig and Murphy 2002; Pfennig 2007; Pfennig and Martin 2009; Grant and Grant 2010). These studies have revealed that plasticity may mediate character displacement through two general mechanisms.

First, plasticity may underlie character displacement when competitively mediated selection promotes the evolution of reaction norms as an adaptive response to interspecific competition (see Box 2.2). Such selection may act on heritable variation in either the tendency to respond to competitors or the manner in which individuals express these responses and thereby propel competitively mediated trait *evolution* (that is, character displacement). Indeed, the genetic regulatory architecture of phenotypic plasticity potentially provides numerous targets on which competitively mediated selection can act (Figure 4.3). Furthermore, experiments have demonstrated that these evolved shifts in the expression of plasticity can satisfy the widely accepted criteria for character displacement (Pfennig and Murphy 2000; for criteria see Table 1.2).

A second general mechanism by which plasticity may promote character displacement involves instances in which an environmentally induced trait is transmitted reliably *across* generations, even in the absence of genetic specification of that trait. Such inherited environmental effects potentially form the basis of an alternative inheritance system on which adaptive evolution can unfold. For example, in many species of animals, learning can influence both an individual's choice of mate (see, for example, Grant and Grant 2008, pp. 79–80; Grant and Grant 2010; Kozak et al. 2011) and its preferred food/habitat (see, for example, Papaj and Prokopy 1989; Price 2008, pp. 129–135). Once a population acquires such learned mate or food preferences, these preferences can be transmitted across generations and even reinforce differences between species, thereby mediating ecological or reproductive character displacement (see, for example, Price 2008, pp. 293–296). Such a situation is illustrated in Darwin's finches, where learning appears to have played a crucial role in mediating reproductive character displacement (Grant and Grant 2010).

Additionally, acquired information or materials can be transmitted via parental effects. Parental effects occur when the phenotype of an organism is determined not only by its genotype and the environment that it experiences, but also by the environment and phenotype of its parents, most commonly those of its mother (in which case, the parental effect is a "maternal effect"; Mousseau and Fox 1998). For example, mothers may differentially endow their eggs or seeds with resources (such as lipids), hormones, or RNA transcripts (Donohue and Schmitt 1998; Mousseau and Fox 1998; Räsänen and Kruuk 2007). Because maternal effects can be triggered by interspecific competition (Allen et al. 2008; Pfennig and Martin 2009), can mediate adaptive phenotypic change

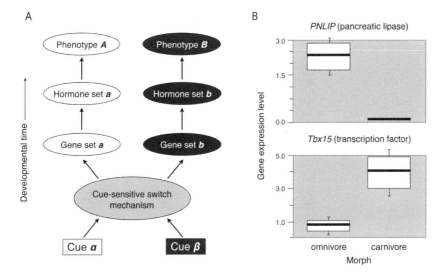

FIGURE 4.3.
The genetic regulatory architecture of phenotypic plasticity potentially provides numerous targets on which competitively mediated selection can act to promote character displacement. (A) In certain forms of plasticity, environmentally induced phenotypes are triggered when specific environmental cues are detected by a cue-sensitive mechanism (for example, a sensory system). Depending upon the cue, this system activates alternative sets of genes and hormonal pathways, which regulate the expression of alternative phenotypes (each pathway is shown in a different color). In species that can respond to heterospecific competitors through phenotypic plasticity, each element inside the ovals could serve as a target of selection. (B) For example, genes that are differentially expressed in alternative, environmentally induced morphs could serve as targets of competitively mediated selection. Shown here are different expression patterns for two genes (*PNLIP* and *Tbx15*) in alternative resource-use morphs of spadefoot toad tadpoles, *Spea bombifrons*. Note that, relative to its expression in the alternative morph, each gene is up-regulated in one morph (that is, the gene product is increased in quantity). Panel A based on West-Eberhard (1992); data in panel B from Leichty et al. (2012).

(Badyaev et al. 2002; Galloway and Etterson 2007), and can be transmitted reliably across generations (Plaistow et al. 2006), maternal effects may play an under-appreciated role in driving character displacement (Pfennig and Martin 2009; Figure 4.4).

In sum, although genetically canalized changes are frequently assumed to be the sole mechanism of character displacement, character displacement can also be mediated by environmentally initiated phenotype shifts; that is, phenotypic plasticity. Additional empirical and theoretical work is needed, however, to determine whether and how each of the above mechanisms contributes to character displacement.

For the remainder of the chapter, we consider some ways in which plasticity may play an important role in character displacement. In particular, we contrast plasticity

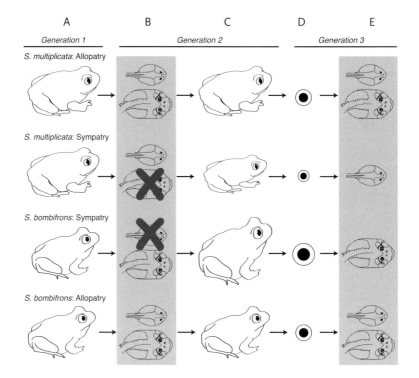

FIGURE 4.4.
A diagram summarizing how an environmentally triggered maternal effect may promote transgenerational plasticity that can mediate character displacement. This particular example is based on spadefoot toads, (A) two species of which (*Spea multiplicata* and *S. bombifrons*) occur in allopatry and sympatry (middle rows) with each other. (B) In allopatric populations, both species produce alternative resource-use morphs: an omnivore morph and a larger, carnivore morph, which is induced by shrimp ingestion. When the two species first come into contact, phenotypic plasticity causes them to diverge in morph production, such that one species (*S. multiplicata*) produces mostly omnivores, whereas the other species (*S. bombifrons*) produces mostly carnivores (in each species, the alternative morph is seldom expressed, as indicated by the gray X). (C) Because *S. multiplicata* produce both morphs in allopatry, but only the smaller omnivore morph in sympatry (*S. multiplicata* are the poorer competitor for shrimp), *S. multiplicata* females from sympatry are smaller and in poorer condition when they mature than *S. multiplicata* females from allopatry. By contrast, because they produce both morphs in allopatry but only the larger carnivore morph in sympatry, *S. bombifrons* females from sympatry mature larger and in better condition than *S. bombifrons* females from allopatry (the carnivore morph is larger because it is able to monopolize the more nutritious shrimp resource). (D) Consequently, in sympatry, the two species diverge in maternal investment: *S. multiplicata* females invest less into offspring by producing smaller eggs, whereas *S. bombifrons* females invest more into offspring by producing larger eggs. (E) Smaller eggs hatch into smaller tadpoles, which tend to become omnivores. By contrast, larger eggs hatch into larger tadpoles, which tend to become carnivores. Thus, by the third generation after encountering the other species, each species may be "epigenetically" canalized to produce an alternative morph. Eventually, the two species may accumulate genetic differences such that sympatric populations of each species become genetically fixed for producing a single morph. Redrawn from Pfennig and Martin (2009) based on data in Pfennig and Martin (2009), Martin and Pfennig (2010b), and Pfennig and Martin (2010).

and genetic canalization in terms of their effects on the rapidity with which character displacement can occur. Although the two mechanisms are not mutually exclusive, they differ in important ways that can influence when and how character displacement occurs.

TEMPO AND MODE OF CHARACTER DISPLACEMENT

Having described how character displacement can occur through either genetically canalized differences or phenotypic plasticity, we now consider how each mechanism affects character displacement's tempo and mode. The speed of divergence is critical, because character displacement is a time-limited process. If character displacement transpires too slowly, a population risks becoming extinct from competitive or reproductive exclusion. Because phenotypic plasticity can promote rapid, widespread, and adaptive divergence, it may be crucial in mediating character displacement. However, genetically canalized differences and phenotypic plasticity are not mutually exclusive mechanisms of character displacement. Below, we consider how character displacement may also proceed through an initial phase in which trait divergence is environmentally induced to a later phase in which divergence becomes genetically canalized in different populations and species.

HOW MECHANISMS DIFFER IN SPEED OF DIVERGENCE

Recall from chapter 3 that the speed of character displacement, like any other form of adaptive evolution, depends partly on the amount of standing variation in the trait(s) under selection (Milligan 1985; Taper and Case 1985; Doebeli 1996; Rice and Pfennig 2007; Barrett and Schluter 2008) or, when standing variation is depleted/initially absent, on the speed with which new variants are introduced into the population. If standing variation is exhausted or absent from the start, and if divergence depends entirely on genetically canalized differences, then new variants must be introduced into the population via mutation, recombination, or gene flow/introgression. Yet, as described in chapter 3, all three processes are likely to either transpire relatively slowly or, ironically, simultaneously counteract character displacement even as they provide the raw material necessary for character displacement to proceed.

This is where phenotypic plasticity potentially becomes crucial in facilitating character displacement. Divergence driven by phenotypic plasticity may promote rapid divergent evolution for at least three reasons. First, because individuals within the same population often harbor genetic variation in the degree to which they respond to environmental cues (see Box 2.2), the expression of environmentally induced phenotypes can reveal "cryptic" genetic variation—genetic variation that is not normally expressed as phenotypic variation (Gibson and Dworkin 2004). Indeed, such hidden reaction norms may represent an evolutionarily significant pool of cryptic genetic variation on which competitively mediated selection can act when it becomes revealed (Schlichting 2008). Therefore, as soon as a population experiences competition, abundant standing variation can be released and exposed to selection, thereby fueling competitively mediated trait evolution (see, for example, Ledón-Rettig et al. 2010). In essence, the time between when a population begins to experience competition and when it begins to express variation on which selection can act to produce an adaptive response is negligible. Rather than being produced over generations (as with genetically canalized traits), new variants that are triggered by the environment are produced over developmental time; that is, over a single individual's lifetime.

Second, environmentally induced changes typically occur in numerous individuals simultaneously. Such widespread recurrence of environmentally initiated change contrasts markedly with the production of new genetic variants, which typically arise in only a few, or even just one, individual (West-Eberhard 2003). Moreover, because (as noted above) these individuals may harbor genetic variation in the form of hidden reaction norms, substantial variation that increases the likelihood and speed of an evolutionary response may be present.

Finally, a third reason why phenotypic plasticity may promote rapid character displacement is that the ability to respond to competition adaptively through induced niche shifts may already exist in many populations, having previously evolved in response to *intra*specific competition. As described in the next chapter, many species respond to intraspecific competition through facultative shifts in traits associated with resource use or reproduction. The presence of such pre-existing plasticity (and cryptic genetic variation) is important because selection can repurpose the underlying genetic pathways and developmental mechanisms and thereby drive adaptive divergence between heterospecific competitors; that is, character displacement. Indeed, as we describe below, empirical evidence exists to suggest that intraspecific variation generated by plasticity may form the basis for interspecific variation during character displacement. Essentially, pre-existing plasticity (and the underlying developmental mechanisms that mediate such plasticity) may enable character displacement to evolve rapidly along the lines of least resistance (*sensu* Schluter 1996).

THE PLASTICITY-FIRST HYPOTHESIS

Above, we suggested that—because phenotypic plasticity can generate rapid, widespread, and adaptive changes in resource-use or reproductive phenotypes—phenotypic plasticity might play a key role in character displacement. But phenotypic plasticity may also contribute to the evolution of genetically canalized differences that mediate character displacement, such as those described above (see "Genetically Canalized Divergence"). Indeed, character displacement may often evolve from an initial phase in which trait divergence is environmentally induced to one in which such divergence is expressed constitutively. As Wilson (1992) put it:

Imagine a case in which two such species have been squeezed together in the same communities long enough for evolution to occur. When they first came into contact, they were elastic and could diverge in their habits enough to lessen competition. The differences were phenotypic, the result of environment and not genes. The compression occurred in traits that were relatively easy to change, most likely by a retreat from parts of the habitat and diet by one or both of the species. As the generations passed, genetic differences arose and hardened the distinction between the two species. (Wilson 1992, p. 174)

This plasticity-first hypothesis is, of course, not the only way in which character displacement might evolve—plasticity might not play any role in some species or populations, whereas in other species, it might play a solitary role by mitigating against selection for any further genetically canalized divergence (Price et al. 2003; Schlichting 2004). Nevertheless, character displacement may generally be mediated *initially* by plasticity, followed by "genetic hardening," for two reasons. First, those populations that initially express plasticity may be more likely to be buffered from extinction while genetically canalized traits evolve that harden the distinction between species (see above). When faced with a new, superior competitor, populations lacking the ability to respond through plasticity may simply undergo reproductive or competitive exclusion before canalized differences evolve (Pijanowska et al. 2007). Second, even in populations where plasticity successfully minimizes competitive interactions with heterospecifics, genetic canalization of such traits may be favored if there are costs to plasticity per se (Price et al. 2003; for discussions of costs of plasticity, see DeWitt et al. 1998; Relyea 2002; Sultan and Spencer 2002; Auld et al. 2010). In other words, plastic traits may evolve more readily (and therefore initially) for the reasons outlined in the previous section, but plasticity may ultimately be replaced with genetic canalization.

The plasticity-first hypothesis rests on the long-standing observation that an induced phenotype can lose its environmental sensitivity over evolutionary time and eventually become constitutively produced, such that it no longer requires the original environmental stimulus to produce it (Baldwin 1896, 1902; Morgan 1896; Waddington 1953, 1957). Waddington (1953) coined the term "genetic assimilation" to refer to this process by which a plastic trait becomes subsumed into the genome as a genetically canalized trait (see recent reviews by West-Eberhard 1989, 2003, 2005; Pigliucci and Murren 2003; Lande 2009; Moczek et al. 2011).

The key to understanding how genetic assimilation comes about is the concept of interchangeability (Whiteman and Agrawal 2009). As noted above, most traits have both genetic and environmental influences, and these two influences are potentially evolutionarily interchangeable, meaning that selection can slide trait regulation anywhere along the continuum from total environmental control to total genetic control (Figure 4.1). Specifically, when genetic variation for the degree of environmental influence is present, selection can act on this variation to promote the evolution of either increased or decreased environmental sensitivity (West-Eberhard 2003). If selection favors the elimination of all environmental influences (that is, if genetic assimilation occurs), the end result is a genetically canalized trait. (As an aside, plasticity can also be lost through *stochastic* processes. Specifically, when a change in the environment results in one trait rarely being produced, then the underlying alleles regulating expression of this "hidden" phenotype [for example, see differentially expressed genes in Figure 4.3B] would not be exposed to selection and would therefore be at greater risk of chance loss; see Masel et al. 2007; Lahti et al. 2009.)

With genetic assimilation, the origin of a new, canalized trait does not require new genes; instead, selection can act on existing genetic architecture and epigenetic interactions (Schlichting and Pigliucci 1998; see, for example, Emlen et al. 2007; Ledón-Rettig et al. 2008; Aubret and Shine 2009). In other words, a plastic trait can be converted to a canalized trait through evolutionary adjustments in *the regulation of trait expression.* Experiments have demonstrated such evolutionary shifts (Suzuki and Nijhout 2006), including the complete loss of plasticity (Waddington 1953), and numerous examples of evolution by natural selection may reflect such genetic assimilation (Matsuda 1987; Hall 1999; Pigliucci and Murren 2003; West-Eberhard 2003; Moczek et al. 2011).

Genetic assimilation may be a relatively common mechanism of character displacement (Pfennig et al. 2010). As indicated by Wilson's quote above, two interacting species may initially respond to each other's presence adaptively through the evolution of facultative adjustments in resource-use or reproductive phenotypes (Figure 4.5A). Over time, as each species facultatively expresses a different subset of the initial phenotypes (Figure 4.5B), the two species may ultimately lose this plasticity and may instead become fixed for a different alternative phenotype (as genetic assimilation promotes fixation of alternative alleles in each species; Figure 4.5C; for empirical examples, see Schwander and Leimar 2011). Thus, under persistent selection for traits that minimize competition or reproductive interactions with heterospecifics, distinctive traits that are initially environmentally induced may eventually become genetically canalized, and thereby mediate character displacement.

This plasticity-first hypothesis has two important ramifications for the ways that character displacement occurs. First, character displacement may not always transpire slowly. Although most theoretical models predict that character displacement should proceed slowly, especially in the early stages of divergence (see, for example, Taper and Case 1985; Doebeli 1996), empirical studies have demonstrated that it is possible for divergence to be rapid—that is, occurring within a few to dozens of generations; see, for example, Fenchel 1975; Diamond et al. 1989; Yom-Tov et al. 1999; Pfennig and Murphy 2003; Pfennig 2003; Hoskin et al. 2005; Grant and Grant 2006). Although such rapid evolutionary responses can occur via shifts in genetically canalized traits—specifically, when selection is strong and standing genetic variation is abundant—the incorporation of plasticity into models of character displacement could increase the chances that character displacement can occur when non-cryptic genetic variation is exhausted or absent from the start.

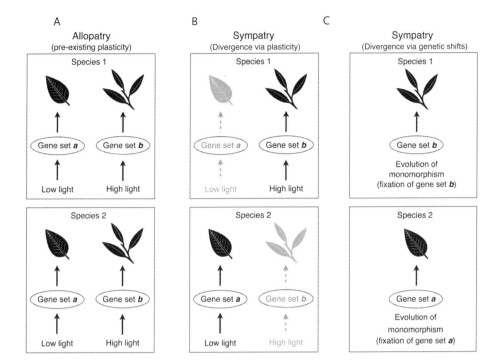

FIGURE 4.5.
Character displacement may evolve from an initial phase in which trait divergence is environmentally induced to a later phase in which divergence is expressed constitutively. (A) Initially, two interacting species may express plasticity in resource-use or reproductive phenotypes. In this diagram, two species of plants facultatively produce different-sized leaves as an adaptive response to different light levels. (B) When they come into sympatry and compete (for example, for light), each species may begin to facultatively express a different subset of the initial phenotypes. Here, species 1 overtops species 2, thereby gaining increased access to light, which triggers the facultative production in species 1 of small leaves only. By contrast, because species 1 shades it, species 2 facultatively produces larger leaves only (in each case, the expressed phenotype—and underlying genes—are shown in black, whereas the unexpressed ones are shown in gray). If individuals within the same population harbor genetic variation in the degree to which they respond to environmental cues, and if these reaction norms evolve in response to competitively mediated selection and thereby minimize competition between species, then such evolved environmentally induced shifts would constitute character displacement. (C) Over time, both species may lose this pre-existing plasticity and become fixed for a different alternative phenotype, possibly because of the loss (through selection or chance) of alleles or gene combinations underlying the non-expressed phenotype. Thus, character displacement may proceed through an initial phase in which trait divergence is environmentally induced to a later phase in which divergence becomes genetically canalized. Essentially, during character displacement, each species may evolve from expressing a wide range of phenotypes to becoming genetically canalized for a narrower range of phenotypes (in this case, as each species evolves from being polymorphic for leaf shape to being monomorphic). Reproduced from Pfennig and Pfennig (2012), with the kind permission of the publisher.

Second, character displacement that is initiated by plasticity and followed by genetic assimilation should generate repeated evolution of the same phenotype that minimizes interspecific competition in multiple, independently evolving sympatric populations. Such parallel character displacement has been documented in several systems (Schluter and Nagel 1995; Hansen et al. 2000; Marko 2005; Matocq and Murphy 2007; Rice et al. 2009; Adams 2010). Of course, the plasticity-first hypothesis does not uniquely predict parallel character displacement. Instead, standing variation in genetically canalized traits could respond similarly to the same selective pressures across different populations (Schluter and Nagel 1995). However, if stochastic processes (mutation, recombination) produce the variation in genetically canalized traits that mediate character displacement, then parallel evolution becomes less likely. Under such circumstances, populations would be expected to evolve *different* ways of responding to competitors. In contrast, with pre-existing plasticity, the same sets of phenotypes are repeatedly revealed when individuals in different, independent populations experience similar selective pressures from competition. Generally, the environment may play a key role not only in exerting parallel selection pressures in different populations, but also in *generating* parallel distributions of traits on which selection acts (West-Eberhard 2003, 2005).

EMPIRICAL TESTS OF THE PLASTICITY-FIRST HYPOTHESIS

If the plasticity-first model of character displacement is correct, then plasticity-mediated shifts in ancestral populations that have not undergone character displacement should resemble the constitutively expressed trait differences observed in derived populations that have undergone character displacement. Thus, testing this prediction requires that one characterize ancestral reaction norms in present-day pre-displacement (for example, allopatric) populations. As we describe below, in several systems, the available data suggest that competitively induced plasticity may have preceded the evolution of canalized genetic differences.

The first such example comes from stickleback fishes of the *Gasterosteus aculeatus* complex. Recall from chapter 3 that certain small coastal lakes in southwestern Canada harbor two closely related species, one of which expresses a distinctive benthic ecomorph and the other of which expresses a limnetic ecomorph (Schluter and McPhail 1992; Figure 4.6A). Each coexisting benthic and limnetic species pair is thought to have arisen from two separate invasions by marine sticklebacks into lakes at the end of the Pleistocene period, 13,000 years ago (Taylor et al. 1997). Presumably, the first such invader evolved an intermediate phenotype, similar to that expressed by sticklebacks in modern-day single-species lakes (Schluter and McPhail 1992). It is thought that, when a subsequent such invasion took place, ecological and reproductive character displacement increased the ecological and phenotypic divergence between the two sets of invaders such that the first invaders evolved a new benthic ecomorph and the second invaders

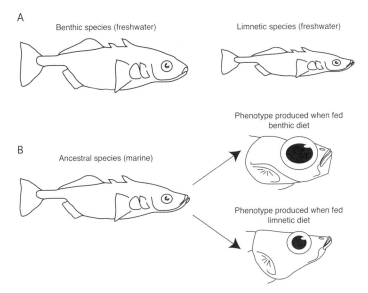

FIGURE 4.6.
Phenotypic plasticity may have mediated character displacement between different species of three-spined sticklebacks (*Gasterosteus aculeatus* complex). (A) In certain lakes, two species have undergone character displacement, resulting in one species expressing a robust benthic ecomorph (left), which feeds on invertebrates on lake margins, and another expressing a thinner limnetic ecomorph (right), which feeds on plankton in open water. Breeding experiments have shown that these phenotypic differences are genetically canalized. (B) When marine sticklebacks (representing the ancestral, pre-displacement population) are reared either on a benthic or on a limnetic diet, however, the resulting environmentally induced changes in head and mouth shape resemble (in direction, but not magnitude) the phenotypic divergence between the derived ecomorphs shown in panel A (in these drawings, shape differences between diet treatments are exaggerated × 4 to highlight the effects of diet on morphology). Drawings in panel B based on Wund et al. (2008).

evolved a limnetic ecomorph, similar to the phenotype of its marine ancestors (Rundle and Schluter 2004 and references therein).

The early stages of this divergence may have been facilitated by phenotypic plasticity. Experiments (Wund et al. 2008) have revealed that diet-induced plasticity is present in modern-day marine sticklebacks, which serve as proxies for the ancestral, pre-displacement populations (see above). More importantly, this diet-induced plasticity mirrors the benthic and limnetic ecomorphs found in modern-day, post-displacement (freshwater) populations (Wund et al. 2008; Figure 4.6B). However, as noted above, other studies have found that phenotypic differences between benthic and limnetic ecomorphs reflect genetically canalized differences.

Therefore, character displacement in sticklebacks may have unfolded from an initial phase in which trait divergence was environmentally induced to one in which divergence became genetically canalized. Moreover, because the ancestral plasticity potentially influences a trait—body size—that is involved in both resource competition and reproductive interactions between species (McKinnon et al. 2004; Rundle and Schluter 2004), plasticity may have facilitated both ecological and reproductive character displacement. More generally, ancestral plasticity in resource-use traits has been detected in numerous species of freshwater fishes that have also undergone character displacement (Robinson and Wilson 1994; Skúlason et al. 1999), suggesting that many species of fishes may fit the plasticity-first scenario.

A second example that appears to conform to the plasticity-first model of character displacement comes from spadefoot toads. In southeastern Arizona, two species (*Spea multiplicata* and *S. bombifrons*) have undergone ecological character displacement (Pfennig and Murphy 2000, 2002, 2003; Pfennig et al. 2006, 2007; Rice et al. 2009; see also Figure 4.4). Specifically, in allopatry, both species produce similar, intermediate frequencies of two, environmentally induced, larval ecomorphs: a large-headed carnivore morph, which preys on fairy shrimp; and a small-headed omnivore morph, which utilizes detritus. In these allopatric populations, carnivores are induced by the ingestion of shrimp (Pfennig 1990). In sympatry, by contrast, each species shows reduced expression of one of the two morphs—that is, they show increased canalization in morph production—with *S. multiplicata* producing mostly omnivores, and *S. bombifrons* producing mostly carnivores (Pfennig and Murphy 2000, 2002). By shifting from producing both morphs in allopatry to producing mostly one morph in sympatry, the two species minimize competition with each other for food (Pfennig and Murphy 2000, 2002; Pfennig et al. 2007; Rice et al. 2009).

This system provides an ideal opportunity for testing the plasticity-first model of character displacement. To see why, recall that the plasticity-first model predicts that plasticity-mediated shifts in ancestral (pre-displacement) populations will mirror the canalized trait differences observed in derived populations that have undergone character displacement. This prediction appears to be fulfilled in spadefoots. Specifically, when allopatric *S. multiplicata* are experimentally exposed to *S. bombifrons*, they facultatively produce mostly omnivores (Figure 1.7B), which is similar to the pattern of morph expression found among *S. multiplicata* in naturally occurring sympatric populations. Conversely, when allopatric *S. bombifrons* are experimentally exposed to *S. multiplicata*, they facultatively produce mostly carnivores (Figure 1.7B), which is similar to the pattern of morph expression found among *S. bombifrons* in sympatry—*S. bombifrons* produce more carnivores in the presence of *S. multiplicata* because *S. bombifrons* are more effective than *S. multiplicata* are at capturing and consuming shrimp, which is the environmental cue that induces the production of carnivores (Pfennig 1990). Moreover, phylogenetic analyses have established that the allopatric condition is ancestral and that the sympatric condition is derived (Rice et al. 2009). Thus, experimentally initiated niche shifts in ancestral populations mirror in direction and magnitude the more highly canalized niche shifts observed in derived populations, thereby supporting the plasticity-first model of character displacement.

Interestingly, the mechanism of canalization appears to differ for the two species. In *S. multiplicata*, the canalized differences in morph production between sympatric and nearby allopatric populations appear to reflect a condition-dependent maternal effect (Figure 4.4; Pfennig and Martin 2009). In *S. bombifrons*, the same phenotypic shifts appear to reflect genetically canalized differences (Pfennig and Martin 2010). Possibly, sympatric populations of these two species differ in the mechanism of canalization because they differ in the length of time that each has been in sympatry. Specifically, because *S. bombifrons* has been expanding its range and continually invading new habitat formerly occupied solely by *S. multiplicata*, populations of *S. bombifrons* on the front edge of this expansion have had long evolutionary contact with *S. multiplicata* (Rice and Pfennig 2008) and therefore have had more time to accumulate genetically canalized differences. By contrast, populations of *S. multiplicata* along this wave front in southeastern Arizona have only recently encountered *S. bombifrons*, and, thus, might not have had sufficient time to undergo *genetic* canalization. In short, sympatric populations of the two species appear to represent different stages of the character displacement process, with sympatric *S. multiplicata* representing an earlier stage than sympatric *S. bombifrons*.

Taken together, these data therefore suggest that character displacement in spadefoots may have evolved through the following sequence of events. First, when the two species initially came into contact, phenotypic plasticity mediated adaptive divergence between them in morph production (as observed when modern-day allopatric populations are experimentally combined; see Figure 1.7). Later, differences in morph production between species (and, within each species, between conspecific populations in allopatry versus in sympatry with the heterospecific) became more canalized developmentally— but not genetically (as in modern-day sympatric populations of *S. multiplicata*, where character displacement is mediated by a maternal effect; Pfennig and Martin 2009). Finally, these phenotypic differences become more canalized both developmentally and *genetically* (as in modern-day sympatric populations of *S. bombifrons*, where character displacement appears to be mediated by genetic shifts; Pfennig and Martin 2010).

The spadefoot system thus supports the plasticity-first hypothesis for character displacement. However, this system also illustrates another important point: canalized differences need not be strictly genetic. Instead, phenotypic differences may undergo "*epi*genetic assimilation," wherein trait expression becomes less sensitive to environmental influences because of inherited environmental effects, such as a maternal effect. Such a process may be an evolutionary precursor to genetic assimilation, a possibility that requires further theoretical and empirical evaluation.

A third possible example comes from *Anolis* lizards from Caribbean islands. On islands where they are sympatric, different species evolve into different ecomorphs, which partition their habitat in different ways, residing (for example) on tree trunks or on twigs (Figure 2.7). Ecomorphs also differ in body size and limb length, and these differences appear to have evolved via character displacement (Losos 2009, pp. 127–129). Recent experiments have revealed that at least two species (*Anolis carolinensis* and *A. sagrei*) are capable of facultatively altering their hind-limb length in response to being reared on broad versus narrow surfaces, possibly mimicking the variation in perch sites experienced by members of different ecomorph classes (Losos et al. 2000; Kolbe and Losos 2005). Although these induced differences are not as great as differences observed among different ecomorph classes (Losos 2009, p. 250), these data suggest that the early stages of character displacement may have been mediated by environmentally induced niche shifts (Kolbe and Losos 2005). Specifically, if, in ancestral populations, different species competed for perch sites such that each species was shunted onto a different type of perch, then induced differences in hind-limb length may have preceded the evolution of the canalized differences observed between different ecomorphs (Figure 2.7).

In addition to the above three examples, recent research has also revealed that beak development in certain birds (a trait that has been implicated in both ecological and reproductive character displacement) may be at least partially environmentally inducible (see, for example, Gil et al. 2008). In this way, ancestral plasticity may have even played a role in instigating the classic case of character displacement—divergence in beak morphology in Darwin's finches (see chapter 1)—as well as in other possible cases of character displacement in birds (see, for example, Reifová et al. 2011).

In sum, studies of character displacement in at least four diverse groups of vertebrates (fish, amphibians, reptiles, and birds) are consistent with the plasticity-first hypothesis. Nevertheless, more work is needed to determine how general this mechanism is for explaining character displacement. Specifically, additional theoretical and empirical work is needed to determine whether competitively mediated plasticity precedes and ultimately promotes the sort of genetically canalized traits described earlier in this chapter. Such studies promise to have implications not only for understanding character displacement, but also for understanding adaptive evolution more generally. Indeed, because many of the ideas that we have discussed in this chapter apply to other forms of trait evolution (such as that stemming from selection to avoid predation; see, for example, Scoville and Pfrender 2010), insights gleaned from studying the mechanisms of character displacement can contribute to a broader understanding of adaptive evolution.

SUMMARY

Despite character displacement's importance in the origins of diversity, relatively little is known of its underlying proximate mechanisms. Two key questions regarding these mechanisms are: (1) what is the source of the phenotypic variation that fuels character displacement? and (2) how do different sources of variation affect the tempo and mode of

BOX 4.1. Suggestions for Future Research

- Identify the proximate mechanisms of character displacement in diverse systems.
- Determine the role that regulatory mutations play in mediating character displacement (for a review of possible mechanisms, see Wittkopp et al. 2004; Wray 2007).
- Use theoretical and empirical approaches to determine whether different proximate mechanisms differ in the ease with which (and, hence, the likelihood that) character displacement occurs.
- Identify what role (if any) transgenerational plasticity (for example, maternal effects, learning) plays in the evolution of character displacement.
- Explore the conditions under which competitively induced plasticity impedes rather than promotes character displacement (see, for example, Price et al. 2003; Schlichting 2004).
- Use experimental or comparative studies to determine whether, when confronted with a novel competitor, populations (or species) consisting of more plastic genotypes undergo character displacement more rapidly (and more readily) than those consisting of less plastic genotypes. Experimental evolution studies with rapidly evolving organisms, such as microbes (see, for example, Tyerman et al. 2008; Bono et al., in press; reviewed in Kassen 2009) may be ideal for addressing this issue.
- Use selection studies (*sensu* Suzuki and Nijhout 2006) to determine whether competitively induced phenotypes can lose their environmental sensitivity over evolutionary time and thereby undergo genetic assimilation.
- Identify additional systems to evaluate the plasticity-first hypothesis for the evolution of character displacement and, ultimately, to determine how common this route is.

character displacement? Character displacement can be mediated either by genetically canalized changes or by environmentally induced shifts. Yet these two mechanisms of trait divergence likely cause differences in the tempo of character displacement. Specifically, character displacement mediated by phenotypic plasticity may generally occur more rapidly than character displacement mediated by genetically canalized changes. However, these two mechanisms are not mutually exclusive, and they likely often act together to determine character displacement's mode. In particular, character displacement may often evolve from an initial phase in which trait divergence is environmentally induced to one in which such divergence is expressed constitutively. Although this plasticity-first hypothesis has increasing empirical support, additional tests are needed. We list some key challenges for future research in Box 4.1.

FURTHER READING

Abzhanov, A., M. Protas, B. R. Grant, P. R. Grant, and C. J. Tabin. 2004. *Bmp4* and morphological variation of beaks in Darwin's finches. *Science* 305:1462–1465. This paper examines the developmental bases of differences in beak morphology in Darwin's finches.

Ortiz-Barrientos, D., B. A. Counterman, and M. A. F. Noor. 2004. The genetics of speciation by reinforcement. *PLoS Biology* 2:2256–2263. This paper reviews the genetic underpinnings of reproductive character displacement.

Pfennig, D. W., M. A. Wund, E. C. Snell-Rood, T. Cruickshank, C. D. Schlichting, and A. P. Moczek. 2010. Phenotypic plasticity's impacts on diversification and speciation. *Trends in Ecology and Evolution* 25:459–467. This paper reviews the various means by which phenotypic plasticity may foster diversification, including its possible role in mediating character displacement.

West-Eberhard, M. J. 2003. *Developmental plasticity and evolution*. Oxford University Press: Oxford, UK. A thought-provoking book that explores how phenotypic plasticity may play a central—and often decisive—role in adaptive evolution.

Wund, M. A., J. A. Baker, B. Clancy, J. L. Golub, and S. A. Foster. 2008. A test of the "flexible stem" model of evolution: Ancestral plasticity, genetic accommodation, and morphological divergence in the threespine stickleback radiation. *American Naturalist* 172:449–462. This paper presents experimental support for the plasticity-first hypothesis during the adaptive radiation of stickleback fish.

5

DIVERSITY AND NOVELTY WITHIN SPECIES

Most species exhibit a striking amount of phenotypic variation. Indeed, in some cases, trait variation between different members of the same species is as great as that normally seen between different species (see, for example, Figure 1.1B, C). Here we consider the role of competitively mediated selection in generating and maintaining such diversity within species.

In explaining this variation, we shift our focus in this chapter to intraspecific competition, which contrasts with previous chapters, where the focus was on *inter*specific competition. Compared with interspecific competition, intraspecific competition is probably more common and frequently stronger (Gurevitch et al. 1992; Dybzinski and Tilman 2009). It is likely more common because individuals generally encounter conspecifics more often than they encounter heterospecifics. Moreover, intraspecific competition is probably also frequently stronger than interspecific competition because conspecifics are typically more similar in resource-use requirements and reproductive strategies than are heterospecifics, thereby making interactants more evenly matched and competition between them intense.

In this chapter we describe how trait evolution stemming from competition within species is the intraspecific analog of character displacement between species. In particular, we describe how trait evolution can arise as an adaptive response to competition for resources or successful reproduction *among conspecifics* through a process known as "intraspecific character displacement" (*sensu* West-Eberhard 2003, p. 397; see also Dayan and Simberloff 2005).

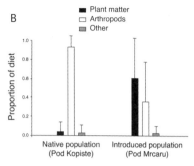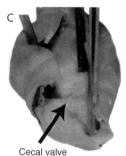

FIGURE 5.1.
Many novel features—especially those involved in resource or mate acquisition—may arise as an adaptive response to intraspecific competition. (A) A possible example comes from populations of Italian wall lizards, *Podarcis sicula*, on a pair of small islands in the middle of the South Adriatic Sea that belong to Croatia. Thirty-six years after being introduced from their native island of Pod Kopiste to the neighboring island of Pod Mrcaru (where they were the only species of lizard present), the translocated lizards' population size expanded dramatically. (B) Possibly as an adaptive response to intense intraspecific competition for food, these lizards expanded their diet to include an underutilized resource: plants (unlike on their native Pod Kopiste, annual plants abound on Pod Mrcaru). (C) Individuals in the introduced population also evolved a novel morphological structure—a cecal valve, which is special muscle in the intestine that slows the passage of food, thereby allowing time for fermentation. Photos in panels A and C reproduced with the kind permission of Anthony Herrel. Graph in panel B based on Herrel et al. (2008).

Evolutionary biologists and ecologists have long recognized that intraspecific competition can promote trait evolution within species. In this chapter, we highlight how intraspecific character displacement impacts diversification in four significant ways. First, intraspecific character displacement often leads to an increase in phenotypic variance within populations. Second, it can favor the evolution of discrete, alternative phenotypes within populations, including phenotypes that are entirely novel (Figure 5.1). Third, intraspecific character displacement may ultimately foster diversity by facilitating speciation and by reducing a lineage's risk of extinction. Finally, by increasing phenotypic variation within species, intraspecific character displacement facilitates character displacement between species.

HOW INTRASPECIFIC CHARACTER DISPLACEMENT WORKS

Long-standing theory maintains that populations facing intense intraspecific competition will tend to evolve to utilize a wider range of resources (Van Valen 1965; MacArthur and Wilson 1967; MacArthur 1972; Roughgarden 1972; see also Darwin 1859 [2009], p. 127; Haldane 1932 [1993], p. 96). Moreover, as we describe below, such populations may also express more diverse reproductive traits. In other words, intraspecific competition promotes niche-width expansion.

To understand why niche-width expansion may represent an adaptive response to intraspecific competition, consider a hypothetical population that exploits a continuously varying resource gradient, such as a gradient of prey size (although we focus here on competition for resources, competition for successful reproduction can occur similarly). If all resource types along the gradient (for example, all prey size classes) are not utilized to the same extent, then those individuals that specialize on underutilized resources (for example, underutilized size classes) should experience less-intense intraspecific competition. Competitively mediated selection should thereby result in individuals being more or less evenly spaced along the resource-use gradient, such that all individuals have equivalent fitness derived from resource acquisition (Van Valen 1965; MacArthur and Wilson 1967; MacArthur 1972; Roughgarden 1972). Such niche-width expansion, arising as an adaptive response to intraspecific competition, may be a common manifestation of intraspecific character displacement.

An effective way of visualizing this process is to imagine the fitness landscape as the upper surface of a sphagnum bog: a floating mat of vegetation, with an uneven upper surface, that can be locally submerged when weight is placed on it (Rosenzweig 1978). In this metaphor, individuals specializing on the most common resources would initially be associated with a high point on the sphagnum bog (fitness) surface. However, as these individuals become more abundant, the weight of their increased numbers would begin to depress the fitness surface, but only locally. By contrast, individuals specializing on underutilized resources may initially have fewer resources available, but—because they would also have fewer competitors with which to share those resources—their overall fitness may equal that of individuals specializing on initially common resources. In the bog metaphor, these rare resource-use phenotypes may initially be associated with a lower point on the sphagnum bog (fitness) surface, but, as long as their numbers remain low, their net weight depresses the local fitness surface less than that of the more common phenotype(s). The result is that individuals may end up with equal fitness across the landscape. This process is driven by negative frequency-dependent selection, in which rare resource-use phenotypes always have a fitness advantage relative to more common phenotypes, because they experience decreased competition.

Niche-width expansion can be achieved in two different ways (Klopfer 1962; Roughgarden 1972). First, niche-width expansion may be achieved through individual specialization, in which a population contains a variety of different phenotypes, each of which specializes on a narrower range of resources than the population as a whole (see, for example, Bolnick et al. 2003). Alternatively, niche-width expansion may be achieved when individuals are generalists that consume a range of resources similar to that utilized by the population as a whole. We will return to the consequences of specialization versus generalization later in this chapter.

Regardless of how it comes about, abundant empirical evidence demonstrates that intraspecific competition can promote niche-width expansion. Below we review the observational and experimental evidence in support of intraspecific character displacement.

INTRASPECIFIC CHARACTER DISPLACEMENT: OBSERVATIONAL EVIDENCE

Empirical support for intraspecific character displacement comes from comparative studies of trait variation in conspecific populations that are sympatric with a heterospecific competitor versus those that are allopatric. Specifically, many studies have documented a distinctive pattern of trait variation known as "character release," where allopatric populations of a given species exhibit greater trait variation than conspecific populations that are sympatric with heterospecifics (see also Figure 1.4). Character release gets its name because it is thought to arise when allopatric populations are "released" from interspecific competition (Wilson 1961). In such a situation, the absence of heterospecific competitors makes ecological opportunity available, which allows populations to respond adaptively to intraspecific competition by increasing the width of their niche.

Character release is often found in populations that colonize isolated archipelagos or similar island-like settings (such as lakes), where individuals typically encounter few heterospecific competitors (MacArthur and Wilson 1967). For example, the Cocos finch, *Pinaroloxias inornata*, is endemic to Cocos Island, a small island 750 km northeast of the Galápagos Islands (it is the only species of Darwin's finch that occurs outside the Galápagos Islands; Grant and Grant 2008). This species has a broader diet, and forages in a greater variety of ways, than any of its Galápagos Island relatives; that is, the Cocos finch displays character release in foraging behavior (Werner and Sherry 1987). Character release has been documented in many species, including numerous species of freshwater fishes that occur in newly formed lakes (Robinson et al. 1993; Robinson and Wilson 1994).

The Cocos finch example is particularly instructive, however, because it illustrates that character release need not occur exclusively (or even at all) in morphological characters. The Cocos finch exhibits a broader range of foraging *behaviors* than do other species of Darwin's finches that co-occur with closely related heterospecific competitors, but Cocos finch populations exhibit relatively low *morphological* variation (Grant 1986). Thus, traits such as behavioral or physiological characters may often exhibit enhanced variation in response to intraspecific competition, even when morphological characters do not.

An additional manifestation of character release is an increase in the degree of sexual dimorphism (that is, the extent to which sexes differ in size or morphology) in populations where heterospecific competitors are absent compared with those where such competitors are present. Sexual dimorphism is widespread, and Darwin (1871) first suggested that it could arise as a means of minimizing competition between the sexes.

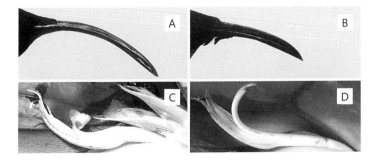

FIGURE 5.2.
Sexual dimorphism may evolve as an adaptive response to competition between the sexes, in which each sex becomes specialized for a separate resource—that is, sexual dimorphism may arise through intraspecific character displacement. For example, the purple-throated carib hummingbird, *Eulampis jugularis*, from the island of St. Lucia exhibits extreme dimorphism in the size and shape of its bill, with (A) the female's bill being longer and more curved than (B) the male's bill. Females drink nectar from (C) the long, curved corolla of *Heliconia bihai* flowers, whereas males feed on (D) *H. caribae* flowers, the corollas of which are straighter and shorter. Each sex feeds most efficiently at the flower species approximating its bill dimension. Photos reproduced from Temeles and Kress (2003), with the kind permission of the authors.

Recent studies have confirmed that, indeed, in the absence of heterospecific competitors, males and females of the same species can lessen competition with members of the opposite sex by diverging in where and how they forage (Figure 5.2; Selander 1966; Dayan and Simberloff 1994; Temeles et al. 2000; Bolnick and Doebeli 2003; Butler et al. 2007; Aguirre et al. 2008; Temeles et al. 2010; Pfaender et al. 2011). As with other forms of character release, sexes generally differ more on islands (Darwin 1871; Simberloff et al. 2000; Butler et al. 2007) or island-like settings, such as newly formed lakes (Aguirre et al. 2008; Pfaender et al. 2011). Presumably the paucity of heterospecific competitors in these environments provides the ecological opportunity that enables males and females to minimize competition with the opposite sex by specializing on different resources (Temeles and Kress 2003).

Before leaving the topic of character release, we must clarify two points. First, character release is generally assumed to derive exclusively from resource competition rather than from *reproductive* competition. Although within-species competition for mates can generate divergent mating tactics (reviewed in Andersson 1994; see also below), little is known of whether and how a release from interspecific competition engenders an expansion of reproductive traits in allopatry as a response to intraspecific competition. Future work on systems where variation in reproductive traits can be characterized in sympatry versus allopatry would be ideally suited for evaluating this issue.

Second, although character release is often interpreted as having arisen via selection, and is therefore thought to be the result of intraspecific character displacement, character release can also result from nonselective processes. For example, in allopatric populations, where there is a lack of directional selection for traits that differ from that expressed by a heterospecific, traits associated with resource use or reproduction may simply accumulate variation through mutation or genetic drift. Thus, as with character displacement between species, inferring intraspecific character displacement solely from observational data is problematic unless there is additional corroborating evidence that competitively mediated selection caused the pattern of trait variation (see chapter 1).

INTRASPECIFIC CHARACTER DISPLACEMENT: EXPERIMENTAL EVIDENCE

Empirical support for intraspecific character displacement does not rest solely on observational evidence. Intraspecific character displacement has also been demonstrated experimentally.

In an elegant experimental demonstration of intraspecific character displacement, Bolnick (2001) gave fruit flies (*Drosophila melanogaster*) a choice of ovipositing in vials with fly food laced with varying concentrations of cadmium (cadmium is normally toxic to flies, but flies can acquire new genetic mutations that allow them to tolerate cadmium). Initially, cadmium-intolerant flies dominated the population, and these flies had greatly reduced performance on high cadmium concentrations. Yet flies preferred to oviposit in vials containing high levels of cadmium, because there were few larvae present in these vials. However, the few cadmium-tolerant flies that were present experienced reduced competition because they could exploit a resource for which there were few competitors: cadmium-laced food. Consequently, these cadmium-tolerant flies were favored and increased in frequency in the population—that is, intraspecific character displacement occurred. Moreover, populations experiencing high levels of intraspecific competition evolved cadmium tolerance more rapidly than populations experiencing lower levels of intraspecific competition. This intraspecific character displacement occurred remarkably rapidly—within four generations—suggesting that competitively mediated selection acted on standing genetic variation for cadmium tolerance (Bolnick 2001). Intraspecific character displacement has also been demonstrated in laboratory populations of viruses, where intense intraspecific competition for hosts led to the evolution of new host-range variants (Figure 5.3).

These experiments are instructive not only in providing direct support for intraspecific character displacement, but also in what they say about intraspecific character displacement's potential ramifications. Namely, by depressing the fitness of individuals to a level at which some would have increased fitness by seeking alternative resources that are less in demand—or even previously avoided altogether—intraspecific competition can favor the utilization of *novel* resources. Such a process may explain the origins

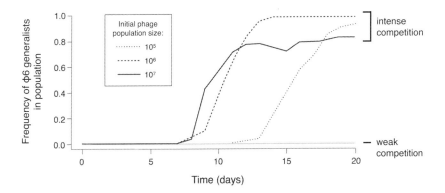

FIGURE 5.3.
Experimentally demonstrated intraspecific character displacement involving the evolution of a host-range shift in the bacteriophage Φ6. This virus normally infects the *phaseolicola* strain of the bacterium, *Psuedomonas syringae*. However, Φ6 readily produces mutants capable of infecting a variety of novel bacterial host strains. When Φ6 were subjected to intense intraspecific competition for *phaseolicola* hosts (by experimentally manipulating the ratio of Φ6 phage to *phaseolicola* hosts), Φ6 evolved, within a matter of days (that is, a few dozen generations), a new generalist morph that expanded its host range to include a novel host: the ERA strain of *P. pseudoalcaligenes*. This novel generalist evolved the most rapidly when the starting population size of Φ6 was the largest (that is, when the starting population was 10^7 Φ6 phage). By contrast, when Φ6 experienced only weak competition (that is, they were given unlimited *phaseolicola* hosts), they did not evolve this generalist morph (solid gray line). Data from Bono et al. (in press).

of novel resource-use traits in numerous natural populations (Figure 5.1; Carroll et al. 1997, 1998; Jones 1998; Aubret and Shine 2009; Martin and Pfennig 2010a).

EVOLUTION OF ALTERNATIVE PHENOTYPES

Above, we focused largely on how intraspecific competition promotes an increased range of continuous traits. Here, we consider how such competition can also promote an increase in phenotypic variance by favoring the evolution of discrete, alternative phenotypes; that is, distinct phenotypes that are expressed in the same sex and at the same life stage within the same population.

Alternative phenotypes have long fascinated evolutionary biologists for various reasons, not the least of which is that they furnish some of the most dramatic examples of diversity within species (for example, see Figure 1.1C). Indeed, alternative morphs frequently exhibit strikingly different behaviors, morphologies, physiologies, and life histories that are often as pronounced as the corresponding differences seen between different species (see, for example, Liem and Kaufman 1984; Benkman 1988, 2003; Gross 1996; Hendry et al. 2006; Calsbeek et al. 2007; Wund et al. 2008). For this reason, some evolutionary biologists have contended that the same factors that generate such diversity within species may also promote diversity between species (West-Eberhard 2003).

Furthermore, from an ecological standpoint, alternative phenotypes may function as separate species (Harmon et al. 2009). Additionally, their existence within a population may alter the outcome of interactions with other species (Bolnick et al. 2011; see also below) and even increase the chances that potential competitors will ultimately coexist (Clark 2010).

Yet the foremost reason why alternative phenotypes are fascinating is that their stable existence is seemingly difficult to explain. How can selection simultaneously favor—and maintain—two or more discrete phenotypes in the same population? What prevents selection from favoring one optimal phenotype in a population and eliminating all others?

We address these questions in our discussion below. Specifically, we describe how intraspecific competition can act as a potent agent of frequency-dependent disruptive selection and thereby maintain two or more discrete phenotypes in the same population.

FREQUENCY-DEPENDENT DISRUPTIVE SELECTION AND THE EVOLUTION OF ALTERNATIVE PHENOTYPES

A population experiences disruptive selection when two or more modal phenotypes have higher fitness than the intermediate phenotypes between them (Figure 5.4; Mather 1953; Rueffler et al. 2006). A widely cited cause of disruptive selection is ecological specialization (Smith and Skúlason 1996). According to this view, disruptive selection arises when individuals with certain phenotypes are better adapted to alternative niches (for example, different resource types) than are individuals with intermediate phenotypes (Levene 1953; Mather 1955; Maynard Smith 1962; Levins 1968). Such ecological specialization, stemming from functional trade-offs, often accompanies disruptive selection (Figure 5.5; for additional examples, see Smith 1993; Robinson et al. 1996; Medel et al. 2003; Bolnick 2004; Calsbeek 2009; Hendry et al. 2009).

Historically, disruptive selection was assumed to be rare (Endler 1986). This is because theory suggested that disruptive selection driven solely by ecological specialization would break down, as the population eventually became fixed for a single morphotype (under the assumption that one morphotype would invariably have higher fitness than the other[s]). However, this viewpoint presupposed a static fitness landscape (Bolnick 2004). If, instead, fitness is frequency dependent—as when intraspecific competition prevails (see above)—then the fitness landscape will be dynamic, such that no one morphotype will be intrinsically better than all others. In such a situation, as any one morphotype becomes more common, it competes more against itself than against

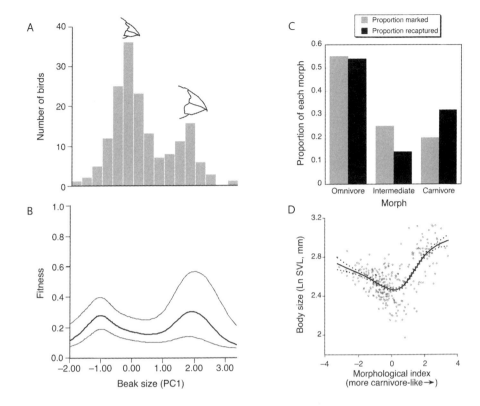

FIGURE 5.4.

Disruptive selection in the wild. (A) As described in chapter 3, the adult population of Darwin's finches (*Geospiza fortis*) at El Garrapatero (Santa Cruz Island, Galápagos archipelago) consists of a small-beaked morph and a large-beaked morph. (B) This population experiences disruptive selection, in which individuals with small or those with large beaks (but not intermediate-sized beaks) are favored. Selection on beak morphology is depicted as a cubic spline (heavy line; the lighter lines denote upper and lower 95% confidence intervals). (C) Similarly, disruptive selection disfavors individuals with intermediate trophic phenotypes in Mexican spadefoot toad tadpoles (*Spea multiplicata*). This bar chart shows the probability of survival for individuals expressing different ecomorphs—omnivores, carnivores, and intermediates—as determined by a mark-recapture experiment in a natural pond in Arizona. Individuals expressing the intermediate phenotype had a lower probability of survival than either carnivores or omnivores. In particular, the proportion of intermediates recaptured was lower than expected, while the proportion of omnivores recaptured was statistically equal to the proportion marked and the proportion of carnivores recaptured was higher than expected. (D) Disruptive selection in this population can also be visualized by a cubic-spline estimate of body size (a fitness proxy) on a composite shape variable of trophic morphology (morphological index). The cubic spline (solid line) is bracketed by the upper and lower 95% confidence intervals (dashed lines). As in panel B, the presence of an intermediate fitness minimum suggests that disruptive selection acts on trophic morphology. Panel A redrawn from Huber and Podos (2006); panel B redrawn from Hendry et al. (2009); panels C and D redrawn from Martin and Pfennig (2009).

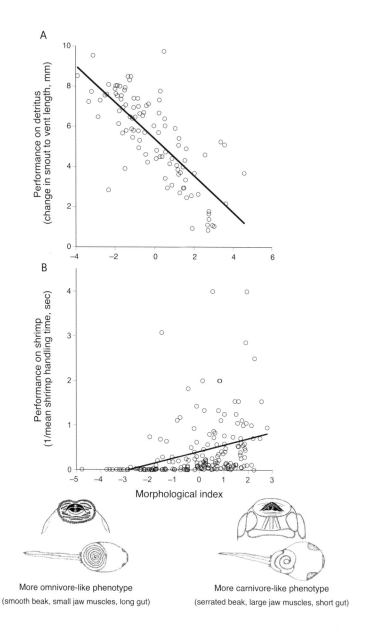

FIGURE 5.5.
Trade-offs in performance on alternative resources disfavors intermediate phenotypes and thereby forms the basis for disruptive selection in many species, such as polymorphic spadefoot toad tadpoles (*Spea multiplicata*). In each panel, an individual tadpole's performance when fed on alternative resources is plotted against its trophic morphology (as measured by a morphological index, in which lower values represent more omnivore-like morphologies and higher values represent more carnivore-like morphologies). (A) When reared on detritus only, the more omnivore-like a tadpole was in morphology, the larger it grew. (B) When offered fairy shrimp, however, the more carnivore-like a tadpole was in morphology, the faster it was able to capture and consume these shrimp. Note that, when compared with the performance of extreme trophic phenotypes, the performance of intermediate individuals was poor with both detritus and fairy shrimp. Reproduced from Martin and Pfennig (2009), with the kind permission of the publisher.

other morphotype(s). Because reduced competition favors rarer morphotype(s), these rare alternatives will be maintained in the population by frequency-dependent selection. In short, frequency-dependent disruptive selection (driven by intraspecific competition) is a plausible mechanism for maintaining alternative phenotypes in the same population (reviewed in Doebeli 2011).

A meta-analysis of empirical studies reveals that disruptive selection may be more widespread than was formerly presumed (Kingsolver et al. 2001). Indeed, several studies of natural populations have documented disruptive selection acting on resource-use traits (Figure 5.4; see also Smith 1993; Medel et al. 2003; Bolnick 2004; Pfennig et al. 2007; Bolnick and Lau 2008; Calsbeek and Smith 2008; Martin and Pfennig 2009, in press; Elmer et al. 2010; Cucherousset et al. 2011). Thus, disruptive selection may be relatively common, suggesting that it could play a general and important role in promoting diversity within species.

As further evidence that intraspecific competition promotes alternative resource-use morphs, a number of studies have also documented frequency-dependent competition for resources. For example, in Africa's Lake Tanganyika, a cichlid fish, *Perissodus microlepis*, has evolved a unique resource polymorphism. This species specializes on eating the scales of other cichlid fish by approaching from behind and plucking scales from its prey's flank. However, the population consists of alternative morphs that attack prey on different sides (Hori 1993). One morph approaches its prey from the left side (this morph's mouth opens to the right), whereas the other morph approaches its prey from the right side (this morph's mouth opens to the left; Figure 5.6A). By sampling the frequencies of the two morphs over time, Hori (1993) established that the frequency of each morph fluctuates around 0.5 (Figure 5.6B), suggesting that they are maintained by frequency-dependent selection. Presumably, the prey's escape behavior accounts for this frequency dependence: as one morph becomes more common, prey become more wary of approaches from that side, and the rarer morph is then favored (Hori 1993). Only when the two morphs are equally abundant does neither have an advantage.

Such frequency dependence has also been demonstrated experimentally. For example, crossbills (*Loxia* spp.) are members of the finch family, so named because their mandibles cross at the tips (Benkman 1996). Crossbills specialize on conifer cones, and their unusual bill shape is an adaptation for extracting seeds from cones. In some individuals, the lower mandible crosses to the right, whereas in other individuals it crosses to the left. Experiments have revealed that when a crossbill forages on a cone previously used by a crossbill of the same morph, it takes longer to extract the seed than when a crossbill follows a crossbill of the opposite morph (Benkman 1996). Thus, frequency-dependent selection, stemming from resource competition for access to seeds, maintains both morphs within the same population. Similarly, frequency-dependent intraspecific competition for resources has also been experimentally demonstrated in polymorphic spadefoot toad tadpoles (*Spea multiplicata*; Pfennig 1992) and tiger salamander larvae (*Ambystoma tigrinum*; Maret and Collins 1997).

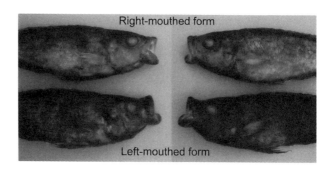

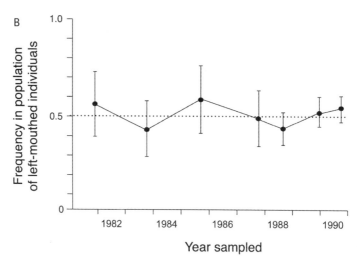

FIGURE 5.6.

Evidence that frequency-dependent competition maintains alternative resource-use phenotypes within a natural population of Lake Tanganyikan scale-eating cichlids, *Perissodus microlepis*. (A) Individuals, which occur as either a right-mouthed or a left-mouthed morph, differ in the direction from which they attack their prey (each form is shown from both sides). (B) The frequencies of these morphs fluctuate around 0.5. These frequencies are stabilized by prey escape behavior. Specifically, when the morph that attacks from the left (the right-mouthed form) becomes common, prey become more wary of a fish approaching on their left side. Consequently, the alternative morph that attacks from the right (the left-mouthed form) is more successful and increases in frequency, until it becomes common, at which point the right-mouthed form is favored. Only when both morphs are equal in frequency does neither have a selective advantage. Photos reproduced from Hori (1993), with the kind permission of the publisher.

The discussion above highlights ways that intraspecific competition can drive frequency-dependent disruptive selection and, ultimately, promote the evolution of alternative phenotypes—that is, intraspecific competition can promote polymorphism. Although we focused on how disruptive selection stemming from resource competition may foster alternative resource-use phenotypes (resource polymorphism), competition for mates may similarly promote alternative mate-acquisition tactics (mating polymorphism). In the next section, we discuss the evolution of resource polymorphism in greater detail, before considering the evolution of mating polymorphism in the subsequent section. Although we treat the two types of polymorphisms separately for the purposes of our discussion, we emphasize how the processes involved in the production of each type of polymorphism are similar.

EVOLUTION OF RESOURCE POLYMORPHISM

Resource polymorphism is the occurrence within a single population of alternative morphs showing differential resource use (Table 5.1). As described above, theory suggests that such polymorphisms evolve as an adaptive response to intraspecific competition for resources (Mather 1955; Maynard Smith 1962). In other words, resource polymorphism results from intraspecific character displacement (see also Smith and Skúlason 1996). However, for intraspecific character displacement to actually promote the evolution of a resource polymorphism, ecological opportunity (in the form of alternative resource types underutilized by other species) must also be present (Smith and Skúlason 1996; Martin and Pfennig 2010a). In particular, because the evolution of a resource polymorphism typically involves the evolution of novel resource-use phenotypes (for example, see Figure 5.1), underutilized resources must be available for these new resource-use phenotypes to acquire and thereby succeed evolutionarily. In the absence of such resources, the evolution of resource polymorphism is unlikely to occur. By contrast, when underutilized resources are present, a population experiencing intense intraspecific competition can, as an adaptive response to competition, expand the range of resources it uses.

As predicted, resource polymorphisms are found most often in environments where intraspecific competition is intense, underutilized resources are present, and interspecific competition is relaxed (the latter two factors combine to increase ecological opportunity). For example, lakes in recently glaciated (10,000–15,000 years ago) regions of the Northern Hemisphere are often species poor. In such lakes, fish tend to occur in either benthic or limnetic habitats, and many species produce discrete, alternative ecomorphs that specialize on these alternative niches (reviewed in Robinson and Wilson 1994; Wimberger 1994; Skúlason et al. 1999). Similarly, alternative resource-use morphs in Galápagos land snails (Figure 5.7) and larval amphibians (Walls et al. 1993; Michimae and Wakahara 2002; Martin and Pfennig 2010a) typically occur in populations where competing species are rare or absent. Indeed, in spadefoot toad tadpoles, experiments

TABLE 5.1. Resource polymorphisms in selected taxa showing the nature of the ecological segregation between alternative resource-use ecomorphs

Organism	Nature of the ecological differences	Source
Viruses		
Lambda bacteriophage (λ Phage, a parasite of the bacterium, *Escherichia coli*)	Reproduction through lysis (in which the host cell is destroyed as it releases phage viruses) vs. lysogeny (in which the phage chromosome is incorporated into the host chromosome and is replicated along with the host chromosome)	Ptashne (1986)
Bacteria		
Pseudomonas fluorescens	Different morphotypes that differ in resource use	Rainey and Travisano (1998)
Fungi		
Mycorrhizal fungi*	Different host plants	Taylor et al. (2004)
Ciliates		
Tetrahymena vorax	Bacterivore vs. carnivore niches	Ryals et al. (2002)
Lembadion bullinum	Noncannibal vs. cannibal niches	Kopp and Tollrian (2003)
Rotifers		
Asplanchna sieboldi	Noncannibal vs. cannibal niches	Gilbert (1973)
Insects		
Blueberry and apple maggot flies (*Rhagoletis pomonella*)*	Different host plants	Feder et al. (1989); Filchak et al. (2000)
Geometrid moth caterpillars (*Nemoria arizonaria*)	Different host plants	Greene (1989)
Leaf beetles (*Neochlamisus bebbianae*)*	Different host plants	Funk (1998)
Gallmaking tephritid flies (*Eurosta solidaginis*)*	Different host plants	Abrahamson et al. (2001)
Walking sticks (*Timema cristinae*)*	Different host plants	Nosil et al. (2002)
Fishes		
Arctic charr (*Salvelinus alpinus*)	Different habitat and dietary preferences	Skúlason et al. (1999)
Sunfish (*Lepomis* spp.)	Benthic vs. limnetic niches	Robinson et al. (1993)

TABLE 5.1. *(continued)*

Organism	Nature of the ecological differences	Source
Fishes *(continued)*		
Numerous species of Nearctic freshwater fishes	Benthic vs. limnetic niches	Robinson and Wilson (1994)
Sockeye salmon (*Oncorhynchus nerka*)*	Different habitat preferences	Hendry et al. (2000)
Three-spined stickleback complex (*Gasterosteus aculeatus*)*	Benthic vs. limnetic niches	Rundle et al. (2000)
Midas cichlid species complex (*Amphilophus* spp.)*	Benthic vs. limnetic niches	Barluenga et al. (2006)
Amphibians		
Tiger salamander larvae (*Ambystoma tigrinum*)	Planktivore vs. cannibal niches	Collins and Cheek (1983)
Tiger salamander larvae (*Ambystoma tigrinum*)	Paedomorph vs. metamorph life histories	Whiteman (1994)
Eastern newts (*Notophthalmus viridescens*)	Paedomorph vs. metamorph life histories	Harris (1987); Takahashi and Parris (2008)
Spadefoot toad tadpoles (*Spea* spp.)*	Omnivore vs. carnivore niches	Rice and Pfennig (2010)
Birds		
Black-bellied seedcracker finches (*Pyrenestes ostrinus*)	Different food niches	Smith (1993)
Blue tits (*Parus caeruleus*)*	Different habitat preferences	Blondel et al. (1999)
Crossbills (*Loxia* spp.)	Alternative resource-use morphs that differ in handedness	Benkman (1996)
Galápagos warbler finches (*Certhidea olivacea* and *C. fusca*)*	Different habitat preferences	Tonnis et al. (2005)
Galápagos medium ground finches (*Geospiza fortis*)*	Different food niches	Hendry et al. (2006, 2009); Huber et al. (2007)

*In cases marked with an asterisk, natural populations that differ in morph expression are (potentially) partially reproductively isolated from each other.

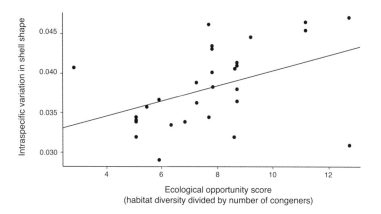

FIGURE 5.7.
Ecological opportunity predicts the degree to which populations express resource polymorphism. For example, the greater the number of local host plant species that Galápagos Island land snails encounter, the greater the variation that they express in shell shape (where different shell shapes are adapted to different host plants). Redrawn from Parent and Crespi (2009).

have established a causal relationship between the level of ecological opportunity that a population experiences (as measured by the range of resources available) and the degree to which resource polymorphism is expressed in that population (Martin and Pfennig 2010a).

Resource polymorphism can evolve even in species-rich communities, where ecological opportunity is seemingly low, if underexploited resources are available that require specialized phenotypic traits on the part of the consumer (Smith and Skúlason 1996; Robinson and Wilson 1998). In such cases, ecological opportunity could actually be high for any species that evolves such specialized traits. Essentially, organisms can create their own ecological opportunity by carving out a novel niche within an already-speciose community. For example, black-bellied seedcracker finches (*Pyrenestes ostrinus*) from Cameroon occur as two distinct morphs: a small-beaked morph and a large-beaked morph (Smith 1993). Normally, both morphs eat soft seeds only. However, when competition for these soft seeds becomes intense, the large-beaked morph is uniquely adapted to utilize harder seeds, which cannot be cracked open by the small-beaked morph (Smith 1993). Similarly, the cichlid *Perissodus microlepis* of Lake Tanganyika described above has evolved a unique resource polymorphism in which alternative morphs attack their prey on different sides (Figure 5.6; Hori 1993). In both of these examples, resource polymorphism has evolved despite the species occurring in species-rich communities, because each species has evolved the unique ability to utilize previously inaccessible resources (in these instances, hard seeds or fish scales).

In the scale-eating cichlids, the alternative resource-use morphs are phenotypic specialists for alternative resources: one specializes on eating scales from the left side of its prey, whereas the other specializes on eating scales from the right side of its prey. Similarly, alternative resource-use morphs that differ in "handedness" in crossbills (*Loxia* spp.) also represent phenotypic specialists for alternative resources. However, it is not necessary for the different resource-use morphs that constitute a particular resource polymorphism to represent alternative phenotypic specialists for different resources. Instead, one resource-use morph may be a specialist for a specific resource, whereas the other morph may be a dietary generalist. In such cases, the dietary specialist is expected to specialize on less-preferred prey (for example, prey that are more difficult to obtain), while retaining the ability to use preferred (generalist) resource(s) when competition for the more desirable resource becomes less intense.

For example, the cichlid fish *Herichthys minckleyi* from the Cuatro Ciénegas valley in Mexico's Chihuahuan desert consist of a molariform morph, which possesses large molar-like pharyngeal jaw teeth used for crushing and eating snails, and an alternative papilliform morph, which possesses small pencil-like pharyngeal teeth used for shredding and eating aquatic plants, arthropods, and detritus (Hulsey et al. 2005, 2006). Liem and Kaufman (1984) demonstrated in the laboratory that the molariform morph feeds on hard snail prey significantly more often than papilliform morphs, but only when densities of alternative arthropod prey are low (as may occur when competition for alternative prey is severe). Thus, in this system, a dietary specialist (the molariform morph) has evolved morphological specializations that provide exclusive access to an alternative, albeit less preferred, resource (snails). Such a pattern is found in many fish species, which appear to have evolved similar phenotypic specializations for utilizing less-preferred resources while at the same time not greatly compromising their ability to use preferred resources (Robinson and Wilson 1998). Thus, counterintuitively, when competition for the preferred resources is relaxed, specialists may reject the very resources that they have evolved traits to use (Robinson and Wilson 1998). In such systems, specialists and generalists coexist within the same population because of temporal variation in the abundance of different resources (Smith and Skúlason 1996; Robinson and Wilson 1998).

An alternative pattern could emerge when specialists monopolize a more profitable prey type and thereby exclude the more generalist consumers from that resource. For example, spadefoot toad tadpoles (genus *Spea*) have evolved alternative resource-use morphs: an omnivore morph, which is a dietary generalist that feeds on plants, detritus, and small invertebrates including fairy shrimp, and a morphologically distinctive carnivore morph, which is a dietary specialist that feeds almost exclusively on fairy shrimp (reviewed in Ledón-Rettig and Pfennig 2011). In this case, the dietary specialist monopolizes the more profitable and preferred resource: tadpoles prefer to eat, and grow better on, fairy shrimp as opposed to plants or detritus (Pfennig 2000a).

When dietary specialists monopolize the more profitable resource, thereby reducing the generalists' access to this resource, fitness trade-offs may explain how specialist and generalist morphs can coexist. Specifically, although dietary specialists may benefit by being able to monopolize a more profitable resource than that used by the generalist, specialists will concomitantly experience more intense competition than the generalist (Paull et al., in press). In such cases, negative frequency-dependent selection maintains both morphs within the same population, such that each morph has, on average, equal fitness (recall the sphagnum bog metaphor at the beginning of this chapter).

Regardless of how alternative morphs partition resources, the proximate mechanisms that generate resource polymorphism can assume different forms. These mechanisms can also shed light onto the selective factors that favor such morphs. Although some alternative morphs are generated by relatively simple genetic mechanisms—for example, beak-size polymorphism in *Pyrenestes* finches (Smith 1993) and handedness in *Perissodus* cichlids (Hori 1993) appear to be determined by alternative alleles at a single locus—in many systems, such as the crossbills described above, morph determination appears to be strongly influenced by the environment (Edelaar et al. 2005). Moreover, in such environmentally influenced systems, the external cues that promote the expression of a particular morph tend to be associated with intraspecific competition or ecological opportunity, as would be expected if these factors favor the evolution of alternative resource-use morphs. For example, many such morphs are induced by tactile cues associated with an increased density of conspecifics (Collins and Cheek 1983; Hoffman and Pfennig 1999) or by the handling and/or ingestion of a resource for which the morph becomes specialized (Gilbert 1973; Bernays 1986; Meyer 1987; Greene 1989; Pfennig 1990; Day et al. 1994; Loeb et al. 1994; Smith and Palmer 1994; Padilla 2001; Michimae and Wakahara 2002).

Additionally, in many species, which resource-use morph an individual expresses depends on its size or condition at the time in development when morph determination occurs. For instance, in tiger salamanders (Maret and Collins 1997) and spadefoot toad tadpoles (Frankino and Pfennig 2001), larger individuals are more likely to develop into a more robust cannibal or carnivore morph. Such condition dependence makes adaptive sense: larger individuals are more likely to succeed at preying on (or competing for) larger food items than are smaller individuals (see, for example, Frankino and Pfennig 2001).

EVOLUTION OF MATING POLYMORPHISM

Mating polymorphism occurs when discrete intraspecific morphs showing differential mate-acquisition tactics are present within the same population. As with resource polymorphism, mating polymorphisms have been documented in numerous taxa (Table 5.2). Indeed, in many natural populations, pronounced, discontinuous differences exist among males in mating behavior and associated morphological structures (reviewed in Andersson 1994; Gross 1996; Shuster and Wade 2003; West-Eberhard 2003).

A common form of mating polymorphism is for males to employ either fighting or

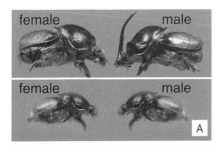
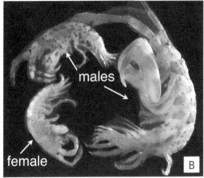

FIGURE 5.8.
Examples of mating polymorphisms. (A) Alternative morphs in the beetle *Proagoderus* (*Onthophagus*) *lanista*, showing variation in body size and, in males, horn size among individuals. (B) Alternative male morphs in marine amphipods, *Jassa marmorata* (clockwise from lower left): female, minor male morph (which resembles females), and major male morph. Photos in panel A by Ben Ewen-Campen, courtesy of Doug Emlen; photos in panel B reproduced from Kurdziel and Knowles (2002), with the kind permission of the authors and publisher.

sneaking as alternative mating tactics. Typically, distinctive morphological structures accompany these different behavioral tactics (Moczek 2005). For example, in dung beetles, some adult males have large horns, whereas others have small horns or lack horns altogether (Figure 5.8A). Large-horned males are more successful at fighting and thereby defending a resource—tunnels underneath dung pads—that attracts females (Emlen 2000). Small-horned or hornless males, by contrast, attempt to sneak matings (Emlen 2000). In some species, sneaker males mimic females and thereby evade other males (Figure 5.8B; see also Table 5.2).

Three lines of evidence suggest that mating polymorphisms typically evolve through intraspecific character displacement. First, mating polymorphisms tend to be restricted to males only (reviewed in Andersson 1994; Gross 1996), suggesting that selection (specifically, sexual selection) favors mating polymorphism (see also chapter 7). Second, selection on such morphs is often negatively frequency dependent, such that rare mate-acquisition phenotypes have a fitness advantage because of decreased competition with more common forms (see, for example, Shuster and Wade 1992; reviewed in Andersson 1994; Gross 1996). Negative-frequency dependence is a hallmark of competitively mediated disruptive selection. Finally, mating polymorphisms are typically most common when competition for mates is intense (see, for example, Radwan 1993; Tomkins and Brown 2004).

As with resource polymorphisms, ecological opportunity is essential for mating polymorphisms to evolve, especially when female choice is involved. To see why, consider

TABLE 5.2. Male mating polymorphisms in selected species
and the nature of the divergence between morphs

Organism	Nature of the divergence	Source
Mites		
Caloglyphus berlesei, Rhizoglyphus echinopus	Fighter males (thickened legs used to kill other males) vs. nonfighter males (nonmodified legs)	Radwan (1993)
Insects		
Numerous species of fig wasps	Nondispersing fighter males (large head and mandibles, no wings) vs. dispersing nonfighter males (small head and mandibles, wings)	Hamilton (1979); Pienaar and Greeff (2003)
Ground-nesting bees (*Perdita texana*)	Fighter males (large body size, large head) vs. nonfighter males (small body size, small head)	Danforth and Neff (1992)
Numerous species of scarab beetles	Major males (large body size, large horn on head, fight for females) vs. minor males (small body size, small or no horn on head, attempt to sneak fertilizations)	Eberhard (1982); Emlen (1997); Moczek and Emlen (2000)
Thrips (*Hoplothrips karnyi*)	Nondispersing fighter males (large head and mandibles, no wings) vs. dispersing nonfighter males (small head and mandibles, possesses wings)	Crespi (1988)
Desert grasshoppers (*Ligurotettix coquilletti* and *L. planum*)	Dominant/territorial males vs. subordinate/non-territorial males	Greenfield and Shelly (1985); Shelly and Greenfield (1989)
Crustaceans		
Isopods (*Paracerceis sculpta*)	Alpha males (large body size and abdominal spines, aggressive, territorial), beta males (intermediate body size, lack spines, mimic female behavior), gamma males (small body size, lack spines, attempt to sneak fertilizations)	Shuster (1989)
Amphipods (*Jassa marmorata*)	Major males (large body size, aggressive, territorial) vs. minor males (small, attempt to sneak fertilizations)	Kurdziel and Knowles (2002)

TABLE 5.2. *(continued)*

Organism	Nature of the divergence	Source
Fishes		
Coho salmon (*Oncorhynchus kisutch*)	Hooknose males (large body size and teeth, aggressive, territorial) vs. jack males (small body size and teeth, attempt to sneak fertilizations)	Gross (1985)
Sunfish (*Lepomis* spp.)	Sneaker males, satellite males, and parental males	Gross (1982)
Lake Tanganyika cichlids (*Lamprologus callipterus*)	Large, nest-building males and dwarf, sneaker males	Schutz et al. (2010)
Amphibians		
Great plains toad (*Anaxyrus (Bufo) cognatus*)	Calling males (which call to attract females) vs. silent "satellite" males (which attempt to intercept females on their way to callers)	Sullivan (1983); Krupa (1989)
Reptiles		
Side-blotched lizards (*Uta stansburiana*)	Orange-throated males (aggressive, defend territories within which live several females), blue-throated males (less aggressive, defend small territories in which they guard a single female), yellow-throated males (do not defend territories, resemble receptive females, attempt to sneak fertilizations)	Sinervo and Lively (1996)
Birds		
Ruffs (*Philomachus pugnax*)	Territorial males (dark-colored ornamental neck ruffs and head tufts, defend small territories to which females are attracted for mating), satellite males (white ruffs and tufts, attempt to sneak fertilizations of females attracted to territories), female-mimic males (no ruff, attempt to sneak fertilizations)	Lank et al. (1995); Jukema and Piersma (2006)

that the evolution of mating polymorphisms typically entails the appearance of a novel mate-acquisition tactic (see Figure 5.8). For such a trait to persist in a population, environmental circumstances must be conducive to individuals expressing the new tactic; for example, they must have available reproductive-trait space that is not already occupied by other species. In the absence of such ecological opportunity—either because of the presence of a similar heterospecific or environmental constraints (see chapter 7)—the new mating tactic is unlikely to arise or be maintained evolutionarily in the population.

As with resource polymorphisms, mating polymorphisms may also be mediated by phenotypic plasticity (for example, Pienaar and Greeff 2003) or genetic polymorphism (Shuster and Wade 1992; Lank et al. 1995; Sinervo et al. 2000). When mating polymorphisms arise through phenotypic plasticity, the cues that induce the plasticity are often associated with intraspecific competition for mates (for example, Radwan 1993). Moreover, which morph any given individual becomes often depends on the individual's status or condition at the point in development when morph determination takes place. For example, in horned beetles, larvae that are fed on a better diet mature into larger adults that develop bigger horns, whereas those fed on a poorer diet mature into smaller adults that develop smaller horns or lack horns altogether (Moczek 2005). Indeed, in most mating polymorphisms, larger individuals generally become a fighter morph, whereas smaller individuals generally become a sneaker morph (Figure 5.8B; reviewed in Gross 1996). As with resource polymorphisms, such condition dependence makes adaptive sense: larger individuals are more likely to succeed at fighting for mates than smaller ones.

INTRASPECIFIC CHARACTER DISPLACEMENT AND SPECIES DIVERSITY

Our emphasis in this chapter has, until now, been on how intraspecific character displacement fosters diversity within species. Ultimately, however, by fostering diversity within species—especially by promoting the evolution of a resource or mating polymorphism—intraspecific character displacement may also enhance diversity *between* species. For example, the evolution of a resource or mating polymorphism may represent an early stage in the speciation process (West-Eberhard 2003, 2005; Mallet 2008; Hendry 2009; Pfennig and McGee 2010; see also chapter 8). Indeed, the phenotypic alternatives that constitute resource or mating polymorphism frequently possess some of the same characteristics that different species possess—including ecological and genetic differentiation and even partial reproductive isolation (Tables 5.1, 5.2). Here, we review the evidence that the evolution of a mating or resource polymorphism is associated with an increase in species diversity. We also describe how such polymorphism may foster species diversity in either of two ways: by increasing the likelihood that new species will form, or by decreasing the chances that existing species will become extinct.

We begin by discussing how the evolution of a mating polymorphism may enhance species diversity. Mating polymorphisms, by their nature, should seemingly have high potential to contribute to diversity. After all, reproductive isolation is a critical element of the speciation process (see chapter 8), and any change in mating between morphs or populations has the potential to promote speciation (West-Eberhard 1979, 1983). Indeed, mating polymorphism can foster speciation in two ways. First, morphs themselves can become reproductively isolated from each other. Second, and somewhat paradoxically, the differential loss of morphs among populations can lead to divergence, and possibly, speciation, between such populations.

Regarding the possibility that alternative morphs can become reproductively isolated from each other, the existence of alternative reproductive tactics within a population could foster reproductive isolation if individuals mate assortatively by morphotype (as might occur with color polymorphisms; reviewed in Gray and McKinnon 2007). However, if the polymorphism were strictly the result of alternative mating tactics in a system where females possessed unimodal/unidirectional preferences—for example, where females preferred males of larger size or better condition—such assortative mating and speciation would be unlikely.

Regarding the possibility of differential evolution and/or loss of polymorphism contributing to speciation, evidence consistent with this possibility comes from side-blotched lizards, *Uta stansburiana*, from the western United States (Corl et al. 2010; for a possible example in birds, see Hugall and Stuart-Fox 2012). In most populations, males employ one of three mating tactics (see Table 5.2), which are specified by alternative alleles (Sinervo et al. 2000). These alternative mate-acquisition morphs are maintained in the same population by negative frequency-dependent selection in a sort of biological "rock-paper-scissors" game (Sinervo and Lively 1996). This polymorphism is geographically widespread, and it has been maintained for millions of years (Corl et al. 2010).

However, eight evolutionarily independent lineages have undergone a reduction in the numbers of male morphs: some populations have two morphs, and others have only one morph (in all such populations, the same morph—a sneaker morph—was lost, indicating that loss was likely caused by selection; Corl et al. 2010). Because the fitness of each morph depends on the frequency of other types in the population, such morph loss changes a population's competitive dynamics (Sinervo and Lively 1996). Indeed, populations that have undergone morph loss are so distinct from each other (in adult size and degree of sexual dimorphism, for example) that they are regarded as separate subspecies and possibly even separate species (Corl et al. 2010). Thus, the loss of mating polymorphism can promote rapid divergence among populations and possibly aid species formation.

As noted above, another outcome of intraspecific character displacement is the evolution of resource polymorphism. As with mating polymorphism, the evolution of resource polymorphism may foster species diversity. Indeed, clades in which resource polymorphism has evolved are more species rich than are their sister clades (those to

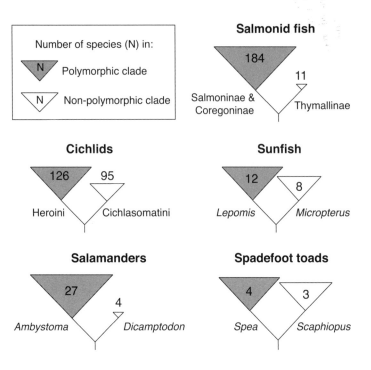

FIGURE 5.9.
Evidence that resource polymorphism is associated with greater species richness in various clades of fish and amphibians. In each case, the clade in which resource polymorphism is known to have evolved (shaded clade) is more species rich than its sister clade, in which resource polymorphism is not known to have evolved (unshaded clade). Reproduced from Pfennig et al. (2010), with the kind permission of the publisher.

which they are most closely related) that lack resource polymorphism (Pfennig and McGee 2010; Figure 5.9).

One way in which the evolution of resource polymorphism may enhance species richness is by enabling a lineage to occupy a substantially new ecological niche. The invasion of a novel niche is often associated with an increase in diversity (Niklas et al. 1983; Bambach 1985), and it may represent a key way in which resource polymorphism drives speciation. Speciation may occur between populations that have evolved such a polymorphism and those that have not if these two different types of populations come to occupy different niches. Additionally, the alternative morphs that constitute a resource polymorphism may also separate into distinct species if they occupy different niches and therefore diverge from one another owing to contrasting selective pressures. Either route could explain why clades in which resource polymorphism has evolved are more species rich (we discuss the potential role of resource polymorphism in speciation in greater detail in chapter 8).

However, not only might resource polymorphism facilitate the formation of new species, it might also maintain existing species by reducing their risk of extinction. Consider that clades in which resource polymorphism has evolved appear to occupy more diverse habitats and possess wider geographical ranges than sister clades lacking resource polymorphism (Pfennig and McGee 2010). Consequently, resource polymorphic species may be less restrictive in their habitat requirements and therefore less likely to become extinct owing to habitat change or loss. Indeed, taxa with wider geographical ranges are generally less likely to become extinct (Jablonski 1986). Any factor that reduces extinction risk should also tend to cause any clade that exhibits that factor to become more species rich, for two reasons. First, reduced extinction allows more species to accumulate in the clade. Second, reduced extinction gives individual lineages more time to diversify. Indeed, certain clades may have more species than other clades simply because they are older and have had more time to accumulate species (McPeek and Brown 2007).

Further research is needed to determine whether resource polymorphism increases species richness by facilitating speciation or by buffering existing lineages from extinction. Both routes, alone or in combination, could explain why clades in which resource polymorphism has evolved are more species rich. Regardless of how resource polymorphism increases species richness, the available data suggest that the evolution of either resource polymorphism or mating polymorphism—deriving from intraspecific character displacement—may play a key role in facilitating diversification, and that species in which such polymorphisms have evolved may be predisposed to diversify.

CHARACTER DISPLACEMENT WITHIN VERSUS BETWEEN SPECIES

At the beginning of this chapter, we noted that, in the absence of heterospecifics, intraspecific competition might become more important in driving trait evolution within populations. However, with the possible exception of populations in island-like settings, populations will rarely encounter intraspecific competition exclusively. Given that most populations likely experience both intraspecific *and* interspecific competition (Connell 1983; Gurevitch et al. 1992; Denno et al. 1995; Kaplan and Denno 2007; Dybzinski and Tilman 2009), how should traits evolve when both forms of competition act together?

Although character displacement within species and character displacement between species are each adaptive responses to competition, the evolutionary outcome of these two forms of character displacement differs. Whereas character displacement within species generates niche-width expansion (as populations adopt a *wider* range of resource-use or reproductive traits), character displacement between species generally promotes niche-width contraction (as post-displacement populations of each species adopt a *narrower* range of resource-use or reproductive traits). Consequently, a tension arises between the opposing effects of intraspecific competition and interspecific competition. Ultimately, an equilibrium may be reached where the strength of selection favoring traits that reduce competition with *conspecifics* (and the resulting effects on trait evolution) matches the strength of selection favoring traits that reduce competition with *heterospecifics* (and its resulting effects on trait evolution; Pianka 2000).

Despite the tension between intra- and interspecific competition, intraspecific character displacement can play a key role in promoting character displacement between species for at least two reasons. First, as we have seen, intraspecific character displacement can favor the evolution of resource or mating polymorphism. The evolution of distinct morphs, prior to interactions between species, can fuel rapid character displacement via differential sorting of the alternative phenotypic variants when a heterospecific competitor *is* encountered (Figure 3.1; Rice and Pfennig 2007). Put another way, when a polymorphic species experiences interspecific competition, it may rapidly become monomorphic and specialize on the resource-use or mating strategy that is the most distinct from that expressed by the heterospecific (reviewed in Schwander and Leimar 2011). Although such a transition from a polymorphic to a monomorphic population may engender greater intraspecific competition and reduce population sizes, it may also be effective at reducing the chances of competitive or reproductive exclusion (see chapter 2).

Second, even when intraspecific character displacement does not favor distinct morphs within species, it maintains (and often increases) both phenotypic and genetic variation in natural populations (see above). Abundant standing variation generated by intraspecific character displacement could therefore increase the chances that character displacement will occur between species (see chapters 3 and 4).

Ultimately, then, intraspecific character displacement can also promote diversity by facilitating character displacement between species. Indeed, as we stressed in chapters 2–4, any factor that facilitates character displacement between species promotes diversity, because lineages that undergo character displacement are less likely to become extinct through exclusion. Moreover, as we discuss in chapter 8, character displacement may promote speciation. Hence, as part of a more general theory for why some groups of organisms are more diverse than others, it is important to consider that taxa that have undergone intraspecific character displacement may be more likely to diversify. Thus, understanding the causes of divergence within species is crucial for illuminating diversity's origins.

SUMMARY

Intraspecific competition can have a profound influence on shaping diversity within species. Intraspecific competition can promote niche-width expansion, sexual dimorphism, and resource and mating polymorphism. Such polymorphisms are of interest in their own right because they provide some of the most dramatic examples of diversity within species. Indeed, the process of diversification and innovation may often begin within species through the evolution of such alternative phenotypes. Thus, contrary to the widespread belief that "without speciation, there would be no diversification" (Mayr

> **BOX 5.1.** Suggestions for Future Research
>
> - Determine the extent to which character release arises via competitively mediated selection as opposed to other selective or nonselective processes.
> - Identify diverse systems to test whether intraspecific character displacement promotes the evolution of novel resource-use or reproductive traits.
> - Use theoretical and empirical approaches to determine whether intraspecific character displacement fosters species diversity, and if it does, the precise mechanisms by which it does so. Specifically, does intraspecific character displacement increase the rate of speciation, reduce the risk of extinction, or both?
> - Use experimental evolution studies to determine whether intraspecific character displacement can trigger adaptive radiation (see chapter 9).
> - Use theoretical and empirical approaches to evaluate the conditions under which mating polymorphism facilitates speciation (see, for example, Corl et al. 2010).
> - Determine whether intraspecific character displacement promotes species diversity by facilitating character displacement between species.
> - Evaluate how factors other than competition—for example, the degree of genetic relatedness among competitors (see Pfennig and Collins 1993; Pfennig et al. 1993; Passera et al. 1996; Pfennig and Frankino 1997; West et al. 2001, 2002)—influences the likelihood that, and the extent to which, a population will undergo intraspecific character displacement.

1963, p. 621), intraspecific character displacement can generate diversity within species that rivals speciation as an important setting for evolutionary innovation and diversification (West-Eberhard 2003). Intraspecific divergence is also potentially important, however, because of its possible role in facilitating character displacement between species and possibly even speciation. In short, intraspecific competition for resources and mates can have far-reaching impacts on the origins of diversity, ranging from promoting diversity and novelty within populations to promoting diversity between species. We list some key challenges for future research in Box 5.1.

FURTHER READING

Davies, N. B., J. R. Krebs, and S. A. West. 2012. *An introduction to behavioural ecology* (4th edn.). Wiley-Blackwell: Oxford, UK. This introductory textbook contains an excellent overview of how intraspecific competition can lead to variable foraging and mating behavior within a population.

Doebeli, M. 2011. *Adaptive diversification*. Princeton University Press: Princeton, NJ. This book reviews the theory on how frequency dependent selection arises and how it maintains alternative phenotypes within a population.

Gross, M. R. 1996. Alternative reproductive strategies and tactics: Diversity within sexes. *Trends in Ecology and Evolution* 11:92–98. This paper presents a comprehensive review of the causes and consequences of mating polymorphism.

Robinson, B. W., and D. S. Wilson. 1994. Character release and displacement in fish: A neglected literature. *The American Naturalist* 144:596–627. This paper examines the evidence for character release in fish and describes why character release may be widespread.

Rueffler, C., T. J. M. Van Dooren, O. Leimar, and P. A. Abrams. 2006. Disruptive selection and then what? *Trends in Ecology and Evolution* 21:238–245. This paper describes the causes and consequences of disruptive selection.

Smith, T. B., and S. Skúlason. 1996. Evolutionary significance of resource polymorphisms in fishes, amphibians, and birds. *Annual Review of Ecology and Systematics* 27:111–133. This paper presents an extensive review of the diversity, ecology, and evolution of resource polymorphisms, with a focus on vertebrates. However, many of the concepts discussed apply more generally to other organisms.

6

ECOLOGICAL CONSEQUENCES

Having focused on the causes of character displacement in the first half of the book, we now explicitly examine some of the consequences of character displacement, starting with its ecological consequences. In this chapter, we consider how the study of character displacement provides a unifying framework for understanding the maintenance, abundance, and distribution of biodiversity.

We begin by discussing how character displacement contributes to species coexistence by promoting niche differences among interacting species. In particular, we describe how competitively mediated selection can promote niche differences, either by causing species to diverge and assume new resource-use or reproductive traits, or by causing each species to become more specialized, such that each utilizes a narrower niche (that is, a more restrictive range of resources or reproductive-trait space). Each of these two non–mutually exclusive processes fosters species coexistence by minimizing competitive interactions between species.

We then discuss character displacement's potential contributions to community organization. Although some authors have questioned competition's role in shaping ecological communities (Andrewartha and Birch 1954; Hubbel 2001)—and, indeed, some patterns in nature (described below) are difficult to reconcile with predictions of competition theory—we review the data suggesting that niche differences promote community diversity.

Next, we consider how character displacement can force one of the two interacting species onto poor-quality resources or reproductive-trait space. Consequently, character displacement can paradoxically fail to prevent extinction.

Finally, we conclude the chapter by considering whether character displacement can influence a species' local distribution and its geographical range. As we describe, there is evidence to suggest that the ability to respond adaptively to competition via character displacement can determine and thereby set the limits of a species' local distribution and possibly its broad-scale geographical range as well.

Our emphasis in this chapter is on community diversity and patterns of species distributions, which are, of course, not the sole foci of ecology. However, species diversity has far-reaching ramifications that range from affecting food-web dynamics to influencing ecosystem functioning (reviewed in Levin 2009). Thus, although we do not offer a broad survey of all possible ecological impacts of character displacement, our goal is to highlight character displacement's fundamental effects on biodiversity, which can have consequences for any process that is tied to diversity.

But first, to lay the foundation for this chapter, we start by discussing how character displacement affects the evolutionary expression of a species' niche.

EVOLUTION OF THE NICHE

Character displacement necessarily causes a species to occupy a new niche. Although niche theory has traditionally focused on features that affect a species' utilization of resources (for example, see Pianka 1976 and references therein), a species' niche also includes environmental features that affect its *reproduction*. Indeed, the competitive exclusion principle (Hardin 1960) can be broadened to state that no two species can occupy a similar ecological or reproductive niche (*sensu* Yoshimura and Clark 1994) or else competitive or reproductive exclusion may occur (Gröning and Hochkirch 2008).

An effective way of visualizing how character displacement (whether ecological or reproductive) leads to the evolutionary occupation of a new niche is to realize that a species' niche can be subdivided into its "fundamental niche" and "realized niche" (Hutchinson 1957; Vandermeer 1972). The fundamental niche is the total range of environmental conditions under which the species could exist in the absence of other, antagonistic species (that is, competitors, predators, parasites; Hutchinson 1957). Essentially, a species' fundamental niche describes where the species *could* live, what it *could* eat, and so on, when these other species are absent. However, other antagonistic species almost always are present, and these heterospecifics (be they competitors, predators, or parasites) can prevent the focal species from occupying its entire fundamental niche and thereby force it into a narrower niche: its realized niche. Consequently, in the presence of such heterospecifics, a species may become more limited in where it *does* live, what it *does* eat, and so on (Figure 6.1).

The classic empirical demonstration of fundamental and realized niches comes from Connell's (1961a, b) research on the barnacles *Chthamalus stellatus* and *Semibalanus balanoides*. Recall from chapter 2 that these species have a stratified distribution on intertidal rocks: *Chthamalus* occurs higher than *Semibalanus* in the intertidal zone.

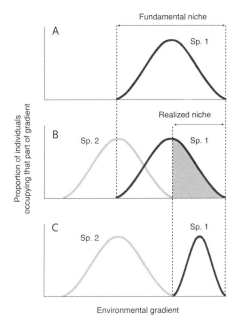

FIGURE 6.1.
How competition can affect a species' niche breadth. (A) The total range of environmental conditions under which a species can live and replace itself when it occurs in the absence of other antagonistic species constitutes that species' fundamental niche. In this case, species 1 initially possesses a broad fundamental niche, as depicted by the broad range of some environmental gradient that it occupies. (B) If a stronger competitor (species 2) invades the habitat of species 1, competition imposed by species 2 can favor those individuals of species 1 (shown in gray) that use a more restricted niche, which would constitute its realized niche. (C) Over evolutionary time, species 1 may assume a narrower niche breadth as an adaptive response to competition from species 2, such that its realized niche becomes its new fundamental niche.

Connell carried out experiments in which he excluded each species separately from intertidal rocks and observed what happened. He found that when *Chthamalus* was excluded, *Semibalanus* did not expand higher up on the rocks, because it apparently could not tolerate the stress of drying out for long periods of time. Thus, *Semibalanus*'s realized niche is similar to its fundamental niche. By contrast, when *Semibalanus* was excluded, *Chthamalus* spread down the rock. Thus, because of competition from *Semibalanus*, the realized niche of *Chthamalus* is much narrower than its fundamental niche.

If a species expresses a realized niche for a sufficiently long period of time, then it may lose the ability to utilize a broader fundamental niche. That is, a population expressing a realized niche is likely to evolve specializations that subsequently render it capable of using only this more restricted niche (see chapter 4). In other words, a species' realized niche could become its new fundamental niche. Essentially, such an evolved niche shift—from a broad, ancestral fundamental niche to a narrower, derived fundamental niche (for example, one that provides a narrower range of resources)—can be viewed as a manifestation of character displacement, as long as this shift occurs in direct response to competitively mediated selection.

In Connell's and other similar examples, one might contend that the ability to distinguish between a fundamental and realized niche is *prima facie* evidence against character displacement having occurred. In particular, if a species can express a wider niche in the absence of competitors (as was the case with *Chthamalus* barnacles in Connell's experiment), then this must mean that the species has undergone no evolution of its niche in response to competition, and, hence, no character displacement. However, the *ability to respond facultatively* to competitors is itself a trait that could evolve (Box 2.2). Thus, the tendency to express a realized niche in the presence of a superior competitor may be a common manifestation of character displacement (see chapter 1 and Table 1.2 for a summary of how to detect character displacement).

Having discussed how character displacement can affect the evolutionary expression of a species' niche, we now turn to what is perhaps the most important ecological consequence of character displacement: it contributes to coexistence within ecological communities by promoting differences between species in resource use, reproductive traits, or both.

PARTITIONING OF RESOURCES AND REPRODUCTION: A REPRISE

In chapter 2, we introduced the concepts of resource partitioning and reproductive partitioning. The former occurs when interacting species subdivide shared resources (Schoener 1974), whereas the latter occurs when species subdivide reproductive-trait space (Littlejohn 1959; Otte 1989; for examples, see chapter 2).

Darwin (1859 [2009]) was the first to suggest that partitioning promotes species coexistence and thereby enhances species diversity. Indeed, Darwin (1859 [2009], p. 127) famously observed that ". . . more living beings can be supported on the same area the more they diverge in structure, habits, and constitution."

But exactly how does partitioning, whether for resources or reproduction, actually promote species coexistence? Moreover, do communities that have greater opportunity for partitioning harbor more species? In this section, we address both questions.

Before proceeding, recall from chapter 2 that partitioning can come about for two evolutionary reasons (Case and Sidell 1983). First, partitioning may involve character displacement, where species come to differ from each other owing to the action of competitively mediated selection (Figure 2.8A). Second, partitioning may involve species sorting, where species that already possess divergent characters upon contact with each other are less likely to undergo competitive or reproductive exclusion than those species that do not differ (Figure 2.8B). Recall also that, of course, these two processes are not mutually exclusive (Figure 2.9).

We begin by considering how competitively mediated selection promotes the partitioning of resources among an assemblage of species. Imagine that numerous species are competing for resources along a single resource spectrum (Figure 6.2A). This spectrum could represent, for example, the range in prey size that predators consume (for example, see Figure 1.8). Imagine further that the overall resource use of each species falls somewhere along this spectrum (for example, each species utilizes a certain range of prey sizes) and that each species' resource use (that is, its resource utilization function; see Figure 2.4) overlaps with that of other species. The intensity of competition

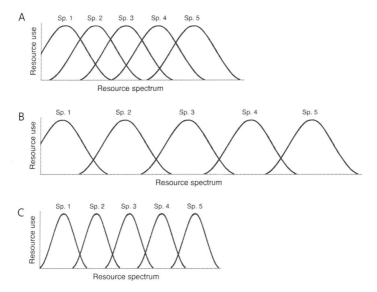

FIGURE 6.2.
How resource partitioning reduces competition between species. (A) Imagine five species sharing a common resource spectrum. (B) When this resource spectrum is broad, selection can minimize overlap between species by promoting divergence between them. (C) Alternatively, when the resource spectrum is narrow, selection can minimize overlap by favoring increased specialization, where each species evolves to utilize a narrower range of resources.

between species increases with increasing overlap in resource use (see Figure 2.5). With complete overlap, one species excludes the other (see Figure 2.2C).

Selection could cause species to minimize overlap in resource use, and thereby enhance their likelihood of coexistence, by promoting character displacement in either of two non–mutually exclusive ways. First, selection could reduce overlap between species by causing species to diverge from one another, such that each species comes to occupy a position along the resource spectrum that overlaps the least with heterospecifics. This selective process can result in different species becoming distributed evenly along the resource spectrum such that overlap between them is minimized (Figure 6.2B). In this case, each species' fundamental niche and realized niche would be equivalent in breadth, but they would occupy different positions along the resource spectrum. Such divergent character displacement is a common response to interspecific competition (Schluter 2000; Dayan and Simberloff 2005).

However, character displacement can occur in the above way only when the resource spectrum is sufficiently broad to enable all competing species to spread out from each other; that is, it can occur only when ecological opportunity is present. Yet in situations where ecological opportunity is low—because, for example, numerous species are already present (Simpson 1953) or alternative resources are scarce (Holmes and Pitelka 1968)—resource partitioning through such divergent character displacement may not be possible (see chapter 3).

In the absence of ecological opportunity, an alternative means of resource partitioning is for each species to become more specialized in resource use (Figure 6.2C; Lack 1954; Mayr 1963). Such specialization can come about when selection reduces variation in resource use within each species by culling those individuals that express the most extreme resource-use traits that overlap with those possessed by heterospecifics (for example, culling the individuals that utilize the smallest and largest prey sizes). Essentially, overlap between species can be minimized if each species specializes on a narrower range of resources. The narrower the range of resources that each species utilizes, the more species can coexist within the community (but see Box 6.1 for an explanation of how *greater* variation within species can promote species coexistence and community diversity). Although the evolution of increased specialization can increase the risk of extinction by making such specialists more susceptible to any change in their environment (Pianka 2000), and by intensifying intraspecific competition (see chapter 5), the narrowing of individual species' resource use by selection reduces the overlap among sympatric species and thereby results in decreased interspecific competition (for theoretical discussions, see Hutchinson 1959; MacArthur and Levins 1967; MacArthur 1972; May and MacArthur 1972; May 1973, 1974; Roughgarden 1974; for an empirical test of the theory, see Werner 1977). In short, increased specialization generally increases the chances that competing species will coexist within the same community.

Essentially, character displacement via increased specialization can be thought of as a mechanism for "species packing" (*sensu* MacArthur 1969). An apt metaphor for visualizing this process was first advanced by Darwin:

> The face of Nature may be compared to a yielding surface, with ten thousand sharp wedges packed close together and driven inwards by incessant blows, sometimes one wedge being struck, and then another with greater force. (Darwin 1859 [2009], p. 67)

In other words, in crowded ecological communities where competition for resources is often severe, competing entities such as species (that is, the "wedges" in Darwin's metaphor) can coexist only when each occupies a specific place along a resource spectrum (that is, a specific realized niche). New species can join the community only by squeezing ever tighter into the available resource spectrum (or by displacing an existing species through exclusion).

Character displacement via increased specialization can promote coexistence and diversity through species packing. Additionally, such character displacement can also transpire rapidly, thereby making exclusion even less likely to occur. Recall from chapter 3 that character displacement (and, therefore, resource partitioning) can proceed through either in situ evolution of novel phenotypes or the sorting of pre-existing

BOX 6.1. Individual Variation and the Coexistence of Species

If the members of competing species are likely to interact with the same individuals in space or time over their lifetimes (as when species are sessile or sedentary), character displacement can actually promote the evolution of *enhanced* trait variation within a species that minimizes competition between species. In particular, because individuals that are more dissimilar in niche use will be less likely to compete (for example, see Figures 2.3–2.6), neighboring individuals that are most dissimilar from one another will perform better than neighboring individuals that are similar (Box 6.1 Figure 1). Thus, trait variants that differ from *local* heterospecifics (as well as conspecifics; see chapter 5) will be selectively favored. In both species, negative frequency-dependent selection could thereby maintain or even enhance trait variation in response to competition. Consequently, at the population level, conspecifics and heterospecifics can display extensive niche overlap, whereas at the individual level sufficient variation may exist to minimize competition and promote coexistence (for a possible example, see Clark 2010).

In short, although species-level differences may be small, individual variation within species (which might have arisen via intraspecific character displacement; see chapter 5) can enable competing species to coexist, depending on how that individual variation is distributed spatially (or temporally) within each species; that is, whether or not the variation in one species complements that in other species can determine whether coexistence is possible.

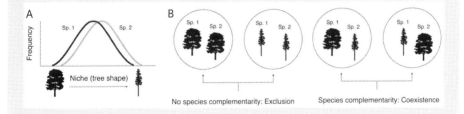

BOX 6.1 FIGURE 1

How individual niche variation *within* species can promote coexistence between species that overlap extensively in niche use, depending on how that variation in niche use is distributed spatially within each species. (A) Species with extensive niche overlap (depicted here as two species of trees that overlap in tree shape) are not expected to coexist stably. However, whether one species excludes the other depends on interactions among *individuals*. (B) When neighboring heterospecific individuals occupy similar niches, competitive exclusion will likely occur. By contrast, when neighboring heterospecific individuals occupy slightly different niches, competition will be less severe and the two species may coexist. Greater variation of traits within species increases the chances that any two individuals in close proximity will differ and therefore experience lower competition. The above scenario will most likely arise in organisms that are likely to interact with the same types of individuals over their lifetimes (for example, in plants or sessile marine organisms).

variation (Figure 3.1). Whereas character displacement that promotes divergent shifts in phenotypes (Figure 6.2B) can occur through either evolutionary route, character displacement that arises via increased specialization *always* involves the sorting of preexisting variation. Because this process merely entails a winnowing of phenotypes that are already present in the population, character displacement that proceeds through this route can transpire in as little as a few generations. Moreover, because the speed of character displacement can determine whether or not exclusion occurs (see chapters 3 and 4), character displacement via specialization can further promote coexistence by simply reducing the likelihood that exclusion occurs.

Regardless of whether character displacement promotes resource partitioning through divergent trait evolution (Figure 6.2B) or through the evolution of increased specialization (Figure 6.2C), both outcomes enhance species coexistence. Indeed, long-standing theory predicts that resource partitioning (whether it stems from character displacement or species sorting) should increase species diversity within communities by promoting coexistence among competitors (Volterra 1926; Lotka 1932; Gause 1934; MacArthur 1972; Schoener 1974; see also Figure 2.3). Empirical support for this general theory comes from both laboratory and field studies. For instance, recall that Gause's (1934) experiments demonstrated that different species of *Paramecium* could coexist stably in laboratory microcosms only when they differed in resource use (see chapter 2). Moreover, as we described above, Connell (1961a, b) showed that two barnacle species coexisted in the wild by partitioning space.

Resource partitioning also appears to enhance species diversity within more complex natural communities. Consider the following classic field study, which motivated much of the theory on partitioning. In this study, MacArthur (1958) asked how five species of warblers could coexist in New England forests, even though they often occur in the same trees. From detailed field observations, MacArthur (1958) found that the five species tended to live in different parts of the same tree; that is, they subdivided the habitat. MacArthur (1958) suggested that these warblers coexisted—even though they used the same habitat and food resources—by feeding and nesting in slightly different microhabitats. More recent experiments have shown that such resource partitioning reduces the costs of competition for nesting and foraging sites (Figure 6.3).

Perhaps more critically, MacArthur and MacArthur (1961) found that 13 different bird communities (from both temperate and tropical regions) exhibit a positive relationship between bird species diversity and foliage height diversity. Thus, the more opportunity there is for species to partition resources (in this case, by foraging at different heights), the greater the species diversity within a community.

Although these differences in foraging height among birds did not necessarily arise via character displacement (such differences could also have arisen via species sorting), these studies suggest that resource partitioning, however it arises, decreases competition and thereby increases community diversity. Presumably resource partitioning that stems specifically from character displacement enhances community diversity.

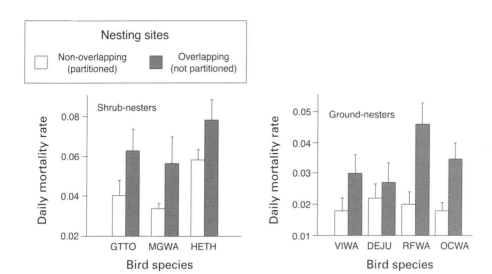

FIGURE 6.3.
Evidence that resource partitioning can minimize costs associated with competition. Seven species of birds that co-occur in the montane forests of Arizona normally do not overlap in nesting microhabitats. When nesting sites were experimentally modified to make nesting microhabitats overlap, those individuals that overlapped in nesting site with a heterospecific experienced higher mortality than did conspecifics that did not overlap in nesting site with a heterospecific. Species abbreviations: GTTO: green-tailed towhee; MGWA: MacGillivray's warbler; HETH: hermit thrush; VIWA: Virginia's warbler; DEJU: dark-eyed junco; RFWA: red-faced warbler; OCWA: orange-crowned warbler. Based on data from Martin (1996).

We have limited our discussion thus far to resource partitioning (primarily because there is an extensive literature on this topic). However, *reproductive* partitioning should be similarly important in maintaining species diversity within communities (reviewed in Gröning and Hochkirch 2008; Pfennig and Pfennig 2009). Indeed, as noted in the previous section, species can coexist stably only if they occupy different reproductive niches; in the absence of such differences, reproductive exclusion may occur (for example, Hochkirch et al. 2007; Kishi et al. 2009). Thus, the evolution of distinctive reproductive niches is no less important for species' coexistence than is the evolution of distinctive niches associated with resource use.

As was the case with resource partitioning, reproductive partitioning can occur when species space themselves out in reproductive trait space (Littlejohn 1959; Otte 1989). Indeed, such a process may even cause different species to utilize completely different sensory modalities to attract and compete for mates. For example, the rapid diversification and coexistence of different species of mormyrid fish from Africa was accompanied by the evolution of a novel communication channel for sexual signaling: a high-frequency electrical channel (Arnegard et al. 2010; see chapter 9).

Alternatively, in the absence of available reproductive trait space, reproductive par-

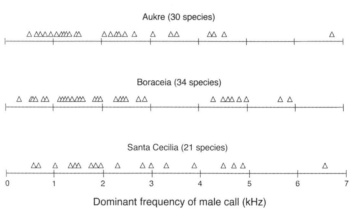

FIGURE 6.4.
An example of reproductive partitioning involving male advertisement calls of hylid treefrog from the eastern Amazon basin. The dominant frequencies of different species are more evenly spaced within each of three species assemblages than expected by chance (within each assemblage, each triangle depicts the dominant frequency of a single species). Based on data in Chek et al. (2003); photo from the public domain.

titioning can occur when each species becomes more specialized in its reproductive traits. For example, in species-rich communities, species may compete for reproductive-trait (or signal) space that would enable individuals to avoid signal masking and species misidentification, both of which decrease the chances of successful reproduction (see chapter 2). In such situations, species' coexistence should be facilitated if each species specializes on a distinctive portion of a reproductive-trait (or signal) spectrum (that is, a spectrum of reproductive-trait [or signal] space that is analogous to the resource spectrum depicted in Figure 6.2).

Essentially, costly reproductive interactions between the different members of an ecological community can be minimized if each species specializes on a narrower portion of the reproductive-trait spectrum. For instance, different species may differ in the range of signals that they each produce or in the timing of signal production. The narrower the range of signal space that each species utilizes for sexual signaling, the more species can be packed into the community. For example, in certain South American frog communities, the vocalizations of the different species are more regularly spaced in acoustic space than expected by chance, suggesting that reproductive partitioning might have fostered species coexistence in these communities by reducing interspecific acoustic competition (Figure 6.4). Likewise, Luther (2009) found that birds singing at the same time during the dawn chorus in an Amazonian rain forest had more differentiated songs in acoustical space than bird species that sang at different times (for other examples of reproductive partitioning in complex communities, see Sueur 2002; Henry and Lucas 2008).

In sum, long-standing theory and empirical studies suggest that when species compete for resources or successful reproduction, they can increase their chances of coexistence by partitioning resources or reproductive-trait space, respectively. Further research is needed, however, to determine the frequency of resource or reproductive partitioning in multispecies communities and to evaluate whether such partitioning is actually essential for species to coexist in complex assemblages where competition may be diffuse (Dayan and Simberloff 2005).

COMMUNITY ORGANIZATION

Above we described how competitively mediated selection can foster species richness within communities by promoting niche differences between species. Left unaddressed, however, is the role of such niche differences in determining the *composition* of species within any given community. Here, we describe the potential contribution of niche differences and character displacement to community organization.

Before proceeding, we must stress two key points. First, niche differences within a community could have come about through species sorting as opposed to character displacement (see chapter 2; Figure 2.8; recall, however, that species sorting and character displacement are not mutually exclusive; see Figure 2.9). Indeed, in many of the examples that we describe below, the origin of niche differences (that is, whether these niche differences arose via species sorting or character displacement) is unknown. One of the key challenges for future research is to determine empirically the degree to which character displacement (and, thus, competitively mediated selection acting in sympatry) accounts for niche differences within, and ultimately the composition of, ecological communities.

Second, ecological communities are complex assemblages that are shaped by numerous factors, including various ecological interactions with other organisms (Table 1.1), but also by environmental constraints and historical contingencies (reviewed in Chave 2009). Thus, the degree to which communities are structured by competitive interactions remains a point of contention (for reviews of this debate, see Lewin 1983; Schluter 2000; Simberloff 2004). Indeed, the "neutral theory of biodiversity" (*sensu* Hubbell 2001) has challenged the notion that niche differences (and, thus, character displacement) are necessary for species to coexist (for critiques of neutral theory, see Dornelas et al. 2006; McGill et al. 2006).

Despite the controversy over competition's role in community organization, numerous studies have indeed found that niche differences exist between species that make up naturally occurring communities (reviewed in Elton 1927; Odum 1959; MacArthur 1972; Pianka 2000; Chase and Leibold 2003; Silvertown 2004). In some cases, niche differences that enable species' coexistence may be present within species rather than between species (Box 6.1). Moreover, field experiments have established the importance of niche differences for maintaining species diversity within natural communities (see, for example, Levine and HilleRisLambers 2009). Given that character displacement is a key (but not exclusive) mechanism for generating niche differences between species (species sorting being the other mechanism), these data therefore suggest that competitively mediated selection can play an important role in determining community composition and organization.

One way to evaluate competition's possible contribution to community composition is to use phylogenetic approaches (Webb et al. 2002; Violle et al. 2011). Such approaches can also suggest whether character displacement, as opposed to species sorting, may have structured a particular community. Specifically, if competition is important in determining the composition of ecological communities, but the reduction of such competition is mediated by species sorting, then species within communities should be more likely to be overdispersed phylogenetically than a same-sized group drawn at random from the regional species pool (Figure 6.5A). That is, the members of a given community should be less closely related to each other than expected by chance. The underlying assumption behind this approach is that, because species within a lineage share the same evolutionary history, they should also be more ecologically similar, have similar niches, and therefore compete more intensely (reviewed in Losos 2008; Wiens et al. 2010). Generally, more closely related species are more ecologically similar (Burns and Strauss 2011) and, possibly because of this, they are more likely to compete with each other, compete more severely, or both (confirmed by Violle et al. 2011; note, however, that competition has also been documented between *distantly* related taxa, such as between ants and rodents [Brown and Davidson 1977]). Closely related competitors should therefore be less likely to coexist in the same community, having either become extinct through exclusion or been prevented from entering the community in the first place. Either process would constitute species sorting.

Empirical studies have found that certain natural communities are indeed overdispersed phylogenetically (for example, Cavender-Bares et al. 2004; Esselstyn et al. 2011). Additionally, some communities are more species rich when they consist of phylogenetically distinct taxa (Maherali and Klironomos 2007). Although character displacement can occur within a community that is phylogenetically overdispersed, character displacement would not likely play a significant role in *structuring* the diversity of such communities. Instead, studies such as those above are more consistent with the hypothesis that species sorting enhances the coexistence of already-dissimilar species within communities.

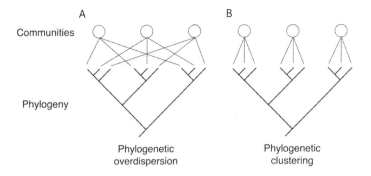

FIGURE 6.5.
Graphic representations of alternative outcomes of phylogenetic tests of character displacement's role in community organization. (A) A pattern of phylogenetic overdispersion—where community members are less closely related to each other than expected by chance—would suggest that competition was important in structuring communities but that such competition was reduced through species sorting. (B) A pattern of phylogenetic clustering—where community members are more closely related to each other than expected by chance—would suggest either that competition was not important in structuring communities or that it was important, but that such competition was reduced through character displacement. Redrawn from Chave (2009).

Other studies, however, have found the opposite pattern, in which communities consist of species that are more phylogenetically *similar* than expected by chance (for example, Webb 2000; Kooyman et al. 2011). Although such phylogenetic clustering (Figure 6.5B) can be explained in terms of closely related species having similar habitat requirements, similar dispersal abilities, or both (Williams 1964; Simberloff 1970), such a pattern is also consistent with character displacement having structured these communities. Specifically, if phylogenetically similar species are more likely to compete with each other, then such species should be more likely to undergo character displacement (see chapter 3). We would therefore not expect communities that have undergone character displacement to be phylogenetically overdispersed (Pfennig and Pfennig 2010; Wiens et al. 2010). Instead, a pattern of phylogenetic clustering would emerge from phylogenetically similar competitors having undergone character displacement rather than having undergone competitive or reproductive exclusion (note, however, that the converse does not apply: evidence of phylogenetic clustering is not conclusive evidence that character displacement has occurred).

Finally, communities may exhibit neither phylogenetic overdispersion nor phylogenetic clustering (for example, see Carranza et al. 2011; Kluge and Kessler 2011). In such communities, character displacement can play a role in promoting species coexistence and diversity without producing a phylogenetic signal (Kluge and Kessler 2011).

To summarize the chapter up to this point, theoretical and empirical studies suggest that by promoting differences between competitors through resource or reproductive partitioning, character displacement lessens costly competitive interactions, reduces the risks of competitive or reproductive exclusion, and thereby increases the chances of species coexistence. Although some have questioned competition's contribution to community organization, abundant data exist to suggest that niche differences foster species diversity and that character displacement can generate such differences. Additional research is needed, however, to determine whether and how character displacement actually does contribute to species richness and composition within communities.

CHARACTER DISPLACEMENT AND DARWINIAN EXTINCTION

Another important ecological consequence of character displacement is that, following its occurrence, one species may monopolize a superior niche at the other species' expense (Pfennig and Pfennig 2005). For example, ecological character displacement results in interacting species utilizing different resources, and these resources may differ in quality (for example, they may differ in nutritional value, search or handling times, or risk of predation or parasitism). When such asymmetries in resources exist, a species that monopolizes the higher-quality resource may even have higher fitness than conspecifics in allopatry (that is, sympatric individuals may be more fit than their conspecifics in allopatry if the former gain access to a new, superior niche). By contrast, a species that is displaced from a high-quality resource will potentially have lower fitness than conspecifics in allopatry (Pfennig and Pfennig 2005).

Reproductive character displacement can have similar effects. For example, following reproductive character displacement in male reproductive traits, one species may monopolize the reproductive trait space (for example, acoustical space that males use to attract females; see chapter 2) that is less "noisy" or in which more reliable information can be conveyed. Consequently, females of this species may experience less difficulty identifying high-quality mates (Pfennig 2000b).

An empirical demonstration comes from spadefoot toads, *Spea bombifrons* and *S. multiplicata*, which have undergone both ecological and reproductive character displacement in response to each other in the American Southwest (Figure 6.6A). Specifically, as a result of ecological character displacement, the tadpoles of one species, *S. multiplicata*, are shunted onto a poorer-quality detritus resource (see Figure 1.7 and chapter 4; Pfennig 2000a; Pfennig and Murphy 2000). Possibly as a consequence, *S. multiplicata* are significantly smaller in adult body size in sympatry than in nearby allopatry (Pfennig and Pfennig 2005; Figure 6.6B). Smaller adult body size is, in turn, associated with lower survival and fecundity (Pfennig and Pfennig 2005; Figure 6.6C). Moreover, because sympatric *S. multiplicata* females are forced to select "average" rather than the most extreme male traits—where these male traits signal male quality—they give up substantial benefits of mate choice (Figure 6.6A; Pfennig 2000b; Pfennig and Pfennig

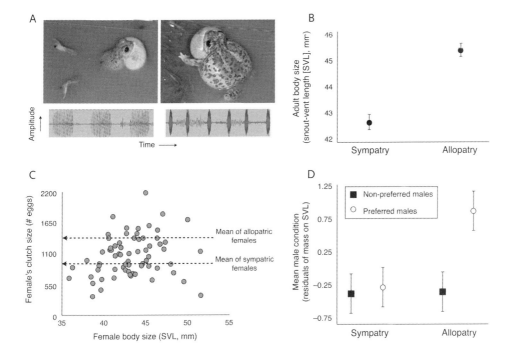

FIGURE 6.6.
As a consequence of character displacement, one species may monopolize a superior niche at the expense of the other species. For example, in spadefoot toads, reproductive character displacement has resulted in (A) male *Spea multiplicata* (left) having a slower call rate than male *S. bombifrons* (right; waveforms showing five seconds of calls for each species are shown below their photos). (B) Possibly because *S. multiplicata* females (unlike *S. bombifrons* females) are unable to use this condition-dependent character to select high-quality mates, *S. multiplicata* adults are significantly smaller in sympatry than in nearby allopatry. (C) Moreover, because there is a significant, positive relationship between female body size and clutch size, smaller adult body size is, in turn, associated with lower fecundity in sympatry. (D) Finally, female mate choice exerts positive, directional selection on male condition in allopatry, but not in sympatry, where they are constrained by the presence of *S. bombifrons*. Specifically, preferred males (that is, the males that females chose as mates) are in significantly better condition in allopatry than are non-preferred males in allopatry, a pattern that does not hold in sympatry. Thus, as a result of character displacement, *S. bombifrons* appears to have monopolized the superior reproductive niche at the expense of *S. multiplicata*. Data from Pfennig and Pfennig (2005) and Pfennig (2008). Photos by D. Pfennig.

2005). These shifts in mate preferences may contribute to the smaller size and fecundity of sympatric *S. multiplicata* (Pfennig 2008).

In contrast to the situation facing *S. multiplicata*, the tadpoles of *S. bombifrons* gain access to the higher-quality shrimp resource following ecological character displacement (see Figure 1.7; Pfennig 2000a; Pfennig and Murphy 2000). Indeed, *S. bombifrons* are significantly larger in adult body size in sympatry than in nearby allopatry (Pfennig and Pfennig 2005), which (as noted above) is associated with increased survival and fecundity in *Spea*. Moreover, because sympatric *S. bombifrons* females are able to select extreme male traits (that is, the fastest-calling males; Figure 6.6A), they reap substantial benefits of mate choice (Pfennig 2000b).

Character displacement may therefore represent a best-of-a-bad-situation (*sensu* Maynard Smith 1982). In particular, character displacement is an adaptive response to competition, but it may also be accompanied by costs: the post-displacement phenotype expressed in sympatry may be associated with lower fitness than the ancestral pre-displacement phenotype expressed in allopatry (Pfennig and Pfennig 2005). Thus, the species that is faced with being shunted into an inferior ecological or reproductive niche following character displacement essentially faces a fitness trade-off between the costs of occupying such an inferior niche and the benefits of avoiding competition with a heterospecific. Generally, character displacement is expected to evolve only when these benefits exceed the costs.

Although one species may be faced with making the best of a bad situation, the other species may actually shift to a new niche that confers *higher* fitness in sympatry than the niche it occupied in allopatry. Yet this raises an important question: if character displacement can create a situation in which one species has higher fitness in sympatry than in allopatry, then why would conspecific populations in allopatry not have occupied that superior niche in the first place?

There are two possible solutions to this problem. First, the superior niche might not be available in allopatry, either because it is not present there or because it is already occupied by another species. Second, allopatric populations might not be able to evolve toward the fitness peak associated with the superior niche in sympatry because of an intervening fitness valley. Yet, in sympatry, the heterospecific's presence may distort the adaptive landscape such that there is no longer a fitness valley (analogous to the sphagnum bog metaphor in character 5), which could explain why the species achieves higher fitness in sympatry than in allopatry.

These asymmetries between interacting species in access to a superior niche following character displacement have at least two major consequences. First, the potential to win exclusive access to high-quality resources or reproductive space can explain why competitively mediated selection may favor increased competitive ability (as opposed to promoting divergence between species in resource use or reproductive traits; see Box 1.1). Indeed, selection may generally favor increased competitive ability, as opposed to divergence onto an alternative, low-quality resource or reproductive trait space, if such resources or reproductive space would not sustain a population (or if alternative resources or reproductive space are not available; see also Price and Kirkpatrick 2009). In such situations, there might be a selective premium on outcompeting the other species, rather than forfeiting a contested resource or reproductive space. However, once one species gains the upper hand, the other species should experience selection for heightened competitive ability (Box 1.1), possibly leading to a co-evolutionary arms race; that is, escalation (see Box 1.1 Figure 1C). Essentially, any given species involved in such an interaction might occupy the superior or inferior niche at different points in evolutionary time. We return to competitively mediated escalation in chapter 9.

A second major consequence of asymmetries between interacting species in access to a superior niche following character displacement is that sympatric populations of one species may be placed at increased risk (relative to the other species and allopatric conspecific populations) of undergoing a Darwinian extinction (*sensu* Webb 2003). By "Darwinian extinction," we mean extinction that results as a consequence of adaptive evolution within a population (Webb 2003; see also Kokko and Brooks 2003). Character displacement can contribute to such enhanced extinction risk when it involves the sort of fitness trade-offs discussed above. Indeed, costs that accrue to individuals in sympatry may reduce population fitness and thereby render sympatric populations more likely to become extinct relative to allopatric populations (Pfennig and Pfennig 2005). For example, if ecological character displacement results in a population being displaced onto a novel resource that is of inferior quality and that therefore supports smaller or otherwise more poorly adapted populations, then such a population may be more susceptible to stochastic extinction events (Pianka 2000). Likewise, displacement onto a more ephemeral resource may make sympatric populations more susceptible than allopatric populations to extinction caused by chance loss of that resource (due to, for example, rapid climate change).

Reproductive character displacement also could engender costs if the displaced phenotypes (for example, male traits or female preferences) are more costly to express than are the pre-displacement phenotypes (Kokko and Brooks 2003). Moreover (and as we describe in greater detail in the next chapter), avoiding heterospecifics may prevent females from selecting high-quality mates and thereby decrease sympatric female fitness relative to allopatric female fitness (Pfennig 2000b; Higgie and Blows 2008; Pfennig 2008). Such trade-offs can reduce female fecundity, rates of reproduction, and even offspring growth or survival (Pfennig and Pfennig 2005). Indeed, if character displacement suppresses condition-dependent sexual selection in sympatry (for example, see Figure 6.6D), sympatric populations may be less able to adapt to changing environments and therefore more susceptible to extinction (*sensu* Lorch et al. 2003).

In sum, character displacement lessens competition, but it can also carry a cost: the post-displacement phenotype that is expressed in sympatry may be associated with substantially lower fitness than the ancestral pre-displacement phenotype that is expressed in allopatry. Therefore, although character displacement is often viewed as a process that minimizes extinction via exclusion, it may paradoxically contribute to higher extinction rates of sympatric populations relative to allopatric populations. In other words, even when populations undergo character displacement, they may still ultimately become extinct.

SPECIES DISTRIBUTIONS AND GEOGRAPHIC MOSAICS

Previously we described how character displacement could determine which species coexist in any given community. These discussions focused on ecological communities as a whole. However, from the perspective of individual species, whether or not character displacement occurs can influence a species' local distribution and its geographical range. In this section, we discuss how character displacement may affect local distributions. In the subsequent section, we consider how it may delimit species' ranges.

To understand how character displacement impacts species' distributions, recall from chapter 3 that populations—not species—undergo character displacement. Moreover, recall that whether or not character displacement occurs in any given population depends on several factors, including the strength of selection for traits that minimize competition, the amount of standing variation in these traits, and the presence of ecological opportunity that enables alternative resource-use or reproductive traits to ultimately succeed. Such factors can vary locally. For example, ecological opportunity may vary even among neighboring populations (Pfennig et al. 2006).

If the facilitators of character displacement vary spatially, then the pattern of species co-occurrence may also become patchy, such that character displacement promotes coexistence in some populations but not in other nearby populations. The resulting species distribution would be a geographic mosaic of habitats in which competitively mediated divergence (and species coexistence) occurred, interspersed with habitats in which exclusion has resulted in the presence of only a single species (Thompson 2005). Indeed, habitats conducive for character displacement may constitute "hotspots" of species diversity that become crucibles of adaptive radiation and further diversification (see chapter 9).

An example in which variation in expression of character displacement promotes such geographic mosaics comes from the spadefoot toads *Spea multiplicata* and *S. bombifrons*. In the southwestern United States, these two species have broadly overlapping ranges. However, on a local scale, these two species co-occur only where ecological opportunity allows ecological character displacement to transpire (Pfennig et al. 2006). In allopatry, the tadpoles of both species display resource polymorphism, producing an omnivore morph that feeds mostly on detritus and a carnivore morph that feeds mostly on fairy shrimp. By contrast, in sympatry, both species are nearly monomorphic: *S. multiplicata* develop primarily into omnivores that outcompete *S. bombifrons* for detritus, whereas *S. bombifrons* develop primarily into carnivores that outcompete *S. multiplicata* for fairy shrimp (see Figure 1.7).

In ponds where detritus and shrimp are both present, the two species undergo character displacement and thereby coexist by utilizing different resources (Pfennig et al. 2006). However, in nearby ponds (which may be less than 1 km away) where only one of these two resources is available, only one species is present—that is, the species that is the superior competitor for the available resource (Pfennig et al. 2006). The result is a patchy distribution of populations where either one or both species is locally present (Pfennig et al. 2006).

As an aside, an alternative pattern of patchy distribution could arise if two species co-occur on a regional scale (as above) but they do not co-occur on a local scale; that is, each species may be present alone in association with a different microhabitat. Such a pattern could arise from differential competitive exclusion (that is, species sorting), in which one species is consistently driven locally extinct in a particular microhabitat because it is outperformed in that microhabitat by the other species (Gröning and Hochkirch 2008). In this way, species sorting could produce a patchy distribution in which species occur in different microhabitats within a broader region of co-occurrence (Gröning et al. 2007). Yet character displacement could also engender such a pattern. Specifically, different species may evolve different habitat preferences as an adaptive response to competitively mediated selection (chapter 2; Price and Kirkpatrick 2009). Distinguishing between the effects of species sorting versus character displacement on species distributions therefore requires detailed investigations into the evolutionary origins of habitat use.

Above, the focus was on spatial variation in the expression of character displacement. Yet the spatial distribution of populations that do and do not experience character displacement will not necessarily be stable in time. Areas in which a population becomes locally extinct can be recolonized, and character displacement can subsequently proceed. Indeed, as we first noted in chapter 3, a population that is declining and moving toward extinction due to competitive or reproductive exclusion could be "rescued" by immigrants from populations that have already undergone character displacement (for experimental evidence that dispersal can promote local adaptation and thereby rescue populations that are experiencing environmental deterioration, see Bell and Gonzalez 2011). In other words, gene flow among populations could enhance genetic variability that promotes local adaptation to competitors (Barton 2001; Case et al. 2005; Bridle and Vines 2007).

CHARACTER DISPLACEMENT AND SPECIES RANGES

In the previous section, we emphasized that whether species are patchily or homogeneously distributed within their range can depend on whether or not character displacement occurs. Indeed, there are many cases in which competitive interactions influence such small-scale patterns of distribution (see examples described earlier in this chapter). However, some authors have questioned whether processes acting at a small scale can also explain large-scale biogeographic patterns. Indeed, species distribution (niche) models generally suggest that climatic tolerances may determine large-scale patterns of distribution and diversity in most species (reviewed in Wiens 2011). Here we consider whether species distributions depend on competitive interactions and whether or not character displacement occurs.

As background, the suitability of habitat at the edge of a species' range is generally thought to limit expansion and therefore set a species' range limits (Sexton et al. 2009). In particular, if a species is adapted to certain environmental conditions, its range may be limited by its inability to tolerate conditions outside this range. Yet, in such cases, what prevents a species from expanding its range by adapting to novel environments?

Whether a species' range expands, contracts, or remains stable is determined by the dynamics of populations at the edge of the species' range (Kirkpatrick and Barton 1997). Because these populations would likely experience environmental conditions that are at the limits of the species' tolerance, such populations are generally smaller and more fragmented than populations in the center of the species' range. These populations are therefore less likely to possess genetic variation that would enable adaptation to conditions outside the species' current range. Moreover, gene flow from the center of the range may continually infuse alleles that are maladaptive at the edge into edge populations (Kirkpatrick and Barton 1997; for example, see Harper and Pfennig 2008). Consequently, only populations that are capable of undergoing local adaptation to edge conditions would persist, and these populations set the limits of a species' range (Kirkpatrick and Barton 1997; Case and Taper 2000; Price and Kirkpatrick 2009; reviewed in Bridle and Vines 2007).

Most studies investigating the causes of species' range limits have focused on abiotic factors. Sexton et al. (2009) recently reviewed empirical studies that evaluated the factors that limit species' ranges. Of 146 studies that focused on abiotic factors, abiotic factors were found to play a role in limiting ranges in 112 studies (77 percent). In contrast, biotic factors were addressed in only 51 studies. Of these, biotic factors were found to influence a species' range in 31 studies (61 percent). Moreover, of these 51 studies, 26 focused on the effects of competition. Although such meta-analyses must be interpreted with caution—because they may be biased by what factors the authors of the original studies chose to investigate—Sexton et al.'s (2009) review does suggest that, although abiotic factors often set species' ranges, competition may also be important.

But how might competition influence a species' range limits, and what role (if any) does character displacement play? Consider that, if, at the edge of its range, a species encounters a heterospecific competitor, then competition would further stress these populations, which are already at their environmental limits. However, whether or not this focal species extends its range into that of the competitor depends on whether or not it can establish viable populations within the competitor's range. This is where character displacement potentially plays an important role in influencing a species' range limits. As described earlier in this chapter, populations that undergo character displacement are more likely to coexist with a heterospecific competitor. Thus, if populations at the edge undergo character displacement, they should be more likely to persist and expand into the competitor's range. In other words, in cases where competition does play a role in setting the limits of a species' range, character displacement can enable range expansion (Case and Taper 2000; Case et al. 2005; reviewed in Bridle and Vines 2007).

Of course, whether character displacement occurs within edge populations depends on the same factors that determine whether a species can adapt to abiotic factors at the range edge (Sexton et al. 2009). Two factors that are likely to be particularly important are the amount of standing variation in edge populations (see chapter 3) and the gene flow from the center of the species' range to the edge. Small populations may lack the genetic variability that enables local adaptation (in this case, local adaptation would be in the form of character displacement). Furthermore, populations at the edge of a species' range may receive an influx (via gene flow) of alleles from the center of the species' range (where the heterospecific is absent) that are maladaptive at the edge (reviewed in Bridle and Vines 2007).

Moreover, ecological opportunity also potentially plays a role in setting a species' range (Price and Kirkpatrick 2009). In particular, for a population to undergo character displacement, exploitable resources or reproductive-trait space must be available (see chapter 3 and the previous section). In the absence of such ecological opportunity, populations will experience competitive or reproductive exclusion (Pfennig et al. 2006). Thus, those edge populations that encounter a competitor but that lack ecological opportunity would be more likely to become extinct. Such populations would thereby set the range limits for that species (Price and Kirkpatrick 2009).

In sum, populations that possess the attributes that facilitate character displacement (see chapter 3) will be more likely to persist in the face of competition (Bridle and Vines 2007). Consequently, as a species extends its range into that of a competitor, the ability to respond adaptively to competition via character displacement can determine and thereby set the limits of the species' range (Case and Taper 2000; Case et al. 2005; reviewed in Bridle and Vines 2007). Character displacement should therefore facilitate wider ranges, whereas its absence may lead to more restricted species ranges.

Future studies are needed to determine the extent to which competition and character displacement influence large-scale biogeographical patterns. Wiens (2011) contends that there are presently no clear-cut examples. However, the paucity of examples may reflect the fact that those who study large-scale biogeographical patterns generally do not consider competition to be a potential explanation for the patterns that they study. Indeed, many such studies do not consider biotic factors at all, possibly because data on abiotic factors (such as climatic variables) are easier to obtain than are data on past and present species interactions. Clearly, additional research is needed to address character displacement's role in species distributions.

SUMMARY

The study of character displacement provides vital insights into patterns of species coexistence, community organization, and species distributions. Character displacement

> **BOX 6.2.** Suggestions for Future Research
>
> - Identify the conditions under which resource or reproductive partitioning unfold via divergent trait evolution (as in Figure 6.2B) versus increased specialization (as in Figure 6.2C).
> - Using experiments and surveys of natural populations, determine whether character displacement increases species diversity within ecological communities.
> - Identify additional cases of reproductive partitioning in natural populations and its causes. Further, determine the degree to which reproductive partitioning affects community organization and species distributions.
> - Using experimental and theoretical approaches, determine whether the spatial distribution of individual variation contributes to species coexistence (for example, see Box 6.1).
> - Gather data from natural populations to determine whether character displacement generally creates asymmetries between interacting species in their access to a superior niche, and if it does, why.
> - Evaluate whether habitats that are conducive for character displacement constitute hotspots of species diversity and/or crucibles of adaptive radiation.
> - Determine the extent to which competition and character displacement influence large-scale biogeographical patterns, such as the limits of a species' range.

fosters species coexistence by promoting differences between species in resource use (resource partitioning) and in reproductive traits (reproductive partitioning). By promoting niche differences, character displacement lessens costly competitive interactions, reduces the risks of competitive or reproductive exclusion, and thereby increases the chances of species coexistence and diversity within communities. Beyond its effects on species coexistence, character displacement has the further potential of altering population dynamics, extinction risk, and, concomitantly, species' ranges. Studying character displacement can therefore reveal how the fitness consequences of interactions between species potentially translate into large-scale patterns of species richness, distributions, and diversity. Thus, the study of character displacement provides a unifying framework for understanding the maintenance, abundance, and distribution of biodiversity. We list some key challenges for future research in Box 6.2.

FURTHER READING

Dayan, T., and D. Simberloff. 2005. Ecological and community-wide character displacement. The next generation. *Ecology Letters* 8:875–894. This paper reviews the theory and evidence for community-wide character displacement.

Haloin, J. R., and S. Y. Strauss. 2008. Interplay between ecological communities and evolution: Review of feedbacks from microevolutionary to macroevolutionary scales. *Annals of the New York Academy of Sciences: The Year in Evolutionary Biology* 1133:87–125. This paper presents an excellent overview of the role of ecological interactions in both micro- and macroevolution. The paper also discusses the interpretation of phylogenetic signal in community assembly.

Hutchinson, G. E. 1957. Concluding remarks. *Cold Spring Harbor Symposium on Quantitative Biology* 22:415–427. This paper first proposed the distinction between an organism's fundamental niche and its realized niche.

Price, T. D., and M. Kirkpatrick. 2009. Evolutionarily stable range limits set by interspecific competition. *Proceedings of the Royal Society B: Biological Sciences* 276:1429–1434. This paper examines how resource competition results in evolutionarily stable range limits.

Webb, C. 2003. A complete classification of Darwinian extinction in ecological interactions. *The American Naturalist* 161:181–205. This paper explores how extinction can occur as a consequence of adaptive evolution within a population.

Wiens, J. J. 2011. The niche, biogeography and species interactions. *Philosophical Transactions of the Royal Society B: Biological Sciences* 366:2336–2350. This paper calls into question whether species interactions play any role in determining large-scale biogeographical patterns, such as determining a species' range limits.

7

SEXUAL SELECTION

In chapter 6, we examined the ecological ramifications of character displacement. For the remainder of the book, we discuss character displacement's evolutionary implications, starting with its implications for sexual selection.

Identifying the factors that can influence sexual selection is critical for understanding the causes of diversification, for at least two reasons. First, traits produced through sexual selection—such as the striking sexually dimorphic features observed in many organisms—provide some of the most dramatic examples of diversity within species (for example, see Figure 5.8). Second, sexual selection may drive speciation (Figure 7.1; reviewed in Andersson 1994; Butlin and Ritchie 1994; Ritchie 2007; Price 2008). In particular, any change in the ways a population of sexually reproducing organisms chooses or obtains mates can promote rapid differentiation from ancestral populations (for example, Fisher 1930 [1999]; Lande 1981, 1982; Higashi et al. 1999; Boul et al. 2007). Because competitively mediated selection can alter sexual traits (even when such selection arises from competition for resources rather than for reproduction; see chapter 3 and also below), understanding how competitively mediated selection and sexual selection interact is vital for explaining diversity's origins.

In this chapter, we begin by considering how character displacement impacts sexual selection by affecting the traits targeted by sexual selection and the strength and mode of such selection. Indeed, when competitively mediated selection generates variation in fitness resulting from either within-sex competition for reproduction or between-sex choice of or ability to find mates, competitively mediated selection *is* sexual selection.

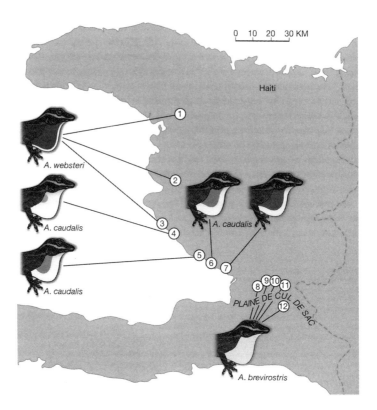

FIGURE 7.1.
Sexual selection typically targets traits that are readily detected by choosy individuals or by same-sex competitors and that provide information about the bearer's quality as a mate. Closely related, sympatric species often differ in traits targeted by sexual selection, suggesting that these traits maintain species boundaries and are therefore involved in speciation. For example, sympatric species of *Anolis* lizards always differ in the coloration of their dewlaps (a colored flap of skin on the throat that males use to attract female mates and deter male rivals). Here, three species of *Anolis* that are distributed along the west coast of Haiti differ in dewlap coloration (the dark shading denotes a vivid orange dewlap, whereas the light shading denotes a light-colored, pale dewlap) but are otherwise nearly indistinguishable in appearance. Note, especially, that adjacent, heterospecific populations differ the most in dewlap coloration, a possible outcome of character displacement. Numbers indicate different populations of each species. Reproduced from Webster and Burns (1973) after Losos (2009), with the kind permission of the author and publisher.

We next describe the consequences of character displacement on sexual selection, especially for diversification. As we will see, when sexual selection and competitively mediated selection act in concert, diversity can be greatly magnified.

Finally, we consider the converse—that is, how sexual selection can affect character displacement. Specifically, we describe how sexual selection can either impede or facilitate character displacement. Ultimately, any trait that enables individuals to simultaneously minimize interactions with heterospecifics while maximizing their mating success should be favored.

We begin, however, with a brief overview of how sexual selection works.

HOW SEXUAL SELECTION WORKS

The idea of sexual selection was initially developed by Darwin (1859 [2009]) to explain an apparent paradox: how could natural selection promote the evolution of traits that seemingly reduce their bearer's survival and that are often expressed in only one sex? The long tail feathers of certain birds are an obvious example (Pryke and Andersson 2005), but such traits are commonly found in many sexually reproducing species (reviewed in Andersson 1994). As Darwin realized, sex provides a solution to this paradox. From an evolutionary standpoint, failing to mate successfully is tantamount to dying young. In both cases, the individual would fail to pass on its distinctive characteristics to future generations. Thus, according to Darwin, any trait that enhances an individual's mating success will be favored, even if this trait also reduces its individual bearer's long-term chances of survival.

Darwin (1859 [2009]) termed the process that favors traits that enhance an individual's mating success "sexual selection." Darwin (1871) further proposed that sexual selection assumes two forms. First, *intra*sexual selection occurs when same-sex individuals compete for mating opportunities. Because females usually invest more into reproduction (Williams 1966; Trivers 1972; Clutton-Brock 1991) and are therefore the limited sex (Bateman 1948), intrasexual selection typically involves males competing for access to females. Second, *inter*sexual selection occurs when one sex chooses members of the opposite sex that bear certain traits. Females are usually the choosier sex (because, again, they typically invest more into reproduction), and they are generally expected to evolve preferences for mates that enhance either their own fitness or that of their offspring (reviewed in Andersson 1994). Thus, variation in mating success is generally greater among males than among females, and this variation reflects the intensity of sexual selection (Wade and Arnold 1980; for empirical tests, see Jones et al. 2000, 2002b, 2005).

Although variation in *mating* success was crucial for Darwin in defining sexual selection, we now know that competition for access to reproduction and mate selection potentially takes subtle forms, some of which (such as competition among sperm) can even occur postmating (Andersson 1994; Eberhard 1996; Wiley and Poston 1996; Jones et al. 2002a). Moreover, much of the work on sexual selection post-Darwin has concerned the evolution of mate choice per se—which is often explained by variation in offspring quality or numbers—rather than variation in mating opportunities (Andersson 1994).

Sexual selection can therefore be defined as the process that arises from variation in fitness resulting from either within-sex competition for reproduction or between-sex choice of mates (Andersson 1994). This more general definition encompasses such phenomena as sperm competition (Parker 1970; Birkhead 2000; Pitnick and Hosken 2010), the evolution of mate choice behavior per se (Lande 1981; Kirkpatrick 1982; Pomiankowski 1987), and cryptic choice that arises postmating (such as the post-copulatory ability of females to discriminate between and differentially utilize the sperm of different males; Eberhard 1996). Natural selection, by contrast, arises when individuals vary in fitness because of variation in survival or fecundity.

In considering how mate preferences evolve—that is, why females adopt preferences for particular traits—it is important to realize that female mate preferences stem from sensory systems that are shaped not only by selection to identify high-quality mates (that is, mates that enhance the female's own fitness [Thornhill 1976] or that of her offspring [Welch et al. 1998]), but also by selection to perform such tasks as foraging successfully and avoiding predation (Ryan 1998; Endler and Basolo 1998). Consequently, males are under strong selection to use traits that are readily perceived by (and maintain the attention of) females. Nevertheless, females will also exert selection on males to use traits that convey reliable information about each male's quality as a mate (Fisher 1930 [1999]; Williams 1966; Zahavi 1975; Thornhill 1976; reviewed in Andersson 1994). Thus, female preferences favor male traits that are readily detected (given the females' sensory system and habitat) and that also provide information about male quality.

Having briefly discussed how sexual selection works, we now consider how character displacement alters patterns of sexual selection. An important point to bear in mind throughout this chapter is that, when competitively mediated selection generates variation in fitness resulting from either within-sex competition for reproduction or between-sex choice of mates, competitively mediated selection *is* sexual selection. Thus, in such cases, character displacement and trait evolution stemming from sexual selection become one and the same. However, competitively mediated selection need not generate such variation in fitness, in which case it would not constitute sexual selection. Nevertheless, the evolution of traits in response to competitively mediated selection—character displacement—can generate new conditions or new trait variants that alter the circumstances under which sexual selection operates. Below, we develop these ideas in greater detail to examine the range of ways in which character displacement can impact sexual selection.

HOW CHARACTER DISPLACEMENT AFFECTS SEXUAL SELECTION

Character displacement can affect the patterns and strength of sexual selection in two general ways (Figure 7.2). First, character displacement often generates shifts in traits associated with reproduction or resource use (see chapter 2). By instigating such shifts, competitively mediated selection can change how sexual traits are produced and perceived, thereby altering the expression and nature of sexual selection (Price 1998). Second, competitively mediated selection (for access to either reproduction or

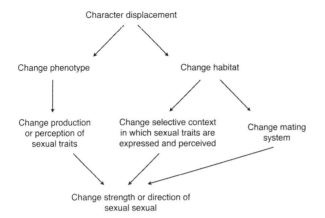

FIGURE 7.2.
Flowchart of the ways in which character displacement may affect sexual selection. Moreover, for interacting species that occur both sympatrically with each other and in allopatry, such effects will arise in sympatry only. Thus, sympatric and allopatric populations may diverge in the strength or mode of sexual selection.

resources) can also cause populations to shift to a new habitat (see chapter 2). A habitat shift can, in turn, modify the selective context in which sexual traits are expressed and perceived and thereby change the pattern of sexual selection in that new context relative to the pre-displacement habitat (Price 1998; see also Endler and Basolo 1998; Boughman 2001). Moreover, habitat shifts can transform the mating system or change the population sex ratio of breeding individuals and thereby alter the strength or mode of sexual selection (reviewed in Emlen and Oring 1977; Andersson 1994; Shuster and Wade 2003; for example, see Karlsson et al. 2010). We discuss each of these two effects separately below.

EFFECTS OF PHENOTYPIC SHIFTS

The impact of competitively mediated selection on sexual selection is most obvious in the context of reproductive character displacement. By causing evolutionary changes in reproductive characters, reproductive character displacement necessarily alters sexual selection in sympatry relative to allopatry. Shifts in any reproductive character—ranging from extravagant secondary sexual traits (for example, see Figure 7.1) to reproductive physiology (for example, the preferential use of conspecific sperm or pollen)—can alter patterns of mate choice and within-sex competition among conspecifics in sympatry relative to allopatry.

Consider, for example, that females often evolve the ability to discriminate conspecifics from heterospecifics better in sympatry than in allopatry (reviewed in Howard 1993). Yet the evolution of such discrimination can rarely occur without also generating new patterns of sexual selection among conspecifics (Ryan and Rand 1993; Pfennig 1998; Ryan 1998; Ptacek 2000; Pfennig and Ryan 2010). In particular, as sympatric females evolve preferences that enable them to better differentiate conspecifics from heterospecifics, sympatric females (but not allopatric females) would select against conspecific males whose sexual traits resemble heterospecific males most closely (Ryan and

Rand 1993; Pfennig 1998). When this occurs, the pattern of sexual selection differs in sympatry relative to allopatry (for example, Pfennig 2000b; Höbel and Gerhardt 2003; Svensson et al. 2010).

The above example illustrates how reproductive character displacement could alter patterns of intersexual selection. In the context of intrasexual selection, changes in male traits brought about by character displacement can alter patterns of male-male competition (if, for example, character displacement results in the loss of male weaponry; Emlen 2008). Moreover, character displacement could alter patterns of sperm competition and sexual conflict among conspecifics in sympatry relative to allopatry (*sensu* Arnqvist and Rowe 2005; see also Palumbi 2009). In particular, selection that minimizes matings with heterospecifics could cause changes in female reproductive morphology and physiology that alter competitive outcomes among conspecific males (or their sperm; Howard 1999; Pitnick and Hosken 2010). For example, male seminal fluid proteins that confer mating success in allopatry may not be successful in sympatry. Regardless of whether by inter- or intrasexual selection, reproductive character displacement causes sexual selection to differ in sympatry relative to allopatry (Pfennig and Ryan 2007).

As with reproductive character displacement, *ecological* character displacement can also alter sexual selection in sympatry relative to allopatry. Ecological character displacement often causes species to adopt novel resource-acquisition traits (chapters 2 and 4). If the adoption of novel resource-use traits also influences mating success (for example, via the production or perception of sexual traits; Etges et al. 2007, 2009), then such shifts would change the strength or mode of sexual selection in sympatry relative to allopatry (Pfennig and Pfennig 2009; see also Maan and Seehausen 2011). Although we focus in the following discussion on how ecological character displacement alters male traits and female preferences, changes in *any* sexual traits in tandem with shifts in resource use can generate differential sexual selection in sympatry relative to allopatry. We begin by exploring the effects of ecological character displacement on male traits.

To illustrate how ecological character displacement can affect the production of male traits, consider the following example. Several species of Galápagos finches have undergone ecological character displacement in beak morphology (reviewed in Lack 1947; Grant 1986; Grant and Grant 2008). These changes in beak morphology can indirectly alter the song that a male produces (Podos 2001), which can change male-male interactions and female mate attraction (Podos 2001; but see Grant and Grant 2010). Thus, by causing shifts in resource-use traits, ecological character displacement can alter the targets of sexual selection between conspecific populations in sympatry versus allopatry (Figure 7.3A–D).

In the example above, male sexual traits were involved in both resource use and mate acquisition. However, selection that lessens resource competition may also promote changes in diet that can affect an individual's ability to acquire dietary components that are, in turn, assimilated into male displays. For instance, males of many animal species incorporate diet-derived pigments (for example, carotenoids) into visual sexual

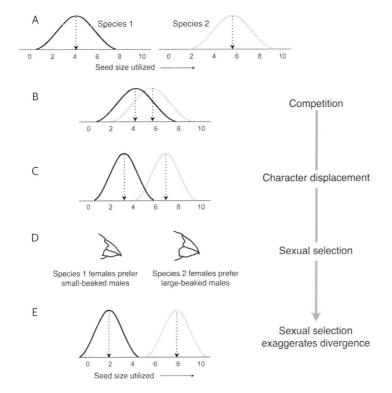

FIGURE 7.3.
How ecological character displacement affects sexual selection. (A) Imagine two seed-eating bird species that initially overlap in the size of seeds that each utilizes when in allopatry (on each graph, the arbitrary units denote different seed-size classes, and the dashed vertical line represents the mean seed size that each population utilizes). (B) When these two species encounter each other, they will compete for seeds of similar sizes. (C) Selection may lead to ecological character displacement, in which one species (species 1) shifts to utilizing small seeds and the other (species 2) shifts to utilizing large seeds. (D) Following character displacement, sympatric females of species 1 may then begin to prefer small-beaked males, because these males will sire offspring that can best utilize smaller seeds. By contrast, sympatric females of species 2 may begin to prefer large-beaked males, because these males will sire offspring that can best utilize larger seeds (these preferences may be based on male song, which is influenced by male beak size and shape; see text). (E) Eventually, these female preferences may promote even greater divergence (in both male traits *and* in resource utilization) than would have occurred by character displacement alone if, for example, females preferred more exaggerated songs in each of the beak types (for example, compare mean seed size utilized in panel E to that in panel C). Although ecological character displacement is illustrated here, the same sequence of events could also unfold under reproductive character displacement such that traits may become more exaggerated beyond what is expected for minimizing reproductive interactions between species.

displays (Andersson 1994; Olson and Owens 1998). If ecological character displacement promotes a dietary change, then sympatric males may differ from allopatric males in their ability to use such pigments in their displays (Grether et al. 1999; Boughman 2007). Likewise, diet can alter the components of chemosensory signals (Ferveur 2005), and differences in such signals can promote divergent behaviors among populations (Smadja and Butlin 2009). Indeed, because male displays are often condition dependent (reviewed in Andersson 1994), shifts in resource use resulting from ecological character displacement can affect the attractiveness of these displays to females (van Doorn et al. 2009).

When males express modified sexual traits following a shift to a novel resource, such traits can become further modified or elaborated via sexual selection in at least three ways. First, intrasexual selection should favor those elements of the male trait that are most effective at mediating competition with other males (Andersson 1994). Second, following ecological character displacement, a newly produced male trait may tap into a hidden preference among females that had not been expressed previously (because such novel male traits did not formerly exist in the population; for example, Basolo 1990; Proctor 1991). In particular, novel environments often reveal hidden genetic variation for traits that had not previously been expressed phenotypically (for example, see Ledón-Rettig et al. 2010 and references therein). The expression of novel male traits may likewise reveal latent female preferences (reviewed in Borgia 2006). Third, and in conjunction with this uncovering of hidden preferences, sympatric females may be under strong selection to identify those aspects of the novel male traits that reliably indicate a male's quality as a fitness-enhancing mate. Novel preferences that identify high-quality males will be favored, and, concomitantly, male traits that reveal quality will be favored.

Shifts to a novel resource can have direct effects on male trait production that indirectly affect female mate preferences. Yet ecological character displacement can also directly modify female preferences when shifts in resource-acquisition traits affect female perception and choice of males. In particular, changes to a female's phenotype, caused by ecological character displacement, can alter whether and how she perceives male traits and thereby modifies her mate preferences. That is, changes in female preference in the post-displacement populations may be a by-product of changes to female sensory systems that arise from shifts to a new resource (*sensu* Endler and Basolo 1998; Ryan 1998). For example, changes to jaw morphology (for foraging) or in visual systems (for detecting prey) could affect females' perception of auditory or visual male traits (Ryan 1998) and, consequently, the likelihood that they would mate with males expressing a particular trait as opposed to some alternative trait.

Ecological character displacement can also directly alter female mate preferences by promoting the evolution of novel preferences for males that can best utilize the new resource (Snowberg and Benkman 2009; van Doorn et al. 2009). Above, we suggested that female preferences might evolve secondarily in response to shifts in male traits that transpire during ecological character displacement. Alternatively, male traits may not initially be impacted by ecological character displacement. Yet those sympatric females may be favored that prefer novel sexual traits that reveal a male's ability to enhance female fitness. For instance, following ecological character displacement, selection would favor sympatric females that prefer males capable of siring offspring that succeed on the new resource (Snowberg and Benkman 2007; van Doorn et al. 2009), and the male traits that indicate this ability may differ from those expressed in allopatry.

As a possible example, recall from chapter 3 that two species of three-spined sticklebacks (*Gasterosteus aculeatus* complex) have undergone character displacement, which has resulted in one species expressing a distinctive benthic ecomorph and the other a distinctive limnetic ecomorph. The benthic ecomorph inhabits the vegetated littoral zone of lakes, whereas the limnetic ecomorph occupies open water. Recall also that the light environments experienced by these two ecotypes differ. In the littoral zone, where the benthic ecomorph forages and mates, red coloration is more difficult to detect (Boughman 2001). By contrast, in open water, where the limnetic ecomorph forages and mates, red coloration is more discernible (Boughman 2001). Finally, recall that only limnetic females use red coloration in mate selection (Boughman et al. 2005).

Red coloration is condition dependent in limnetics, but not in benthics or their anadromous ancestor (Boughman 2007). These results therefore suggest that mate preferences for red in females—and the condition dependence of red coloration in males—arose following divergence of the benthic and limnetic sticklebacks into alternative habitats (Boughman 2007). Indeed, the move into the limnetic habitat may have revealed an underlying bias for red coloration in the limnetic populations, which subsequently favored the condition dependence of red coloration in males (Boughman 2007). Thus, following ecological character displacement, limnetic females may have evolved preferences for males that were capable of siring offspring that could succeed on the limnetic habitat (Boughman 2007).

In sum, by inducing shifts in an organism's phenotype, character displacement—whether reproductive or ecological—changes both the production and perception of sexual traits, and thereby alters the targets and patterns of sexual selection in sympatry relative to allopatry. However, inducing changes in sexual traits is not the only way in which character displacement can affect sexual selection. As the stickleback example above illustrates, character displacement can also change the habitat (and, thus, the selective context) in which mating takes place. It is this topic that we turn to next.

EFFECTS OF HABITAT SHIFTS

As noted in the previous section, when character displacement alters the habitat that individuals use for reproduction, it can change the environmental and selective context in which individuals compete for reproductive opportunities. Consequently, competitively mediated selection can transform the strength and mode of sexual selection in post-displacement populations relative to pre-displacement populations. Generally, the habitat in which mating takes place selects for the evolution of male traits (and female preferences for those traits) that enhance both the detection and discrimination of those traits (Endler 1992; Endler and Basolo 1998; Ryan 1998).

Because an organism's habitat critically affects the attenuation and perception of its sexual traits (reviewed in Wiley 1994; Bradbury and Vehrencamp 1998; Boughman 2002; Gerhardt and Huber 2002), shifts in habitat use (such as those that may be brought about by character displacement) can favor the expression and evolution of novel sexual traits (for example, McNett and Cocroft 2008). Indeed, selection could even favor the use of alternative sensory modalities in both males and females if the post-displacement habitat is no longer favorable for a signaling system used in the pre-displacement habitat. For example, females may switch to relying on olfactory signaling if visual signals are no longer discernible. Because females often rely on multiple cues in mate choice (Candolin 2003; Hebets and Papaj 2005), such switches to an alternative sensory modality may often occur readily (Hankison and Morris 2003).

In addition to physical differences in habitat, predation and parasitism may also alter patterns of sexual selection following character displacement (*sensu* Price 1998; Maan and Seehausen 2011). Because male traits that are conspicuous to females are often also apparent to predators and parasites (for example, Slagsvold et al. 1995; Wagner 1996; Lehmann et al. 2001), males of many species typically face a trade-off between expressing traits that simultaneously maximize mate attraction and localization and minimize predation and parasitism (reviewed Endler 1991; Andersson 1994). Consequently, when predation or parasitism is high, then sexual selection may favor less-exaggerated sexual signals (reviewed in Endler 1991; Houde 1997). Indeed, in extreme situations, conspicuous male traits may be lost altogether if predation or parasitism is intense. For example, male field crickets (*Teleogryllus oceanicus*) from Hawaii have lost the ability to make mating calls because of pressure by the parasitic fly *Ormia ochracea*, which uses the male call to locate hosts (Zuk et al. 2006). Moreover, if females face a high risk of predation or parasitism by expressing strong mating preferences, they may simply become less choosy and sexual selection thereby is weakened (Jennions and Petrie 1997). Thus, when character displacement causes a population to shift to a new habitat, the post-displacement population may become exposed to new predators or parasites, which in turn favors male traits and female preferences that minimize the risk of predation or parasitism.

The above discussion highlights how divergence in habitat use owing to character displacement can instigate changes in the targets and mode of sexual selection. Changes in habitat or resource use can also exert more subtle effects on sexual selection. Specifically, such changes can alter both the mating system (that is, the way in which the members of a population obtain mates, including how many mates a typical individual has; Krebs and Davies 1993) and the operational sex ratio (that is, the ratio of fertilizable females to sexually active males at any given time; Emlen 1976; Emlen and

Oring 1977). Sexual selection's strength depends on both factors (reviewed in Emlen and Oring 1977; Andersson 1994; Shuster and Wade 2003). For example, the farther the operational sex ratio deviates from 1:1, the greater the strength of sexual selection (Emlen 1976; Emlen and Oring 1977; for example, Olsson et al. 2011; but see Weir et al. 2011). Both factors, in turn, depend on, and vary with, the environment (Emlen and Oring 1977; for example, Karlsson et al. 2010; Olsson et al. 2011). These factors can therefore be influenced by changes in habitat or resource use, such as those that might arise via character displacement.

Consider, for example, that ecological character displacement may cause a population to shift from a resource that is distributed uniformly to one that is distributed patchily (see chapter 6). In such cases, the post-displacement population would likely transform from a system in which no given male could monopolize resources (and, hence, gain exclusive access to females) to one in which a few males may be able to do so. In the derived, post-displacement population, the strength of sexual selection would be greater than in the ancestral, pre-displacement population (*sensu* Emlen and Oring 1977; Krebs and Davies 1993). Therefore, by altering the mating system, the operational sex ratio, or both, character displacement can impact the strength of sexual selection in a given population.

An example of a shift in mating system in response to a novel habitat can be seen in the freshwater isopod, *Asellus aquaticus* (Karlsson et al. 2010). Owing to changes in water turbidity, two Swedish lakes have experienced major ecological change. Isopods that invaded these lakes have undergone rapid, parallel morphological and behavioral changes (in approximately 40 isopod generations), resulting in a novel stonewort ecomorph that contrasts with their ancestral reed ecomorph. In the novel habitat, the stonewort ecomorph spends significantly less time mate guarding, and female survivorship during mate guarding is reduced (Karlsson et al. 2010). These differences could affect the level of sexual conflict among the different ecomorphs. These differences could also alter the strength of sexual selection on male behaviors, and this change in sexual selection will depend on the fitness consequences of guarding versus searching for mates (Karlsson et al. 2010). In the case of *A. aquaticus,* the evolution of these different ecomorphs appears to reflect alternative responses to different predation pressures (Karlsson et al. 2010). Nevertheless, competitively mediated selection could generate similar differentiation that ultimately impacts mating systems and the resulting strength of sexual selection in different habitats.

Thus, the effects of character displacement on sexual selection can be direct, as when character displacement alters sexual traits used in competition for successful reproduction. But the effects of character displacement on sexual selection can also be indirect and subtle, as when character displacement changes the environmental context in which sexual selection occurs. Moreover, these different ways in which character displacement can affect sexual selection are not mutually exclusive.

IMPLICATIONS OF THE EFFECTS OF CHARACTER DISPLACEMENT ON SEXUAL SELECTION

Above we described how character displacement could affect sexual selection. Here, we highlight two implications of these effects. First, we discuss how diversity in sexual traits can be greatly enhanced when character displacement and sexual selection reinforce each other. Second, we describe how character displacement's effects on sexual selection can alter the underlying fitness consequences of sexual selection in sympatry relative to allopatry.

Regarding the first implication—that diversity in sexual traits can be greatly magnified when character displacement and sexual selection augment each other—consider that sexual selection has long been viewed as a key driver of diversification in sexual traits among populations and species (Andersson 1994; Butlin and Ritchie 1994; Coyne and Orr 2004; Ritchie 2007; Price 2008). One explanation for how sexual selection drives trait diversity is that chance differences can arise among populations in female preferences (owing to drift or founder effects, for example). Such preferences could then cause male traits to diverge between population or species (Lande 1981; Kirkpatrick 1982).

Trait diversity among populations and species could also arise as a consequence of local adaptation to different habitat types (Bradbury and Vehrencamp 1998; Boughman 2002). In particular, features of the environment can affect the kinds of preferences and traits that are most likely to evolve via sexual selection. In such cases, sexual selection should favor different traits in different habitat types. The converse of this prediction is that *similar* sexual traits should evolve in similar habitat types by sexual selection. Although sexual selection could also produce variation within a habitat (for example, by favoring polymorphisms in mate choice or mate choice based on genetic compatibility; Jennions and Petrie 1997; Tregenza and Weddell 2000), in general, sexual selection is thought to promote similar sexual traits within a habitat type.

For example, different species of birds converge in song attributes when they occupy similar habitat types, but diverge when they occupy different habitat types (Morton 1975; Wiley and Richards 1978; Wiley 1991). Indeed, songs of different forest-dwelling species are often similar to each other—regardless of where they are found in the world—but their songs tend to differ from those of grassland dwellers (for example, Nottebohm 1975; Cardoso and Price 2010; see also Tobias et al. 2010; Kirschel et al. 2011). Presumably, such differences arise because the physical properties of alternative habitats differentially affect song transmission (Wiley and Richards 1978; Wiley 1991). Thus, local adaptation to physical features of different habitat types can promote diversity in sexual traits between populations or species.

Local adaptation to predators and parasites can also promote similar patterns of diversity in sexual traits. For example, male Trinidadian guppies (*Poecilia reticulata*) from different streams have independently converged on the use of orange coloration to attract females (reviewed in Houde 1997). However, because these bright orange male displays also attract the attention of predators, males in populations with high predation risk have evolved to become less colorful (Houde 1988). Thus, guppies in different streams with similar levels of predation express similar sexual traits, but guppies in streams that differ in predation level have diverged (reviewed in Endler 1991; Houde 1997). In sum, local adaptation to either physical features of different habitat types or to the presence of predators and parasites can explain why sexual traits are similar among populations and species that occupy similar habitats but different across different habitats (Andersson 1994).

If sexual selection generally favors similar sexual traits within a habitat type, how does diversity in sexual traits arise *within* habitats? This is where character displacement plays a role—species within the same habitat will be under selection to diverge from one another in sexual traits (see chapter 2; for example, Grant and Grant 2010). In other words, whereas sexual selection acting alone tends to generate trait diversity *between* habitat types (see above; see also Figure 7.4A), character displacement acting alone generates trait diversity *within* habitat types (Figure 7.4B). When these two processes act in concert, however, they can promote trait diversity in sexual traits both between *and* within habitats (but see Cardoso and Price 2010). Thus, sexual selection and character displacement acting together generate diversity beyond that which either process would likely generate when acting alone (Figure 7.4C).

Character displacement, acting in conjunction with sexual selection, can therefore underlie divergent sexual traits between species in similar habitats. Yet diversity is not enhanced only by the tendency for each process to generate orthogonal patterns of divergence. In particular, sexual selection can augment differences that derive initially only from character displacement.

To demonstrate how sexual selection can augment differences created by character displacement, consider divergence in sexual traits in a population in which sexual selection is absent versus in one in which sexual selection is present. In a population in which sexual selection is absent, once a species' (or an ecomorph's; see chapter 5) sexual traits differ sufficiently from those of its competitor, character displacement should cease to promote divergence. This is because selection for divergence lessens as competition between species declines with increasing differentiation (chapter 2).

By contrast, in a population in which sexual selection is acting, divergence in sexual traits can be greatly enhanced. This is because sexual selection would continue to elaborate traits involved in reproduction (for example, because sexual selection can generate a "runaway" process in which male traits and female preferences reinforce each other's evolution; Fisher 1915; Lande 1981; Kirkpatrick 1982). Sexual selection thereby promotes the evolution of ever more extreme male traits. Consequently, when character displacement and sexual selection act together, character displacement may promote the initial divergence in sexual traits, but sexual selection could then cause this divergence to become ever more extreme—more extreme than needed to differentiate between conspe-

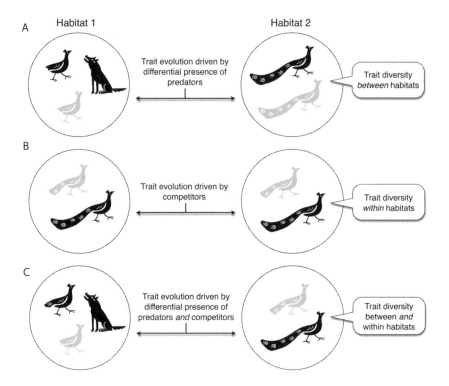

FIGURE 7.4.
A diagram comparing how selection is expected to promote reproductive trait diversity—within and between habitats—when sexual selection and character displacement act separately rather than jointly. In this diagram, the focal reproductive trait is a male bird's tail feathers; birds within a habitat shaded differently are different species (birds in different habitats are also different species). (A) When habitats differ in abiotic factors or the presence of "exploiters" (that is, predators and parasites, depicted here by the wolf symbol), sexual selection promotes trait diversity *between* habitats but similar traits are favored *within* habitats where no competition between species is occurring. (B) When individuals confront competitors, but habitats are otherwise similar in abiotic factors or in the presence of "exploiters," character displacement promotes trait diversity *within* habitats but sexual selection favors similar traits *between* habitats. (C) Sexual selection and character displacement acting together promote trait diversity both between *and* within habitats, thereby generating greater overall levels of trait diversity than when either process acts alone.

cifics and heterospecifics (Figure 7.3E; Lande 1982; Pfennig and Ryan 2010). Thus, when character displacement and sexual selection reinforce each other, divergence in sexual traits can be greatly enhanced (for a possible empirical example, see Pfennig 2000b).

A second major implication of character displacement's impact on sexual selection is its effect on the underlying fitness consequences of sexual selection. Because character displacement can alter the strength and pattern of sexual selection in sympatry versus allopatry, the resulting fitness consequences of sexual selection may also change (Gerhardt 1994; Pfennig 2000b; Higgie and Blows 2007). For example, a trait that was a reliable indicator of a male's quality as a mate (that is, of his ability to provide the female and/or her offspring with fitness benefits) in allopatry may be unreliable in sympatry (*sensu* Schluter and Price 1993; Price 1998). Thus, following character displacement, sympatric females may be under new selective pressures to identify alternative quality-revealing traits or even novel fitness benefits (Schluter and Price 1993; for example, if parasites are prevalent in sympatry but not allopatry, then sympatric females may be under new selective pressures to identify resistant males; Hamilton and Zuk 1982).

Moreover, females may actually face fitness trade-offs between choosing high-quality conspecifics on the one hand and avoiding heterospecifics on the other hand (Figure 7.5; Pfennig 1998; see also Ryan and Rand 1993). In particular, if reproductive character displacement favors the evolution of preferences that ensure mating with the correct species, the resulting preferences that evolve via character displacement may not be those that also enable females to select high-quality conspecific mates (Pfennig 1998, 2000b, 2007; Higgie and Blows 2007, 2008).

Consider, for instance, that sexual selection theory generally predicts that females should prefer males with more elaborate or costly traits that are indicative of male condition and his quality as a mate (Andersson 1994; Bradbury and Vehrencamp 1998). If, however, heterospecifics possess elaborate traits, character displacement should promote the evolution of preferences for less-exaggerated traits, because such traits would be the most dissimilar from those of the heterospecifics (Ryan and Rand 1993; Pfennig 1998, 2000b; Rosenthal et al. 2002; Higgie and Blows 2008; Pryke and Andersson 2008). Yet, by adopting such preferences to avoid costly heterospecific interactions, females may concomitantly forego information about a prospective conspecific mate's ability to convey additional fitness benefits (Pfennig 1998, 2000b; Higgie and Blows 2007, 2008). Therefore, although allopatric females may be able to do so, sympatric females may not be able to reap the benefits of "mate-quality recognition"; that is, sympatric females may be unable to identify males that can provide additional fitness benefits (Figure 7.5; Pfennig 2000b, 2008).

For instance, female Mexican spadefoot toads (*Spea multiplicata*) from southeastern Arizona compromise on mate quality to ensure conspecific matings (Pfennig 2000b). Because male call rate is a condition-dependent character in many frog species—and fast call rates are a reliable indicator of male quality in many frog species—female frogs are often under directional selection to choose more rapidly calling males (discussed in Pfennig 2000b). As predicted, female *S. multiplicata* from populations where they are the only species of *Spea* present (and where, therefore, there is no risk of costly hybridization) prefer fast call rates (Figure 7.6A), and their mate choice preferences result in the direct fitness benefit of increased fertilization success (Figure 7.6B) and better-quality offspring (Pfennig 2008). By contrast, in populations where *S. multiplicata* are sympatric with Plains spadefoot toads (*S. bombifrons*)—a species with which there is

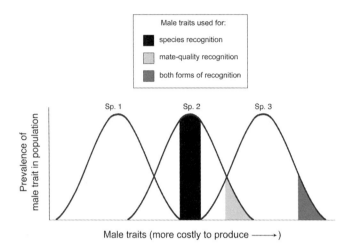

FIGURE 7.5.
When high-quality males resemble heterospecifics, females may be unable to identify both conspecific mates (species recognition) and high-quality mates that can confer fitness benefits to the female or her offspring (mate-quality recognition). In this figure, females of three sympatric species (species 1, species 2, and species 3) are faced with having to select the highest-quality mate (that is, mates that produce the most extreme—and hence costly—traits) while simultaneously avoiding mismatings with the other two species. For species 2, species and mate-quality recognition conflict, because females of this species risk mismatings with species 3 (when they engage in mate-quality recognition) or with relatively low-quality conspecifics (when they engage in species recognition). By contrast, for species 3, these two forms of recognition reinforce each other: by preferring the highest-quality males, females of species 3 also minimize the risk of mismatings with the other two species. Redrawn from Pfennig (1998).

the risk of costly hybridization and whose males also call faster—female *S. multiplicata* prefer slower, average call rates (Figure 7.6A), and they fail to receive the fitness benefits of mate choice that accrue to allopatric females (Figure 7.6B; Pfennig 2000b, 2008).

Character displacement does not always generate fitness trade-offs in choosing among mates, however (Pfennig 1998). For example, if males with the most elaborate characters are also the most dissimilar from heterospecifics (as would be the case for species 3 in Figure 7.5, for example), then character displacement may actually reinforce the evolution of female preferences for exaggerated, condition-dependent male traits. For example, female *S. bombifrons* spadefoot toads from southeastern Arizona are generally under directional selection to prefer the fastest-calling males, both as a means of avoiding mating with the more slowly calling *S. multiplicata* males when hybridization is costly, and also as a means of selecting the highest-quality conspecific males (Pfennig

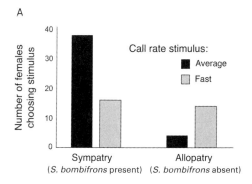
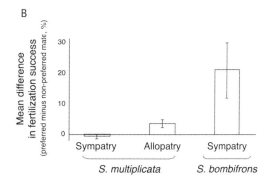

FIGURE 7.6.
Effects of heterospecifics on mate preferences and on the fitness consequences of these preferences. Male spadefoot toads (*Spea multiplicata* and *S. bombifrons*) use calls to attract their mates. (A) Female *S. multiplicata* from sympatric populations prefer average call rates over fast call rates as a means of avoiding hybridization with faster-calling *S. bombifrons* males. However, females from allopatric populations prefer fast call rates over average call rates because faster call rates indicate male condition in anurans. (B) As a consequence of these shifts in mate preferences, *S. multiplicata* females from allopatry choose good-condition mates that provide them with enhanced fertilization success. In sympatry, by contrast, *S. multiplicata* females do not engage in condition-dependent mate choice and do not get enhanced fertilization success from their preferred mates relative to a non-preferred mate. This difference between sympatric and allopatric female *S. multiplicata* is attributed to the way in which they trade off species and mate-quality recognition (that is, *S. multiplicata* is essentially in the position of species 2 in Fig. 7.5). In support of this notion, female *S. bombifrons*, which do not face the trade-off, because preference for fast-calling males reliably identifies conspecifics that are also in good condition, choose good-condition mates that provide them with enhanced fertilization success (that is, *S. bombifrons* is essentially in the position of species 3 in Fig. 7.5). Redrawn from Pfennig (2000b).

2000b). In this species, females reap substantial fitness benefits of mate choice (Figure 7.6B).

When trade-offs in mate choice do arise owing to character displacement, selection may favor the resolution of such trade-offs by promoting the evolution of preferences for multiple traits that enable females to avoid heterospecific interactions while simultaneously assessing conspecific quality (Pfennig 1998; Hebets and Papaj 2005). Thus, character displacement can not only contribute to divergence in male traits, but it can also promote the evolution of multiple or complex traits for discriminating high-quality conspecifics from heterospecifics (reviewed in Pfennig 1998; Gerhardt and Huber 2002; Hebets and Papaj 2005). Indeed, character displacement can even promote the evolution of the differential valuation of those traits in mate choice in sympatry versus allopatry. That is, traits used in species recognition could be valued more highly by females in sympatry than in allopatry (Gerhardt 1994).

As discussed above, character displacement's effects on sexual selection can enhance diversity of sexual traits and also alter the underlying fitness consequences of sexual selection. Both outcomes—acting alone or together—can increase the likelihood of speciation. Specifically, because mate choice plays a critical role in reproductive isolation (reviewed in Andersson 1994; Coyne and Orr 2004), population differences in patterns of sexual selection may cause these populations to diverge in sexual traits to such a degree that they no longer recognize individuals from alternative populations as acceptable mates (Howard 1993; Hoskin et al. 2005; Jaenike et al. 2006; Pfennig and Ryan 2006, 2010). Such an outcome becomes especially likely when fitness trade-offs such as those described above preclude gene flow between divergent populations. We discuss this point in greater detail in chapter 8.

HOW SEXUAL SELECTION AFFECTS CHARACTER DISPLACEMENT

Having discussed character displacement's impacts on sexual selection, we now describe how sexual selection may affect character displacement, by either facilitating or impeding it.

Sexual selection can facilitate character displacement when the direction of sexual selection and competitively mediated selection are concordant. Indeed, selection will most strongly favor a trait that minimizes competitive interactions with heterospecifics when that trait also confers a mating advantage. In other words, selection for traits that minimize competitive interactions with heterospecifics is stronger when such traits are *also* favored by sexual selection, or when competitively mediated selection constitutes sexual selection (for example, see Figure 7.3E and species 3 in Figure 7.5).

Conversely, sexual selection may impede character displacement. For instance, in the example of bird beak evolution illustrated in Figure 7.3, if the derived, post-displacement song were *less* attractive to females than the ancestral, pre-displacement song, then sexual selection would oppose character displacement. In such cases, sexual selection may even preclude character displacement in the first place. Indeed, sexual selection may slow character displacement to such a degree that competitive or reproductive exclusion occurs instead.

Generally, sexual selection will impede character displacement whenever traits that minimize heterospecific interactions also *decrease* an individual's ability to compete for, or attract, mates. Thus, any change in phenotype or habitat use that reduces an individual's ability to acquire high-quality mates will generally be disfavored by sexual selection, even if these traits also minimize resource competition or reproductive interactions with heterospecifics.

Ultimately, however, any sexual trait that enables individuals to simultaneously minimize interactions with heterospecifics while maximizing their mating success should be favored. For instance, in the bird example in Figure 7.3, any female that evolved a novel preference for a modified song produced by a male that has undergone character displacement may accrue the indirect fitness benefit of having offspring with higher fitness owing to reduced resource competition or reproductive interactions with a hetcrospecific. Once again, if female preferences ultimately reinforce selection favoring a reduction in interactions with heterospecifics, then sexual selection will facilitate character displacement (van Doorn et al. 2009). Thus, although sexual selection may initially impede character displacement, over time, novel preferences or traits could arise that would be strongly favored if they minimize the conflict between sexual selection and character displacement (*sensu* Schluter and Price 1993). Whether such a resolution arises before exclusion occurs depends on the amount of standing variation, the strength of sexual selection versus selection for traits that minimize competition, the genetic architecture of the traits involved, and the level and nature of any gene flow among populations (see chapter 3).

In sum, sexual selection can either facilitate or impede character displacement, depending on whether or not traits that minimize heterospecific interactions also enhance or decrease an individual's mating success. Indeed, in some cases, sexual selection may even prevent character displacement from occurring. Yet the degree to which this is so remains an open, and relatively unstudied, question.

A CAUTIONARY NOTE: PROCESS VERSUS PATTERN

In this chapter, we highlighted ways that character displacement can generate different patterns of sexual selection that, in turn, promote diversity in sexual traits. Yet it is important to keep in mind that the mere existence of divergence in sexual traits between populations in sympatry with another species versus in allopatry should not be construed as sufficient evidence for reproductive character displacement (recall chapters 1 and 3). For example, we described above how ecological character displacement could generate different patterns of sexual selection in sympatry versus allopatry and how sexual traits could then diverge as a consequence of these different patterns of sexual selection. However, such divergence of reproductive characters in sympatry versus allopatry would not constitute reproductive character displacement for the simple reason that selection for traits that minimize reproductive interactions between species would have played no role in promoting such divergence. Determining whether divergence in reproductive characters has arisen from reproductive character displacement per se requires experiments and observations aimed at specifically evaluating whether such divergence has arisen as an adaptive response to deleterious reproductive interactions between species (see chapter 1).

For instance, in the sticklebacks described above, divergence in sexual traits between limnetic and benthic species could have hypothetically arisen as a by-product of ecological character displacement altering patterns of sexual selection (rather than as an outcome of reproductive character displacement). As it turns out, both forms of character displacement appear to have occurred in this system (reviewed in Rundle and Schluter 2004). Studies that controlled for the effects of ecological character displacement established that sympatric limnetic and benthic species pairs have increased reproductive isolation (relative to similar ecomorphs from allopatry) in a manner consistent with the

> **BOX 7.1.** Suggestions for Future Research
>
> - Identify the specific means by which character displacement alters sexual selection.
> - Use comparative methods to determine whether sexual selection and character displacement, acting together, can explain why some taxa and communities are more diverse than others.
> - Determine whether character displacement alters mating systems, the operational sex ratio, and, consequently, the strength and mode of sexual selection.
> - Evaluate when and how character displacement alters the underlying fitness consequences of mate choice.
> - Determine the conditions under which character displacement generates selective trade-offs between different fitness components involved in reproduction.
> - Evaluate whether character displacement's effects on sexual selection can initiate divergent trajectories of preference and trait evolution.
> - Using both theoretical and empirical approaches, identify when sexual selection promotes—or impedes—character displacement.

operation of reproductive character displacement (Rundle and Schluter 1998, 2004). Thus, in these sticklebacks, ecological divergence is thought to have initially generated differences in traits that are also used for mating, and subsequently divergence in these same traits was enhanced to minimize costly reproductive interactions between species (Rundle and Schluter 1998, 2004; Boughman et al. 2005). Such studies should be conducted in other systems where the causes of reproductive trait diversity are unclear.

In sum, divergence in reproductive characters does not necessarily constitute reproductive character displacement (see chapters 1 and 3 for similar cautionary points). Instead, such diversity may reflect the effects of character displacement on sexual selection.

SUMMARY

Sexual selection may account for much of the diversity in secondary sexual characters found in sexually reproducing organisms. Moreover, sexual selection may even promote speciation. Explaining the factors that affect sexual selection is therefore crucial for understanding diversity's origins. Character displacement can affect sexual selection by generating shifts in either an organism's phenotype or the habitat in which mating takes place. In so doing, character displacement also alters the underlying fitness effects of mate choice in post-displacement populations. Such changes can enhance divergence between conspecific populations in sympatry with a heterospecific and those in allopatry. Consequently, when character displacement and sexual selection act together, each process can enhance the diversifying nature of the other. Not only can character displacement affect sexual selection, but sexual selection can also affect character displacement by either impeding or facilitating it. In particular, sexual selection may impede character displacement if it precludes the expression of traits that reduce heterospecific interactions. Sexual selection may facilitate character displacement if traits that minimize heterospecific interactions enhance an individual's ability to compete for, or attract, mates. Understanding the relationship between character displacement and sexual selection is therefore vitally important for illuminating diversity's origins. We provide some suggestions for future research in Box 7.1.

FURTHER READING

Andersson, M. 1994. *Sexual selection*. Princeton University Press: Princeton, NJ. This book provides an excellent, general overview of sexual selection.

Endler, J. A., and A. L. Basolo. 1998. Sensory ecology, receiver biases and sexual selection. *Trends in Ecology and Evolution* 13:415–420. This paper reviews, and distinguishes among, the different models for explaining how female sensory systems affect the expression of female mate choice.

Pfennig, K. S. 1998. The evolution of mate choice and the potential for conflict between species and mate-quality recognition. *Proceedings of the Royal Society B: Biological Sciences* 265:1743–1748. This paper discusses how individuals may often face a conflict between avoiding reproductive interactions with other species and selecting a high-quality conspecific mate.

Price, T. 2008. *Speciation in birds*. Roberts and Company Publishers: Greenwood Village, CO. Using birds as the focus for discussion, this book highlights the many ways that speciation occurs and contains a comprehensive review of sexual selection's role in diversification.

Ritchie, M. G. 2007. Sexual selection and speciation. *Annual Review of Ecology Evolution and Systematics* 38:79–102. This article reviews the degree to which there is support for the notion that sexual selection promotes diversification.

Ryan, M. J. 1998. Sexual selection, receiver biases, and the evolution of sex differences. *Science* 281:1999–2003. This paper describes how female mate preferences are shaped not only by selection to identify high-quality mates, but also to forage successfully and avoid predation.

van Doorn, G. S., P. Edelaar, and F. J. Weissing. 2009. On the origin of species by natural and sexual selection. *Science* 326:1704–1707. This paper presents theory demonstrating that ecological character displacement can promote the evolution of novel preferences for mates that can best utilize the new resource.

8

SPECIATION

Understanding how species arise ("speciation") is essential for explaining the origins of diversity. Until this point in the book, we have only briefly mentioned the possible contribution of character displacement to speciation. Here, we explicitly consider character displacement's role in the formation of new species.

Specifically, we discuss how character displacement potentially promotes speciation via three routes: (1) by finalizing speciation following contact between already divergent populations (Figure 8.1 at the arrow labeled a); (2) by initiating speciation between conspecific populations in sympatry with a heterospecific, which have undergone character displacement, versus those in allopatry, which have not undergone character displacement (Figure 8.1 at the arrows labeled b); and (3) by initiating speciation between the members of a population when, as discussed in chapter 5, *intraspecific* character displacement promotes the evolution of divergent strategies for acquiring mates or resources (Figure 8.2).

As background, we begin by reviewing what species are, how they remain distinct, and how new species arise. We then turn to character displacement's role in speciation. As we will see, although character displacement's contribution to diversification has traditionally focused on its role in maintaining and enhancing differences between *already existing* species, it may play an equally crucial role in generating *new* species.

WHAT ARE SPECIES?

Before describing how new species arise, we must first define the term "species" itself. As it turns out, numerous definitions of species have been proposed (Coyne and Orr

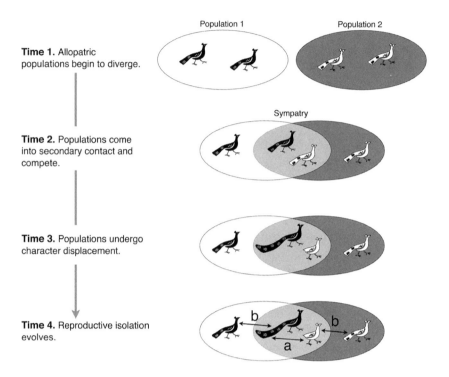

FIGURE 8.1.
Two routes by which character displacement potentially promotes speciation. Both routes begin when separate populations that have been evolving in allopatry (Time 1) contact each other (Time 2) and then undergo character displacement (Time 3). White and black birds are from different populations; character displacement is indicated by divergence in tail length, which signifies here a trait involved in reproduction. In the first route (indicated by the arrow labeled "a" and shown at Time 4), character displacement finalizes speciation when competitively mediated selection promotes differences between populations in traits associated with resource use or reproduction, such that individuals from separate populations no longer interbreed. In the second route (indicated by the arrows labeled "b" and also shown at Time 4), as an indirect consequence of character displacement between newly formed species (as occurs in sympatry during the first route and indicated by the arrow labeled "a"), sympatric and allopatric populations of each species also diverge. This divergence then favors the evolution of reproductive isolation between these populations.

2004; Wilkins 2009). The definition that we adopt here—the biological species concept—states that "species are groups of interbreeding natural populations that are reproductively isolated from other such groups" (Mayr 2001, p. 166). Put simply, under the biological species concept, populations are considered to be separate species if they do not interbreed regularly when sympatric in nature, or, if they do interbreed, they fail to produce fertile offspring.

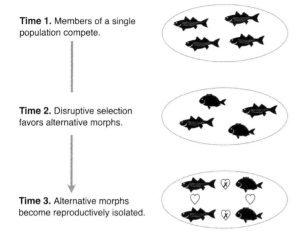

Time 1. Members of a single population compete.

Time 2. Disruptive selection favors alternative morphs.

Time 3. Alternative morphs become reproductively isolated.

FIGURE 8.2.
The steps by which intraspecific character displacement—driven by disruptive selection—potentially promotes speciation among the members of the same population (in this figure, different fish shapes represent alternative ecomorphs, open hearts indicate that individuals prefer to mate with each other, and hearts with an × inside indicate that individuals avoid mating with each other).

As first noted by Mayr (1942, 1963) and Dobzhansky (1937, 1940)—two early proponents of the biological species concept—reproductive isolation (that is, a lack of gene flow) is important in defining species because—in the absence of gene flow between groups of organisms—different patterns of selection, random genetic drift, and mutation will send those groups on separate evolutionary trajectories and thereby cause them to evolve independently of each other. By contrast, when groups of organisms exchange genes via interbreeding, each group tends to lose its evolutionary independence and also its distinctive characteristics.

Under the biological species concept, speciation is tantamount to the evolution of reproductive isolation. However, reproductive isolation is a graded condition, in that the amount of gene flow—and therefore, the degree of reproductive isolation—varies for different interacting groups or species (Grant 1971; Mallet 2008; Nosil 2008). Indeed, there is a continuum in nature that ranges from randomly mating populations to partially reproductively isolated populations or ecomorphs, to species that hybridize, and finally to completely reproductively isolated species (Mallet 2008). The existence of such a continuum implies that the evolution of reproductive isolation (and, hence, speciation) is not an event but rather a process (Coyne and Orr 1989, 2004; Harrison 1998; Grant and Grant 2008; Mallet 2008; Harrison 2010). Moreover, the continuous nature of the speciation process means that it is often not possible to pinpoint precisely when a new species arises (Ridley 1996).

Because reproductive isolation is crucial for delineating species, identifying the factors that preclude gene flow between populations, and evaluating why these factors evolve, is crucial for understanding the speciation process. In the next section, we examine the factors that prevent gene flow between sympatric species. We then discuss how these factors evolve.

HOW ARE SPECIES' BOUNDARIES MAINTAINED?

Under the biological species concept, species are groups of individuals that do not exchange genes with each other, precisely because they have evolved features that prevent gene exchange (Mayr 1942). These features—dubbed "isolating mechanisms" (*sensu* Dobzhansky 1937)—are "biological properties of individuals which prevent the interbreeding of populations that are actually or potentially sympatric" (Mayr 1970, p. 56). Isolating mechanisms can be behavioral, morphological, physiological, genetic, or biochemical attributes of individual organisms.

Generally, isolating mechanisms can be grouped into two broad, non–mutually exclusive categories that differ depending on whether or not mixed-population (also termed "hybrid") zygotes are formed. Prezygotic isolating mechanisms are features that prevent gametes from different populations from uniting and creating a zygote (reviewed in Coyne and Orr 2004). These include traits that preclude mating in the first place. For example, as described in chapter 2, individuals may be active at different times or places, they may possess divergent mating behaviors or mate preferences, or—in the case of plants—they may use different pollinators to transfer their gametes. Prezygotic isolating mechanisms also include traits that prevent zygotes from forming, even though mating takes place and gametes were exchanged or released. Examples of such traits are incompatibilities between gametes (Palumbi 1998; for example, see Galindo et al. 2003), mechanical isolation (when, for example, male and female reproductive structures do not match, where the male's genitalia are a special "key" that can open only a conspecific female's "lock"; Eberhard 1996), or cryptic female choice, such as may occur when females preferentially use gametes of their own population (Howard 1999).

Postzygotic isolating mechanisms, by contrast, prevent hybrid offspring from surviving to adulthood or, if they do survive, prevent them (or their offspring) from reproducing. Essentially, postzygotic isolating mechanisms are features that cause hybrids to have low fitness. Reduced hybrid fitness can stem from two sources. First, hybrids may suffer reduced fitness because of *intrinsic* factors that are internal to the organism. For example, once different alleles accumulate in different populations, alleles at different loci may be incompatible with one another, having never previously been tested together in the same genome (Bateson 1909; Dobzhansky 1934; Muller 1939, 1940, 1942; reviewed in Coyne and Orr 2004). Such negative interactions ("Dobzhansky-Muller incompatibilities") can cause hybrid inviability and sterility (Orr 1995; Orr and Turelli 2001). Second, hybrids may suffer reduced fitness because of *extrinsic* factors in the environment. For instance, hybrids may produce an intermediate phenotype that is poorly adapted to either parental population's competitive environment, thereby making hybrids inferior competitors for resources or mates (Rice 1987; Hatfield and Schluter 1999; Rundle and Whitlock 2001; Rundle 2002; Pfennig and Rice 2007; Fuller 2008;

Svedin et al. 2008). Generally, if environmentally mediated maladaptation limits gene flow between species, it constitutes ecologically dependent postzygotic isolation.

Before leaving the topic of isolating mechanisms, we must stress two important points. First, although isolating mechanisms must be capable of being inherited in order to evolve, their inheritance need not be based on any changes in DNA sequence. Instead, isolating mechanisms may be inherited epigenetically. For example, prezygotic isolation in animals may arise when individuals *learn* to avoid another species (Grant and Grant 2008, 2010; Kozak et al. 2011). Price (2008, pp. 293–296) describes an example from brood-parasitic *Vidua* finches (based on the work of Payne and colleagues; for example, see Payne et al. 2000) in which speciation may have been caused by an isolating mechanism that was inherited entirely through learning (in this case, when the young of brood-parasitic species learn the songs of their foster parents' species). Additionally, mating incompatibilities—and, potentially, intrinsic postzygotic isolation—may arise when individuals differ in whether or not they are infected with certain cytoplasmically inherited endosymbionts (Werren 1997; Rokas 2000; Jaenike et al. 2006; Adachi-Hagimori et al. 2011). Thus, because epigenetically inherited traits may mediate environmentally induced reproductive isolation, they are therefore potential isolating mechanisms.

Second, most species are not isolated from other species by just a single isolating mechanism (Coyne and Orr 2004). Indeed, most species do not interbreed with other, sympatric species because of the existence of multiple isolating mechanisms. Different isolating mechanisms typically come into play in a temporal sequence, like a series of hurdles (for an example from natural plant populations, see Ramsey et al. 2003). Thus, even if isolating mechanisms that act early in an organism's life cycle fail to prevent gene exchange, later-acting mechanisms can further reduce gene flow to negligible levels (Coyne and Orr 2004).

Having introduced the concept of isolating mechanisms, we next describe how isolating mechanisms evolve and, ultimately, how new species form.

THE EVOLUTION OF ISOLATING MECHANISMS

As described above, speciation involves the evolution of isolating mechanisms that prevent populations from exchanging genes with each other when sympatric. But how do these isolating mechanisms evolve in the first place, and what is the biogeographical configuration of populations that evolve isolating mechanisms (Figure 8.3)?

Darwin (1859 [2009]) suggested that speciation could arise when groups of individuals within a population diverge as a result of selection to lessen competitive interactions (see chapter 1). This idea was later challenged by those who argued that *physical* isolation was necessary to prevent gene exchange, which would then allow populations to evolve independently and thereby become different species (reviewed in Coyne and Orr 2004;

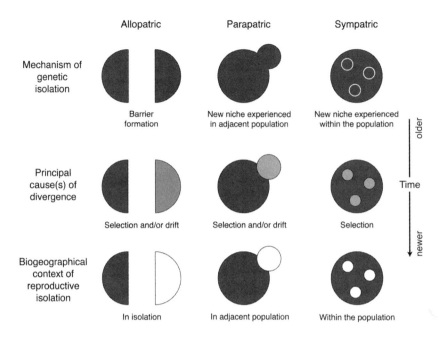

FIGURE 8.3.
Successive stages of three modes of speciation, which differ in the biogeographical context of reproductive isolation. Note, however, that these modes are often not as distinct as is sometimes portrayed.

Polechová and Barton 2005). Indeed, Mayr (1942) presented compelling evidence that speciation often occurred when populations were physically isolated from each other. Moreover, many researchers noted that population differentiation could arise from various selective—and nonselective—processes other than competition. For instance, Mayr (1942) suggested that physically separated populations might diverge purely through chance events, such as random genetic drift. Mayr's ideas have been particularly influential, so they are worth exploring in greater detail.

According to Mayr (1942, 1963, 1970, 2001), speciation typically begins when populations become separated by the establishment of some sort of physical barrier to gene flow. This barrier may be a body of water or an uninhabitable stretch of land, such as a desert or mountain range. Essentially, it could be any physical feature that prevents populations from encountering each other and exchanging genes. Once physically separated populations cease to exchange genes, they will inevitably begin to accumulate genetic and phenotypic differences. These differences may arise when selection favors alternative alleles in different populations (Box 8.1). In addition to (or instead of) selection, physically separated populations may come to differ owing to the differential action of chance events. For example, differences may arise via genetic drift (such as founder effects or population bottlenecks; Mayr 1963; Carson and Templeton 1984) or polyploidization (Ramsey and Schemske 1998; Wood et al. 2009; but see Mayrose et al. 2011).

Regardless of how differences between populations come about, some of these differences may, purely as an incidental by-product, cause the populations to be reproductively isolated from each other (Mayr 1942, 1963, 1970, 2001; Coyne and Orr 2004; Futuyma 2009; Harrison 2010; Sobel et al. 2010). For instance, if these differences cause differential choice of mates, habitats, and/or resources, then the two populations would be prevented from interbreeding via prezygotic isolating mechanisms (for example, individuals may possess divergent mating behaviors or mate preferences, or they may be active at different times or places, as described in the previous section). Indeed, laboratory experiments have demonstrated that independently evolving populations can evolve prezygotic isolating mechanisms purely as a side effect of being separated (Rice and Hostert 1993). Alternatively, physically separated populations may evolve postzygotic isolating mechanisms if they accumulate differences that cause reduced fitness owing to hybridization, as described in the previous section.

This model of speciation—in which physically separated populations evolve reproductive isolation as an incidental by-product of the differences that accumulate between them when they are physically separated from one another—constitutes the allopatric model of speciation (Mayr 1942). Note, however, that allopatry is not the only biogeographical context in which speciation can theoretically occur (Figure 8.3). A new species could also evolve either in a geographically contiguous population (parapatric speciation; Mayr 1942; Endler 1977; Gavrilets 2004) or within the geographical range of its ancestor (sympatric speciation; Maynard Smith 1966; Felsenstein 1981; Dieckmann et al. 2004; Gavrilets 2004; Bolnick and Fitzpatrick 2007; Nosil and Rundle 2009). Although the allopatric model of speciation is widely regarded as speciation's most common mode (Mayr 1942, 1963, 1970, 2001; Coyne and Orr 2004; Grant and Grant 2008; Price 2008; Futuyma 2009; Harrison 2010; Sobel et al. 2010), a growing number of researchers maintain that debates over the biogeographical setting of speciation are unproductive (see, for example, Butlin et al. 2008).

Regardless of the biogeographical setting, any difference between populations will constitute an isolating mechanism if it precludes gene flow when members of those populations have the opportunity to interbreed (for example, during "secondary contact," when populations come into sympatry after initially evolving in allopatry). Thus, whether a difference between populations is actually an isolating mechanism is tested when these populations come into contact. This also means that, although a physical barrier can be instrumental in the evolution of isolating mechanisms (by fostering independent evolution of populations), physical separation, by itself, does not constitute an isolating mechanism: unless individuals have evolved features that prevent gene exchange, they would exchange genes with each other if they came into contact. Thus opportunities for interbreeding, such as secondary contact, become the proving grounds for the efficacy of isolating mechanisms in preventing gene exchange between

BOX 8.1. Selection and the Evolution of Reproductive Isolation

Selection is often thought to play a general and decisive role in promoting divergence and reproductive isolation between populations (reviewed in Schluter 2000, 2009; Coyne and Orr 2004; Gavrilets 2004; Funk et al. 2006; Sobel et al. 2010; Nosil 2012). There are two main models for speciation via selection (Schluter 2009).

Under the "mutation-order" model of speciation (*sensu* Schluter 2009), speciation arises as a consequence of the chance occurrence and fixation of different, incompatible alleles among populations adapting to similar (that is, uniform) selection pressures (Mani and Clarke 1990). Reproductive isolation evolves because different mutations become fixed in different populations, merely because of chance variation in the order in which different mutations arise in each population, even though all of these mutations would be advantageous in the environments experienced by all populations.

Perhaps the best-documented instances of mutation-order speciation come from situations in which reproductive isolation has evolved as a by-product of conflict resolution between different genetic elements within the same genome (Schluter 2009). For example, a mutation that somehow increases its representation in the gamete pool by distorting meiosis (through "meiotic drive") exerts selection on other loci to restore a fair meiosis. Because these distorter and restorer mutations are unlikely to be the same in different populations (irrespective of the environment), the mismatch between the distorter in one population and the restorer in the other can result in reduced fitness of population hybrids and, thus, reproductive isolation (Schluter 2009).

By contrast, under the "ecological speciation" model (*sensu* Schluter 2000, 2009; Rundle and Nosil 2005; Nosil and Rundle 2009; Nosil 2012), divergent selection among environments drives the evolution of reproductive isolation. Specifically, when different populations experience dissimilar ecological circumstances, selection acts in contrasting directions in these different populations (Box 8.1 Figure 1A, B). There are three main forms of divergent selection (Nosil and Rundle 2009). First, divergent selection can arise as a result of divergent *natural* selection when different populations experience, for example, different soil chemistries (Macnair and Gardner 1998), climates (Kozak and Wiens 2006), resources (Funk 1998), competitors (Pfennig et al. 2007), parasites (Buckling and Rainey 2002a), or predators (Langerhans et al. 2007). Second, divergent selection can arise when the mode, direction, or strength of *sexual* selection differs between populations (Fisher 1930 [1999]; West-Eberhard 1983; for a possible empirical example, see Boul et al. 2007). For instance, populations might diverge in mate preferences owing to differences in either abiotic environments (for example, the amount of background noise) or biotic environments (for example, the presence of predators or competitors; see chapter 7). Third, divergent selection can even occur within a single population when the population experiences disruptive selection because of competition for resources or mates (reviewed in Dieckmann et al. 2004), which may ultimately result in the evolution of alternative morphs (for examples, see chapter 5).

Regardless of its cause, the differences that arise between populations (or between morphs) owing to divergent selection can reduce or prevent gene flow when these populations (or morphs) have the opportunity to interbreed, thereby
(continued)

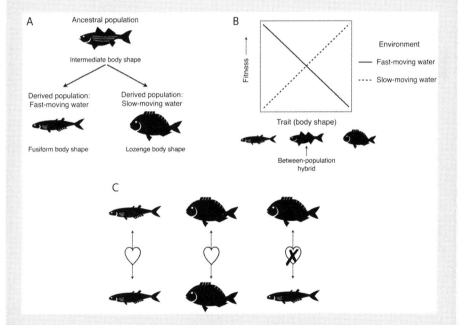

BOX 8.1 FIGURE 1
Ecological speciation arises when barriers to gene exchange evolve between populations as a result of ecologically based divergent selection. (A) For example, an ancestral population of fish might experience divergent selection on body shape when it occupies two different environments: fast-moving water and slow-moving water. (B) If between-population hybrids perform poorly in both environments, postzygotic isolating mechanisms may prevent gene flow between populations. (C) Additionally, if individuals prefer mates with a similar body shape, prezygotic isolating mechanisms might also prevent gene flow.

BOX 8.1. *(continued)*

reproductively isolating them (Box 8.1 Figure 1B, C). Speciation by this route becomes more likely if a strong association exists between the gene(s) that confer local adaptation and the gene(s) that confer reproductive isolation (Box 8.1 Figure 2; Coyne and Orr 2004; Schluter 2009; Nosil and Rundle 2009; Nosil 2012).

As support for ecological speciation, numerous studies have found that divergent selection has promoted differences between populations in traits such as body size or coloration that influence mate preferences (for example, Rundle et al. 2000; Jiggins et al. 2001; Lowry et al. 2008; reviewed in Nosil 2012). The ecological speciation model is also supported by laboratory experimental evolution studies (Rice and Hostert 1993), by comparative studies (Funk et al. 2006), and by instances of parallel speciation, in which traits that determine reproductive isolation evolve repeatedly in independent populations that inhabit similar environments (Schluter 2009; Nosil 2012).

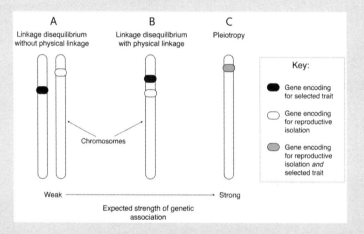

BOX 8.1 FIGURE 2

For ecological speciation to occur, there must be an association between the gene(s) that confer local adaptation and the gene(s) that cause reproductive isolation. Reproductive isolation can evolve through indirect selection, when linkage disequilibrium arises between genes under selection and those conferring reproductive isolation. Such linkage disequilibrium can occur whether genes are (A) on different chromosomes or (B) on the same chromosome. (C) Alternatively, reproductive isolation can evolve through direct selection, when the genes under selection are the same as those conferring reproductive isolation (that is, pleiotropy). The nature of the genetic mechanism is important, because different mechanisms differ in the expected strength of the genetic association between the gene(s) that confer local adaptation and the gene(s) that cause reproductive isolation, and, hence, the likelihood that reproductive isolation will evolve. Redrawn from Nosil and Rundle (2009).

previously separated populations. There are four possible outcomes when populations come into contact.

First, if the members of the two populations are similar enough to compete for shared resources or to interfere with each other's ability to reproduce successfully, then one population may drive the other locally extinct through competitive or reproductive exclusion (see chapter 2). In this case, the barriers to gene flow are not necessarily tested, because one of the two populations is driven extinct.

Second, if reproductive isolation is incomplete, the populations may successfully interbreed with each other and thereby lose any distinctive characteristics that had begun to accumulate (for examples in which previously separated populations merged, see Rhymer and Simberloff 1996; Seehausen et al. 1997; Grant and Grant 2008; Behm et al. 2010). Generally, populations may either go extinct or fuse when they have not diverged greatly (Hendry 2009; Nosil et al. 2009; see also chapter 2).

Third, populations coming into contact may be sufficiently diverged that they do not exchange genes and therefore remain distinct. As described above, differences that accumulate between physically separated populations may, as an incidental by-product, also cause reproductive isolation. In this third outcome, speciation is already complete when populations come into contact. Nevertheless, selection may still favor enhanced differentiation between these species to minimize competitive interactions between them. Thus, character displacement may cause such new species to become further differentiated, but it would have played no role in speciation per se.

Finally, populations that come into contact may be divergent but still capable of exchanging genes (Barton and Hewitt 1989; Arnold 1997). Yet, depending on a number of factors—such as the fitness consequences of interactions between the populations—selection may favor further divergence that reduces gene flow between them (Servedio and Noor 2003; Coyne and Orr 2004). That is, the evolution of isolating mechanisms per se can be favored by selection (Dobzhansky 1951). In this way, isolating mechanisms can themselves evolve via either reproductive character displacement or ecological character displacement, acting either separately or together. For the remainder of the chapter, we examine how isolating mechanisms may arise and be enhanced by character displacement.

CHARACTER DISPLACEMENT'S ROLE IN SPECIATION

As indicated above, and as described in detail below, character displacement may complete the process of speciation by promoting divergence between previously separated populations that have come into contact and potentially interbreed (Figure 8.1 at arrow labeled a; note, however, that character displacement—and hence, speciation—is not guaranteed to occur in such situations; for a discussion of the general factors that facilitate or impede character displacement, see chapter 3). Although we begin by describing character displacement's role in finalizing speciation, character displacement's contribution to speciation is not limited to the end stages of speciation. Indeed, character displacement may *initiate* speciation, as we describe in subsequent sections.

HOW CHARACTER DISPLACEMENT FINALIZES SPECIATION

Under the allopatric speciation model outlined above (Mayr 1942; Coyne and Orr 2004), the process of speciation unfolds when populations that are physically isolated from each other diverge to such a degree that they no longer successfully interbreed upon contact. Character displacement potentially plays a crucial role in finalizing speciation under these circumstances (Figure 8.1 at arrow labeled a). In particular, when partly divergent populations come into contact, selection may favor traits (in either or both populations) that lessen competitive interactions between the populations. This selection could thereby accentuate any pre-existing differences between these populations until they no longer exchange genes (Coyne and Orr 2004; Grant and Grant 2008; Price 2008). This process, in which character displacement finalizes speciation, could unfold regardless of whether incipient species reduce each other's access to resources or whether they reduce each other's access to successful reproduction. In other words, both *ecological* character displacement and *reproductive* character displacement can contribute to the completion of speciation.

For example, incipient species that compete with each other for a shared resource (and consequently risk competitive exclusion; see chapter 2) may undergo ecological character displacement and thereby diverge in traits associated with resource use. More importantly, if the alternative resource types onto which each population is shunted following ecological character displacement occur in separate habitats (for an example, see Rundle and Schluter 2004), then post-displacement populations may cease to encounter each other (or they may encounter one another less frequently). Essentially, ecological character displacement can promote allotopic distributions between populations, in which populations occupy different microhabitats within a broader zone of sympatry (see chapter 6). Consequently, these populations may cease to exchange genes and thereby accumulate further differences (see above). Such a process could accentuate reproductive isolation until these separate populations become distinct species. In this way, ecological character displacement would finalize the process of speciation following contact between already-divergent populations.

The above route is not the only way in which ecological character displacement may finalize speciation. Divergence arising as an adaptive response to resource competition may also incidentally cause divergence in *reproductive* traits that thereby generates reproductive isolation between the two populations. In particular, as we noted in chapters 3 and 7, divergence in reproductive traits is not divorced from divergence in ecological traits (Rice and Hostert 1993; Pfennig and Pfennig 2009; Sobel et al. 2010). Indeed, divergence in ecological traits is unlikely to occur without also potentially altering both male and female traits involved in reproduction. Moreover, divergence in resource-use traits can even generate selection for reproductive isolation (chapter 3).

For example, recall that selection stemming from competition for resources among different species of Darwin's finches has led to divergence in resource use and, consequently, changes in beak size and morphology (Figure 1.6). As we described in chapters 3 and 7, these changes in beak size and morphology are associated with a concomitant shift in the male's song, which is directly involved in species recognition (reviewed in Podos and Nowicki 2004; Grant and Grant 2008; Price 2008). Thus, divergence in resource-use traits may promote incidental divergence in reproductive traits, thereby reducing interbreeding between incipient species.

The above discussion focused on how ecological character displacement could finalize speciation by causing changes in habitat or resource use that also happen to reduce interbreeding between incipient species. Yet, in contrast to these *incidental* effects of ecological character displacement on speciation (incidental in the sense that, under this scenario, selection does not favor the evolution of reproductive isolation per se), reproductive character displacement could itself complete the process of speciation. In particular, if incipient species interbreed and produce hybrids of low fitness, then the process of reinforcement can finalize speciation by favoring the evolution of reproductive traits that promote reproductive isolation (Dobzhansky 1940; Howard 1993; Servedio and Noor 2003; Coyne and Orr 2004). In other words, selection could favor the evolution of prezygotic isolating mechanisms via reproductive character displacement.

Recall from chapter 1 that reinforcement is the process by which pre-existing reproductive isolation between species is enhanced ("reinforced") owing to selection against hybrids between populations or species. Dobzhansky (1951) first fully elaborated the reinforcement hypothesis, and he developed it as a means of explaining how speciation could be completed. As he put it:

> Assume that incipient species, A and B, are in contact in a certain territory. Mutations arise in either or in both species which make their carriers less likely to mate with the other species. The nonmutant individuals of A which cross to B will produce a progeny which is adaptively inferior to the pure species. Since the mutants breed only or mostly within the species, their progeny will be adaptively superior to that of the nonmutants. Consequently, natural selection will favor the spread and establishment of the mutant condition. (Dobzhansky 1951, p. 208)

Thus, once populations have begun to diverge in allopatry, the accumulation of genetic and ecological differences would likely cause any between-population hybrids to be at a selective disadvantage, because of either intrinsic or extrinsic factors (see above). Consequently, individuals would be under strong selection to evolve traits that reduce the chances of mating with members of the alternative population. These features could include mating behaviors that more readily identify conspecifics, but they could also include more subtle traits that facilitate assortative mating, such as conspecific sperm precedence and reduced offspring production by females that mated with heterospecifics (see discussion above on prezygotic isolating mechanisms).

According to Dobzhansky (1951), because the evolution of such traits would reduce costly mismatings between incipient species, they would be expected to be favored and therefore increase in frequency within each population. Moreover, because these traits should also foster assortative mating within each incipient species, hybridization should decline (Dobzhansky 1951). Ultimately, as a consequence of divergence in reproductive traits between the interacting species, the two species would no longer exchange genes and reproductive isolation would become complete (Dobzhansky 1940). In other words, by exaggerating or building upon existing isolating mechanisms, reinforcement solidifies species differences.

Recent theory and data suggest that reinforcement can and does occur in natural populations (Coyne and Orr 1989, 2004; Howard 1993; Servedio and Noor 2003). Yet how common (or essential) reinforcement is during speciation remains unclear (Howard 1993; Coyne and Orr 2004; Bank et al. 2011).

One problem with evaluating the reinforcement hypothesis is that its critical prediction—that the incidence of hybridization should decline over time—is difficult to test, because long-term data of hybridization frequency are generally not available. In cases where this prediction has been tested, the evidence is mixed as to whether the incidence of hybridization actually decreases, and therefore reproductive isolation increases, over time. In a hybrid zone between two species of crickets (*Allonemobius fasciatus* and *A. socius*), for example, no such decline was observed (Britch et al. 2001). Instead, hybridization rates appear to be stable, despite other data suggesting that reinforcement has occurred in this system. In spadefoot toads (*Spea multiplicata* and *S. bombifrons*), by contrast, a decline in hybridization was observed in conjunction with additional evidence of reinforcement (Pfennig 2003). Likewise, in a population of *Bufo* toads, hybridization between *B. americanus* and *B. woodhousii* declined significantly following secondary contact (Jones 1973). In the latter two systems, the decline in hybridization was rapid (25 and 30 years respectively, which amounts to less than 20 toad generations), suggesting that the effects of reinforcement on substantially reducing gene flow can be both rapid and detectable. That species' mating signals can evolve rapidly, and in a predictable direction—that is, away from another closely related species, as Dobzhansky (1940) first suggested—has also been confirmed by laboratory experiments (Higgie et al. 2000; Matute 2010). Thus, reinforcement potentially finalizes reproductive isolation between species, but further studies are needed to evaluate this possibility fully.

Reinforcement is commonly detected by evaluating mating behaviors and sexual signals. However, reinforcement can also arise through the evolution of postmating, prezygotic isolation. For example, on the African island of São Tomé, the widespread fruit fly species, *Drosophila yakuba*, hybridizes with an endemic species, *D. santomea* (Matute and Coyne 2010). However, *D. yakuba* from this hybrid zone have evolved at least 11 distinct reproductive barriers, each of which partially isolates *D. yakuba* from *D. santomea,* and each of which acts at a different point in *D. yakuba*'s life cycle (Matute and Coyne 2010). Laboratory experiments have shown that both behavioral (that is, premating) and gametic (that is, postmating, prezygotic) isolating mechanisms can evolve after only four generations (Matute 2010). Thus reinforcement can favor the evolution of barriers to gene flow that act *after* mating, but *before* fertilization, takes place. Generally, however, reinforcement should favor the evolution of premating isolation, because premating mechanisms prevent individuals from suffering the costs of poor mate-choice decisions (costs, for example, such as loss of reproductive effort; Mayr 1970).

By promoting the evolution of traits that enhance reproductive isolation, reproductive character displacement (specifically through reinforcement) potentially drives speciation toward completion. Yet, as described previously, reproductive isolation may also arise as an incidental consequence of selection to reduce resource competition rather than to avoid interbreeding per se. In sum, both ecological and reproductive character displacement—operating separately or together—may play a key role in increasing reproductive isolation between populations that have already begun to diverge for other reasons.

HOW CHARACTER DISPLACEMENT INITIATES SPECIATION

Above we described how character displacement could finalize the process of speciation. That character displacement can also *initiate* speciation has received much less attention. Yet, as an incidental by-product of promoting divergence between species, character displacement may instigate speciation by driving the evolution of divergent traits between populations in sympatry with a heterospecific competitor versus those in allopatry (Figure 8.1 arrows labeled b; Howard 1993; Jaenike et al. 2006; Pfennig and Ryan 2006, 2007; Pfennig and Rice 2007; Price 2008; Pfennig and Pfennig 2009). In particular, because individuals in sympatry with a heterospecific will experience a different selective environment than the one experienced by conspecifics in allopatry, conspecific populations in these two types of environments are expected to diverge in traits associated with resource use, reproduction, or both. In other words, as with any other source of divergent selection (see Box 8.1), differential exposure to a competitor can drive divergence between populations. Such competitively mediated selection may initiate the evolution of reproductive isolation between conspecific populations in sympatry versus in allopatry via two non–mutually exclusive routes.

First, as noted above, divergent selection may initiate speciation through the evolution of postzygotic isolating mechanisms, stemming from either extrinsic or intrinsic causes (Rice and Pfennig 2010). Regarding the former, offspring created by matings between sympatric and allopatric parents may express phenotypes that are maladaptive in either parental environment (Rice and Hostert 1993; Hatfield and Schluter 1996, 1999; Rundle 2002; Pfennig and Rice 2007; Svedin et al. 2008). For example, individuals produced by matings between sympatric and allopatric populations may express intermediate resource-acquisition phenotypes that make them competitively inferior in either sympatry or allopatry (for example, Pfennig and Rice 2007). Similarly, such individuals may also display reproductive phenotypes (for example, mating behaviors) that are inappropriate for either selective environment (*sensu* Hatfield and Schluter 1996; Vamosi and Schluter 1999; Svedin et al. 2008; van der Sluijs et al. 2008). Consequently, such population hybrids would not fare well when competing for reproduction with pure offspring of either population. Regarding intrinsic causes, isolating mechanisms may arise if populations evolving independently of each other in divergent competitive environments accumulate alleles that are incompatible with genomes from the alternative environment (see above). Thus, as a by-product of character displacement in sympatry, offspring of matings between sympatric and allopatric populations could possess genetic incompatibilities that make them less fertile or less likely to survive than pure offspring of either population (for example, see Pfennig and Rice 2007).

A second, non–mutually exclusive, way in which character displacement *between* species can initiate speciation between sympatric and allopatric populations *within* species is the evolution of prezygotic isolating mechanisms. Reproductive character displacement would be especially likely to generate such isolation between sympatric and allopatric populations. In particular, female preferences or male traits may become so divergent that females in sympatry may fail to recognize allopatric males (or vice versa) as acceptable mates (Pfennig and Ryan 2006, 2010). Consequently, populations in sympatry and allopatry may become reproductively isolated from each other (Howard 1993; Hoskin et al. 2005; Jaenike et al. 2006; Pfennig and Ryan 2006, 2010). Likewise, ecological character displacement between species may generate prezygotic isolation within species between sympatric populations and allopatric populations if shifts in habitat or resource use preclude mating (reviewed in Rundle and Schluter 2004; see also the discussion above on how ecological character displacement promotes divergence between incipient species). Moreover, both ecological and reproductive character displacement can engender different patterns of sexual selection in sympatry versus allopatry, thereby producing divergent mating behaviors that reproductively isolate sympatric and allopatric populations (see chapter 7).

An example in which reproductive character displacement may contribute to reproductive isolation between conspecific populations that are in sympatry with a heterospecific versus those that are allopatric comes from fruit flies, *Drosophila subquinaria*, from boreal forests of Canada (Jaenike et al. 2006). Where this species occurs with a heterospecific, *D. recens*, *D. subquinaria* females discriminate against *D. recens* males owing to selection against hybridization (Jaenike et al. 2006). By contrast, allopatric *D. subquinaria* females do not discriminate against *D. recens* males—that is, reproductive character displacement in female discrimination has occurred (Jaenike et al. 2006). However, sympatric females also discriminate against conspecific males from allopatry

(Jaenike et al. 2006), suggesting that reproductive character displacement could contribute to premating isolation between sympatric and allopatric populations.

Although we focus above on divergence between sympatry versus allopatry, sympatric populations may similarly diverge to the point of being reproductively isolated from one another if they undergo character displacement differently (for example, if they evolve different traits or express different values of the same trait). Such would be the case if populations differed in the factors that facilitate character displacement (see chapter 3). Alternatively, sympatric populations could diverge from one another if different populations undergo character displacement with different heterospecifics (Pfennig and Ryan 2006; see also Lemmon 2009).

A possible example of reproductive character displacement promoting reproductive isolation in this way comes from the green-eyed treefrog, *Litorina genimaculata*, in Australia (Hoskin et al. 2005). This species has undergone reproductive character displacement in different sympatric populations, such that male calls are more divergent from the same heterospecific in one sympatric region than in another sympatric region (Hoskin et al. 2005). Consequently, females in the different sympatric populations select against males from alternative sympatric populations. In this way, reproductive character displacement has promoted reproductive isolation between sympatric populations of the same species that differ in the degree to which they have diverged in response to the same heterospecific.

Character displacement can instigate speciation between sympatric and allopatric populations only if divergent traits that evolve in sympatry do not spread back into allopatry via gene flow (or, in the case of sympatric populations, gene flow among them must be prevented). But what factors prevent this spread? Although gene flow between sympatric and allopatric populations may often be precluded simply by virtue of the geographical distance between such populations (for example, Hoskin et al. 2005), gene flow may also be precluded by *selective barriers that arise as a consequence of character displacement*. These selective barriers stem from fitness trade-offs, in which traits (for example, resource-acquisition traits, female mate-choice preferences, and male signals) that are favored in sympatry are disfavored in allopatry, and vice versa (see chapters 6 and 7; see also Paterson 1978, 1982; Barton and Hewitt 1985; Sanderson 1989; Pfennig and Pfennig 2005). By precluding the spread of traits between sympatry into allopatry—through the evolution of prezygotic isolating mechanisms, postzygotic isolating mechanisms, or both—fitness trade-offs resulting from character displacement foster local adaptation and possibly initiate speciation between such populations.

Character displacement's role in initiating speciation between sympatric and allopatric populations is thus potentially twofold. First, character displacement promotes the evolution of alternative resource-use and reproductive traits in sympatric populations relative to allopatric populations that may contribute to reproductive isolation between sympatry and allopatry. Second, character displacement causes selective trade-offs between sympatric and allopatric populations. These trade-offs reduce migrant fitness and gene flow between sympatry and allopatry, thereby exaggerating any reproductive isolation already caused by the evolution of divergent reproductive and resource-use traits in sympatry relative to allopatry described above (see also Price 2008, pp. 65–73).

In sum, populations in sympatry with a heterospecific competitor will necessarily experience a different selective environment than conspecific populations in allopatry, precisely because of the differential presence of heterospecifics. Sympatric populations that undergo character displacement differently may also diverge from each other. Character displacement therefore generates adaptive trait divergence between different populations, and individuals that fail to express the trait(s) best suited for the prevailing selective environment will be selected against. In this way, character displacement may initiate the evolution of reproductive isolation between conspecific populations and thereby play a critical role in the early phases of speciation.

HOW INTRASPECIFIC CHARACTER DISPLACEMENT INITIATES SPECIATION

In chapter 5, we introduced the concept of *intraspecific* character displacement—trait evolution that arises as an adaptive response to resource competition or deleterious reproductive interactions among conspecifics. We also noted that, when conspecifics compete, competitively mediated disruptive selection can promote the evolution of resource polymorphism or mating polymorphism, where alternative morphs show differential resource-use or mate-acquisition tactics, respectively (for example, see Tables 5.1, 5.2). Further, we also presented evidence in Table 5.1 suggesting that such phenotypic alternatives may constitute incipient species. Here, we describe in greater detail how resource or mating polymorphism may foster speciation, thereby pointing to a role for intraspecific character displacement in instigating species formation (Figure 8.2; see also Maynard Smith 1966; Rosenzweig 1978; Seger 1985; Wilson 1989; Dieckmann and Doebeli 1999; Doebeli and Dieckmann 2000; Friesen et al. 2004; Doebeli 2011).

We begin by discussing how *resource* polymorphism may promote speciation. The evolution of a resource polymorphism potentially initiates speciation because it typically also leads to physical separation, divergence, and possibly even complete reproductive isolation between alternative resource-use morphs (Smith and Skúlason 1996; Pfennig and McGee 2010). Below, we describe how the evolution and elaboration of a resource polymorphism simultaneously fosters each of these key elements of the speciation process (Coyne and Orr 2004).

First, the evolution of a resource polymorphism tends to promote physical separation between alternative ecomorphs. Such separation occurs because alternative resource-use morphs typically differ in the locations and times that they seek their separate resources. For example, many freshwater fishes occur as alternative benthic and limnetic ecomorphs (Table 5.1), which occupy different habitats. Specifically, the limnetic morph forages in open water, whereas the benthic morph forages on the lake margin (reviewed in Robinson and Wilson 1994; Wimberger 1994; Skúlason et al. 1999; Robinson and Parsons 2002). Similarly, in many species of herbivorous insects, different groups of individuals, despite potentially occurring sympatrically with each other, evolve preferences for feeding and ovipositing on alternative host plants (Chew and Robbins 1984; Thompson and Pellmyr 1991; Tilmon 2008). Both examples illustrate how different groups of conspecifics can become physically separated from each other following the evolution of a resource polymorphism.

Once different resource-use morphs become physically isolated from one another, divergence often ensues because each ecomorph is expected to adapt to unique selective pressures that are not experienced by the alternative morph (Schluter 1993; Barluenga et al. 2006; Nosil 2007; Sobel et al. 2010). Essentially, selection should favor morph-specific traits that improve a morph's ability to survive and reproduce in its particular niche (Figure 8.4; see also Figure 5.5). For example, in herbivorous insects, different host races tend to diverge in suites of traits that vary in parallel with each other (Drés and Mallet 2002; Singer and McBride 2010). These host races may diverge in numerous traits, including mouth-part morphology (as in soapberry bugs; Carroll et al. 1997, 1998, 2003); ovipositor morphology (as in goldenrod gall midges; Dorchin et al. 2009); cryptic coloration (as in *Timema cristinae* walking sticks and larch budmoths; Sandoval 1994; Emelianov et al. 1995); clutch size (as in pipevine swallowtail butterflies; Fordyce and Nice 2004); dispersal tendency (as in soapberry bugs and pea aphids; Carroll et al. 2003; Frantz et al. 2009); and diapause/emergence timing (as in *Rhagoletis pomonella* and *Eurosta solidaginis* flies; Smith 1988; Craig et al. 1993; Filchak et al. 2000; Dambroski and Feder 2007).

Generally, morph-specific adaptations involve traits that enhance the utilization of each morph's particular resource (see references above). However, as individuals come to occupy different niches, they would likely experience different pressures from numerous agents of selection. For instance, different resources/habitats may be associated with different light/thermal environments or different parasitism/predation regimes (for example, see Figure 8.4). Consequently, alternative morphs would be under selection to accumulate numerous, morph-specific traits, some of which may not be directly related to resource-use per se. Additionally, as described in chapter 7, ecological separation between alternative morphs can modify the selective context in which *sexual* traits are produced, expressed, and perceived and thereby change the mode, direction, and strength of sexual selection in one morph relative to the other (*sensu* Endler and Basolo 1998; Boughman 2001), which could further fuel divergence between morphs. In short, once a resource polymorphism evolves, each ecomorph is expected to evolve an entire suite of traits that increase that morph's chances of success in its distinctive niche.

As these morph-specific traits accrue, morphs would also be expected to accumulate genetic differences that contribute to genetic divergence between them. Indeed, morphs may become more genetically distinct over evolutionary time, even in situations where

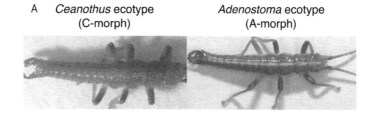
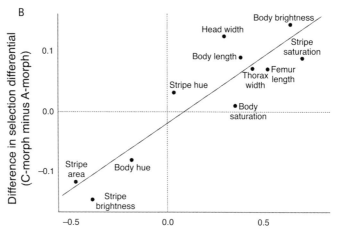

FIGURE 8.4.
Evidence that divergent selection, acting on alternative resource-use morphs, can favor the evolution of morph-specific adaptations. *Timema cristinae* walking sticks are wingless insects inhabiting southwestern North America. Individuals mate on the host plants on which they feed, and they use two host species that differ strikingly in foliage and general morphology: *Ceanothus spinosus* (Rhamnaceae), which is relatively large, tree-like, and broad-leaved; and *Adenostoma fasciculatum* (Rosaceae), which is small and bush-like and exhibits thin, needle-like leaves. (A) *Timema* occur as alternative ecomorphs that are defined by the host species on which they are found: individuals using *Ceanothus* (the "*Ceanothus* ecomorph" or "C-morph") tend to exhibit bright bodies, lack a dorsal stripe, and have relatively narrow heads and long legs. By contrast individuals using *Adenostoma* (the "*Adenostoma* ecomorph" or "A-morph") tend to have smaller and duller bodies, large and bright dorsal stripe patterns, and relatively wide heads and short legs. (B) Divergent, host-associated selection caused by predation acts on these two morphs. When predation is present, the direction and magnitude of divergent selection is positively correlated with the direction and magnitude of trait divergence observed between ecomorphs in nature. Reproduced from Nosil (2007), with the kind permission of the author and publisher.

morph determination is initially influenced by the environment. Because phenotypic alternatives often exhibit different patterns of gene expression (for example, see Figure 4.3), selection could act on this variation to enhance genetic differences between morphs (Pavey et al. 2010). Eventually, such selection may lead to genetic assimilation, where each morph becomes expressed constitutively and characterized by a unique set of alleles (see chapter 4).

Finally, after (or while) alternative morphs accumulate ecological and genetic differences, selection may favor the evolution of assortative mating by morphotype and therefore reproductive isolation between morphs (Emelianov et al. 2001, 2003; Nosil 2007). Assortative mating may evolve, because when disruptive selection promotes the evolution of resource polymorphism, offspring created by between-morph matings may often have an intermediate phenotype (assuming that resource-use traits are inherited in an additive fashion). Such intermediate phenotypes will confer lower fitness than the extreme phenotypes expressed by the offspring of within-morph matings. For example, when insects from different host races are crossed and their offspring are raised on both host plants, these between host-race "hybrids" often have lower fitness than individuals raised on their own host plant (for example, see Via et al. 2000; Nosil 2007). In such cases, selection would be expected to favor individuals that mate assortatively by morphotype. Additionally, different patterns of sexual selection among ecomorphs (patterns that result, for example, from the different perceptive abilities of females of each morph) may foster assortative mating (Boughman 2001; van Doorn et al. 2009). The evolution of such assortative mating may eventually promote complete reproductive isolation between ecomorphs (for examples in which assortative mating may have evolved between ecomorphs, see Table 5.1).

Note, however, that the evolution of a resource polymorphism does not ensure that assortative mating will evolve. Indeed, cases exist in which assortative mating by ecomorphs was expected, but not found (see, for example, Slabbekoorn and Smith 2000; Lee et al. 2010; Raeymaekers et al. 2010). At least two factors may preclude the evolution of assortative mating by ecomorph, and, hence, impede reproductive isolation from evolving via intraspecific character displacement.

First, most sexually reproducing organisms are likely to be subject to both natural and sexual selection simultaneously, and the arguments above presuppose that these two forms of selection will reinforce each other (Maan and Seehausen 2011). Yet, this need not be the case (Andersson 1994). Specifically, sexual selection may favor random mating with regard to a prospective mate's ecomorph, thereby precluding the evolution of assortative mating by ecomorph. For example, in many species, females preferentially mate with males that have larger, more intense, or more exaggerated sexual displays such as color patterns, ornaments, vocalizations, or behaviors (see chapter 7). If these sexual displays do not also reveal a male's ability to compete for one resource type as opposed to another, then sexual selection would tend to preclude the evolution of assortative mating by ecomorph.

A second key factor influencing whether ecomorphs will become reproductively isolated from each other is the presence or absence of a genetic mechanism by which divergent selection on ecological traits can be "transmitted" to the genes that cause reproductive isolation (reviewed in Nosil and Rundle 2009). There are two main mechanisms that allow this to occur.

In the first mechanism, genes that underlie ecomorph-specific traits are physically different (that is, reside at different loci) from those that cause reproductive isolation. In this case, reproductive isolation is said to evolve by "indirect selection," because selection acts on genes causing reproductive isolation only to the extent that they are nonrandomly associated (that is, in linkage disequilibrium) with the genes that are directly under divergent selection (Box 8.1 Figure 2A, B; note, however, that strong linkage can prevent the proper associations from being formed in the first place; Servedio et al. 2011). In the second mechanism, the two sets of genes are the same; that is, the same locus encodes for both the ecomorph that an individual expresses *and* the tendency to prefer the same ecomorph as a mate (Box 8.1 Figure 2C). In this case, reproductive isolation is said to evolve by "direct selection," because the genes causing reproductive isolation are themselves under selection.

There are several possible examples in which the traits associated with resource use appear to affect sexual signaling, and therefore, potentially, mate acquisition. For instance, in Darwin's finches, divergent morphological adaptation (in this case, changes in beak shape and size) to alternative resources coincides with changes in a character involved in mate acquisition—song characteristics (see chapters 3, 4, 5, and 7). Beak morphology is therefore considered a "magic trait" because it is subject to divergent natural selection and it also mediates assortative mating (Maan and Seehausen 2011; Servedio et al. 2011). Other possible examples of magic traits include (but are not limited to) those that arise from: (1) instances in which habitat isolation evolves as a direct consequence of selection on genes affecting habitat choice (as in limnetic versus benthic fishes; see above and also chapter 3); (2) instances in which a single gene could have pleiotropic effects by affecting both coloration and perception of—and preference for—that coloration, as in mimetic passion-vine butterflies (genus *Heliconius*; Kronforst et al. 2006; see chapter 4); (3) adaptations to different pollinators in plants (see chapter 2); and (4) temporal isolation caused by differences in flower timing (Nosil and Rundle 2009). Generally, reproductive isolation between ecomorphs is more likely to evolve when the traits under divergent selection in each ecomorph also cause reproductive isolation pleiotropically (Dieckmann and Doebeli 1999; Kondrashov and Kondrashov 1999; Servedio 2008).

The evidence to date in support of resource polymorphism's possible role in speciation is largely circumstantial. One such line of evidence comes from a comparative study that found that clades in which resource polymorphism has evolved are more species rich than their sister clades in which resource polymorphism has not evolved (Figure

5.9; Pfennig and McGee 2010). However, because these differences in species richness may reflect differences in extinction rates (in addition to, or instead of, differences in speciation rates), more direct tests are needed. Organisms with short generation times that shift hosts and mate on their host, such as certain microbes (Duffy et al. 2007) and herbivorous insects (see references above), may prove to be especially useful in determining whether the evolution of a resource polymorphism can indeed initiate speciation.

Intraspecific character displacement could also initiate speciation through the evolution of a mating polymorphism—alternative phenotypes within the same population that differ in mate acquisition tactics (Andersson 1994; see Table 5.2; Figure 5.8). As with resource polymorphism, once a mating polymorphism evolves in a population, the stage is set for speciation to potentially unfold. In particular, if assortative mating based on morphotype evolves, such assortative mating essentially serves as a prezygotic isolating mechanism that generates reproductive isolation between morphs. Moreover, when assortative mating prevents genetic exchange between alternative mating morphs, genetic differences may accumulate that further contribute to isolation between these morphs. How, then, might assortative mating evolve?

When male mating polymorphisms arise as a consequence of polymorphism in female preference, assortative mating is essentially axiomatic (see, for example, Boul et al. 2007). By contrast, when male mating polymorphisms are driven by male-male competition, the evolution of assortative mating will depend on how and whether female preferences for the alternative morphs are expressed. In particular, the fitness consequences of female mating decisions and the nature of the alternative male mating tactics will likely dictate whether assortative mating evolves.

Consider, for example, situations where males occur either as satellite phenotypes or as phenotypes in which the male actively seeks to attract mates (for example, see Table 5.2; Figure 5.8). In such systems, females are generally expected to prefer the latter type of male as mates, because such males are generally thought to provide enhanced fitness benefits to the females and/or their offspring (see chapter 7). In such systems, assortative mating based on the alternative male phenotypes—and speciation—is unlikely to unfold. Yet, even in such systems, assortative mating can arise if females face high search costs or severe competition with other females for access to these preferred mates (Fawcett and Johnstone 2003; Härdling and Kokko 2005). In such situations, females that cannot afford search costs (or that are poor competitors) may do better to mate with satellite males; alternatively, some females may be avoided by preferred males (Simmons 1994; Thomas et al. 1999). By contrast, females that can afford search costs (or that are superior competitors) will likely do better to mate with the non-satellite (territorial) males. These kinds of dynamics can foster assortative mating (Thomas et al. 1999), but determining whether or not they contribute to reproductive isolation requires additional study.

A more likely scenario in which assortative mating could arise is when males adopt alternative phenotypes for attracting females (for example, different male color morphs). When males possess such alternative phenotypes, females may subsequently undergo disruptive selection on mate preferences. Disruptive selection is expected to arise because the mating success of *sons* may be enhanced if a female mates assortatively based on her own phenotype or that of her father (for example, large females should mate with large males). Matings across type would produce intermediate phenotypes among the female's offspring, and these intermediate phenotypes would likely be disfavored by disruptive selection (see chapter 5). Moreover, if mate competition results in a population becoming subdivided into different microhabitats (for example, those consisting of different light environments; see chapters 3 and 7), then females may be under selection to evolve preferences that facilitate mate assessment in their particular microhabitat; for instance, they may prefer more conspicuous males (Chunco et al. 2007; Gray et al. 2008; Elmer et al. 2009). Thus female preference polymorphisms—and, thereby, assortative mating—may arise via disruptive selection that is generated by divergence in male traits.

For example, in certain species of cichlids, male-male aggression appears to generate disruptive selection for alternative male color morphs (Dijkstra et al. 2008). Males of one color are potentially more aggressive toward males of the same color than they are toward males of an alternative color (Dijkstra et al. 2008). Female cichlids, in turn, preferentially mate with males of their own type. Although assortative mating in this particular system has likely arisen from local adaptation to microhabitat variation in light environment (Seehausen et al. 2008), such a process could also transpire via male-male competition. Clearly, more research needs to be conducted on the role of mating polymorphisms in speciation (see also chapter 5).

SUMMARY

According to the widely used biological species concept, species are groups of individuals that do not interbreed regularly and successfully with other such groups when sympatric in the wild. Typically, new species form during an extended process that unfolds gradually over time. This process may begin when populations initially evolve independently of each other and end when the members of these populations have the opportunity to interbreed, but possess, or subsequently evolve, barriers to gene exchange. Selection potentially plays a general and decisive role in promoting divergence and reproductive isolation between such populations. Character displacement, in particular, potentially plays a key role in speciation by (1) finalizing boundaries between incipient species that diverged in allopatry; (2) initiating divergence between populations in sympatry and allopatry with heterospecific competitors; and (3) initiating and finalizing differentiation of alternative morphs via intraspecific character displacement. In short, character displacement's contribution to diversification is not limited to maintaining and enhanc-

> **BOX 8.2.** Suggestions for Future Research
>
> - Identify when, and how, character displacement has played a role in speciation.
> - Determine the degree to which character displacement's contribution to speciation is to finalize versus initiate the evolution of reproductive isolation.
> - Evaluate the extent to which ecological character displacement, as opposed to reproductive character displacement, promotes speciation.
> - Determine whether, as a consequence of character displacement, selective barriers generally exist between sympatric and allopatric populations and identify the nature and strength of any such barriers.
> - Using organisms in which disruptive selection has favored the evolution of a resource polymorphism, determine whether individuals mate assortatively by resource-use phenotype (for a possible example, see Barluenga and Meyer 2004, 2010; Barluenga et al. 2006; Elmer et al. 2010; for counter examples, see Slabbekoorn and Smith 2000; Lee et al. 2010; Raeymaekers et al. 2010).
> - Determine whether, and under what conditions, the evolution of a resource or mating polymorphism results in speciation.

ing differences between *existing* species; it may play an equally crucial role in generating *new* species. We list some key challenges for future research in Box 8.2.

FURTHER READING

Coyne, J. A., and H. A. Orr. 2004. *Speciation*. Sinauer: Sunderland, MA. This book provides a thorough overview of species concepts and speciation.

Hoskin, C. J., M. Higgie, K. R. McDonald, and C. Moritz. 2005. Reinforcement drives rapid allopatric speciation. *Nature* 437:1353–1356. This paper presents an empirical example in which character displacement has promoted reproductive isolation between conspecific populations.

The Marie Curie SPECIATION Network. 2012. What do we need to know about speciation? *Trends in Ecology and Evolution* 27:27–39. This paper suggests where speciation research should be directed in the coming years.

Nosil, P. 2012. *Ecological speciation*. Oxford University Press: Oxford, UK. This book is the definitive treatment of ecological speciation. Readers seeking a concise overview of the subject should consult the paper by Rundle and Nosil described below.

Polechová, J., and N. H. Barton. 2005. Speciation through competition: A critical review.

Evolution 59:1194–1210. This paper presents a comprehensive review of competition's role in speciation.

Rundle, H. D., and P. Nosil. 2005. Ecological speciation. *Ecology Letters* 8:336–352. This paper examines the role of ecological factors, especially biotic interactions, in the formation of species.

Sobel, J. M., G. F. Chen, L. R. Watt, and D. W. Schemske. 2010. The biology of speciation. *Evolution* 64:295–315. In this paper, the authors argue that ecological adaptation is the major driver of reproductive isolation, and that geographic isolation caused by adaptation to different habitats plays a major, and largely neglected, role in speciation.

9

MACROEVOLUTION

Macroevolution is large-scale evolutionary change, ranging from the origin of species and major new features (such as novel traits or even new body plans) to long-term evolutionary trends (Stanley 1979; Erwin 2000; Levinton 2001; Gould 2002; Futuyma 2009). Beginning with Darwin, many evolutionary scientists have held that macroevolution reflects the long-term accumulation of small-scale evolutionary change occurring within species; that is, macroevolution is the sum of *micro*evolution over time (Mayr 2001). From this perspective, the processes that cause microevolution (such as selection, drift) are also thought to propel macroevolution. Yet, because macroevolutionary change typically unfolds over long periods of time (and, hence, unlike microevolution, cannot normally be observed), this connection between microevolution and macroevolution is difficult to verify.

Selection's role in macroevolution has been particularly difficult to establish (Stanley 1979; Erwin 2000; Gould 2002). Moreover, to the degree that selection is thought to drive macroevolution, there is no consensus over what agents of selection account for macroevolution (Benton 2009). Indeed, whereas biotic interactions are viewed as being important over short time scales, abiotic agents—such as climate change, tectonic events, and asteroid impacts—may dominate over longer time scales (Benton 2009).

In this chapter, we consider the role of competitively mediated selection in macroevolution. Although some have questioned competition's contribution to macroevolution (Gould and Calloway 1980; Benton 1987, 1996; Stanley 2008), and others have cautioned against extrapolating organism-level processes over long time scales (for

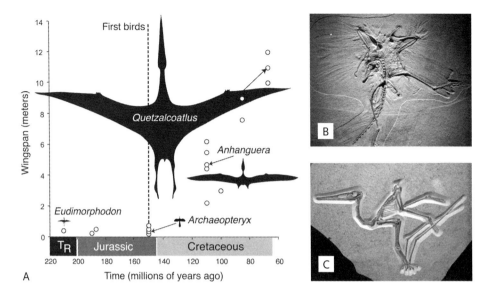

FIGURE 9.1.
Many evolutionary lineages show a trend toward increased size over long periods of time ("Cope's rule"). Such a trend may be a manifestation of character displacement. For example, (A) from the time that pterosaurs (flying reptiles) first appeared during the late Triassic period (T_R) to the time that they went extinct during the end-Cretaceous mass extinction, the largest members of this group increased in body size by a remarkable 3,000 percent (each circle denotes the average wingspan of the largest known pterosaur at a particular time period; silhouettes for representative genera are drawn to scale). Much of this increase occurred during the Cretaceous period, shortly after birds appeared (denoted by the dashed line; the wingspan and silhouette of one of the first known birds, *Archaeopteryx*, is also shown drawn to scale). This dramatic increase in body size among pterosaurs may have been driven, at least in part, by competitively mediated selection imposed on pterosaurs by birds. Fossils of (B) *Archaeopteryx lithographica* and (C) *Pterodactylus kochi* (a small pterosaur) have been found together in the Solnhofen formation of Germany (150 million years old). The fact that early birds and pterosaurs apparently lived together, were similar in body size, and had similar teeth suggests that they may have competed with each other for prey (for example, see Wellnhofer 2009, p. 149). Panel A reproduced from Kingsolver and Pfennig (2007), with the kind permission of the publisher; panels B and C photographed by D. Pfennig at the Houston Museum of Natural Sciences.

example, Benton 2009), we describe how competitively mediated selection—acting on individual organisms—potentially explains certain macroevolutionary phenomena (for example, Figure 9.1).

We begin our discussion below by considering the evidence for competition in the fossil record before outlining some general approaches that can be used to evaluate the importance of competition in driving macroevolution. We then explore character displacement's role in instigating adaptive radiation—the evolution of exceptional, and adaptive, ecological and phenotypic diversity in a rapidly multiplying lineage (*sensu*

Schluter 2000; Losos and Mahler 2010). Because adaptive radiation may account for much of the Earth's biodiversity (Simpson 1953; Mayr 2001; Wilson 2010), understanding the factors that promote adaptive radiation has long been a central goal of evolutionary biology. Finally, we consider competitively mediated selection's possible contribution to two widespread trends in macroevolution: the evolution of increased body size and increased complexity.

COMPETITION IN THE FOSSIL RECORD

Above we noted that some have questioned competition's contribution to macroevolution (Gould and Calloway 1980; Benton 1987, 1996; Stanley 2008). However, several lines of evidence from the fossil record—described below—point to a possible role for competitively mediated selection in macroevolution (see also Sepkoski 1996, 2003).

First, recall from chapter 2 that interspecific competition can result in the local extinction of a species through the process of competitive exclusion. Although competitive exclusion of a single species is difficult to infer in the fossil record, there are plausible cases of such exclusion involving entire clades (such exclusion is termed "competitive displacement"; Figure 9.2; see also Steele-Petrovic 1979; McKinney 1995; but for counter arguments, see Gould and Calloway 1980; Benton 1983). Thus, over macroevolutionary time scales, competition may explain why some clades declined in abundance and diversity of species when others flourished.

The case for competition having played a role in macroevolution does not, however, rest exclusively on the existence of competitive exclusion in the fossil record. Indeed, even if competitive exclusion were found to be rare, it would not necessarily mean that competition has played no role in macroevolution. Rather than experiencing competitive exclusion, many groups may have instead undergone character displacement. Is there evidence of character displacement in the fossil record?

As described in chapter 1, the most compelling evidence for character displacement is to actually document a species undergoing trait evolution in direct response to the presence of a heterospecific competitor. We presented one such direct demonstration of character displacement in Figure 1.6. As it turns out, a remarkably similar pattern has been deduced from the fossil record (Figure 9.3A). Moreover, recall that character displacement is frequently inferred among living species from a pattern of trait overdispersion within a community (for example, see Figures 1.8 and 6.4). Although trait overdispersion can arise for evolutionary reasons other than competitively mediated selection (see chapter 1), such a pattern has also been identified in the fossil record (Figure 9.3B). In short, the two main outcomes of competitively mediated selection—character displacement and competitive exclusion—can be inferred from fossils, pointing to a possible role for competition in macroevolution and suggesting the need for further investigation.

Additionally, resource partitioning has been deduced from the fossil record (recall

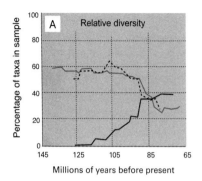 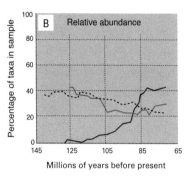 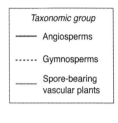

FIGURE 9.2.
A possible example of competitive displacement from the fossil record. Over the past 125 million years, nonflowering plants and flowering plants have lived together and, potentially, competed for space, light, water, and nutrients. The decline in both (A) relative diversity and (B) relative abundance of two major clades of nonflowering plants—gymnosperms (conifers) and spore-bearing vascular plants (ferns)—is mirrored by an increase in both measures for the flowering plants (angiosperms). Based on data in Lupia et al. (1999).

that when resource partitioning arises from trait evolution in response to competitively mediated selection, it constitutes ecological character displacement). Although such partitioning has been inferred from fossil sea urchins (Nichols 1959), trilobites (Thomas and Lane 1984; Fortney and Owens 1999), and dinosaurs (Lyson and Longrich 2011), one of the strongest cases comes from crinoids (sea lilies).

Crinoids are among the most abundant—and most diverse—fossils from Paleozoic strata (542 to 251 million years ago; Hess et al. 1999). Their typical body plan consists of a stalk that is attached to the sea floor on one end and to the feeding part of the animal at the other end, which is held in the water column (see Figure 9.4A). Observations of living crinoids have revealed that they feed by using their feather-like "arms" to filter suspended particles from the seawater. Data from the fossil record reveal that potentially sympatric species differed in the number and characteristics of their arms, which may have enabled different species to specialize on different-sized food particles (Figure 9.4). Perhaps more significantly, different species differed in stalk length, meaning that they fed at differing heights above the sea floor (Figure 9.4; see Ausich 1980 and references therein).

As it turns out, many living sessile invertebrates appear to partition resources by feeding at different heights in the water column (Cooper 1988). Such "tiering" also appears to be widespread among fossils of sessile organisms (Ausich and Bottjer 1982). Interestingly, tiering increased dramatically during the Ordovician period (488 to 444 million years ago), when stalked organisms, such as crinoids, greatly expanded in height (Ausich and Bottjer 2003)—stalks of some Paleozoic crinoids were over a meter long, and later forms may have been over 20 meters long. These increased levels of tiering may have been an adaptive response to heightened levels of competition (Ausich and

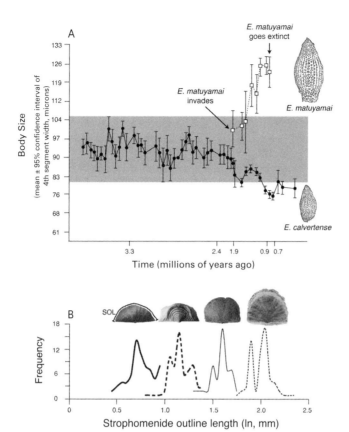

FIGURE 9.3.
Evidence of character displacement from the fossil record. (A) Radiolarians (planktonic diatoms) accumulate in large numbers on the ocean floor, where their morphological evolution can be traced by sampling sediment cores (where different portions of the core correspond to different time periods, which can be assigned absolute dates using magnetic field reversals in the sediment). Using this approach, character displacement has been inferred between *Eucyrtidium matuyamai* and *E. calvertense* from the North Pacific sea floor. At this site, these two species came into sympatry about 1.9 million years ago, when *E. matuyamai* invaded. Before then, *E. calvertense* had not undergone any net change in body size for at least 2 million years. However, following the invasion, *E. calvertense* became significantly smaller and *E. matuyamai* became significantly larger. (The gray bar marks the upper and lower 95% confidence intervals on the estimate of mean size of *E. calvertense* prior to the invasion by *E. matuyamai* to illustrate subsequent changes in the mean; compare the pattern in this graph with that shown in Fig. 1.6.) (B) Character displacement has also been inferred among co-occurring species of brachiopods from the upper Ordovician (450 million years ago) of Indiana. Specifically, different species differ more in strophomenide outline length (SOL) than expected by chance (SOL distributions of four species are shown here). SOL is a proxy for the length of the lophophore, the brachiopod feeding organ. Additionally, the SOL is a measure of a brachiopod's "footprint." Thus, similarity between brachiopod species in SOL indicates the degree to which species compete for both food and space (compare the general pattern in this graph with that in Fig. 1.8). Panel A reproduced from Kellogg (1975) and panel B reproduced from Hermoyian et al. (2002), with the kind permission of the publishers.

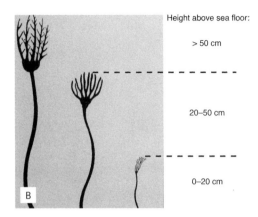

FIGURE 9.4.
A possible example of resource partitioning involving fossil crinoids (sessile marine echinoderms). (A) Sixty-three different species, representing a 340-million-year-old near-shore marine community, have been found in a remarkable state of preservation near Crawfordsville, Indiana, United States (this slab contains *Hypselocrinus hoveyi* [left] and *Sarocrinus nitidus* [right]). (B) This exceptional preservation has revealed that different species may have partitioned food resources in two ways. First, different species differed in the length of the stalk that elevated their feeding apparatus above the sea floor, suggesting that they obtained food at different height zones (living crinoids feed passively on suspended food particles). Second, within a single height zone, species differed in the number and characteristics of their feathery "arms," which may have enabled them to specialize on different-sized food particles. Either or both features could explain why so many crinoid species shared this one ecosystem. Photo in panel A by D. Pfennig; panel B redrawn from a display at the Smithsonian National Museum of Natural History (Washington, DC).

Bottjer 2003). Competition presumably intensified during the Ordovician, given that taxonomic diversity increased dramatically during this period (Sepkoski 1984, 1993; Alroy et al. 2008). Moreover, the same competitive pressures that favored tiering may have been at least partly responsible for the rapid diversification of new body plans during the early Paleozoic era (Ausich and Bottjer 2003).

Finally, another pattern in the fossil record is that of "incumbent replacement," in which a major taxonomic group diversified only after an ecologically similar group had gone extinct. Some researchers contend that incumbent replacement is widespread in the fossil record, and that this implies that competition is of little importance in macroevolution (Gould and Calloway 1980; Benton 1996). According to this line of reasoning, the existence of incumbent replacement suggests that other agents of selection—such as climate change, tectonic events, and asteroid impacts—drove many taxonomic groups extinct, clearing the way for other groups to diversify (Gould and Calloway 1980; Benton 1996). However, finding that a taxonomic group diversified only *after* another ecologically similar group had become extinct suggests that the latter's presence prevented the former from diversifying, possibly because of competitive interactions (Jablonski 2008).

Thus, to the degree that incumbent replacement is widespread in the fossil record, it too points to a possible role for competition in macroevolution.

In sum, although competition is clearly more difficult to deduce from fossils than from living organisms, competition appears to have been instrumental in shaping the ecology and evolution of extinct communities and lineages, just as it does in living communities and lineages. But the relative importance of competition (as opposed to, for instance, abiotic factors) in driving macroevolution requires further study. In the next section, we consider two approaches that have been used to tackle this problem.

METHODS FOR STUDYING MACROEVOLUTION: REPLAYING THE TAPE OF LIFE

A major reason for uncertainty about the drivers of macroevolution is that it is seemingly impossible to "replay the tape of life" (Gould 1989, pp. 45–51). Here, we briefly discuss two ways of confronting this problem.

First, evolutionary biologists have increasingly turned to rapidly evolving microbial populations to study experimental evolution in the laboratory (reviewed in Rainey et al. 2000; Kassen 2009). For example, Rainey and Travisano (1998) used the bacterium *Pseudomonas fluorescens* to study the causes of adaptive radiation. They established that heterogeneously distributed resources caused different starting colonies of these bacteria to diversify repeatedly—and independently—into the same three morphotypes, which differed in resource use (Figure 9.5). Moreover, they showed that these different morphotypes experienced frequency-dependent selection, which, as noted in chapter 5, is a hallmark of competitively mediated selection. Such studies therefore demonstrate that adaptive radiation is a deterministic process and that competitive interactions can play a direct role in driving it (see also Saxer et al. 2010 and Figure 5.3).

Additionally, nature has performed similar "experiments." For example, replicated adaptive radiations occur when the same ecomorphs evolve repeatedly in evolutionarily independent lineages. Essentially, replicated adaptive radiations are a series of miniature adaptive radiations. A spectacular example comes from *Anolis* lizards on the Greater Antilles, which have diversified into the same suites of ecomorphs on different islands (Figure 9.6). Such a pattern is compelling evidence of selection's role in promoting trait evolution (Clarke 1975; Endler 1986; Schluter and Nagel 1995). Moreover, the fact that evolutionarily independent lineages responded in the same way to similar ecological circumstances (such as the presence of few competitors and abundant and diverse resources) suggests that these circumstances favored adaptive radiation (Losos 2009).

These two approaches challenge the notion that macroevolution is entirely stochastic, and that if we "replayed the evolutionary tape" and started it again from the same point, the outcome would be very different (Gould 1989, 2002). Evolution is often deterministic, and studying cases in which it has produced the same outcome in evolutionarily independent lineages can offer important insights into selection's role in macroevolution.

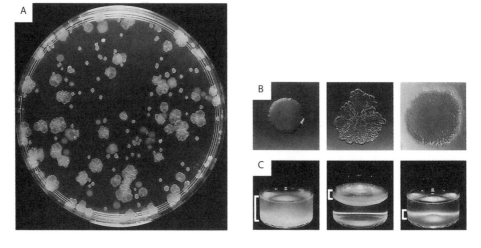

FIGURE 9.5.
An experimental approach for studying adaptive radiation in the bacterium *Pseudomonas fluorescens*. (A) When reared in microcosms in which resources were distributed heterogeneously (and when ecological opportunity was thereby enhanced), bacterial cells repeatedly diversified into three distinct morphotypes, clearly visible as different-shaped colonies on a petri dish. (B) Most phenotypic variants could be assigned to one of three morph classes (left to right): ancestral "smooth" cells; "wrinkly spreader" cells; or "fuzzy spreader" cells. (C) These three morph classes were maintained in the same population by frequency-dependent selection (suggesting that competition played a key role), and they showed marked niche preferences (the niche preference of each morph class is denoted by the brackets to the left of each microcosm). Reproduced from Rainey and Travisano (1998), with the kind permission of the publisher.

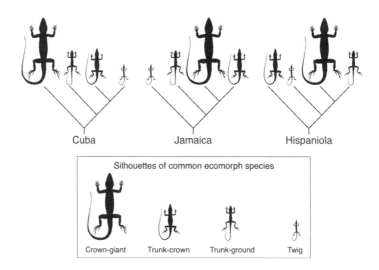

FIGURE 9.6.
Evidence of replicate adaptive radiations in *Anolis* lizards in the Greater Antilles. Generally, when different species of *Anolis* occur together, they tend to occur as different ecomorphs that utilize different microhabitats (see Fig. 2.7). Phylogenetic analyses reveal that the same four ecomorph classes have evolved independently on different islands (only three islands are shown here). Data from Losos et al. (1998); silhouettes reproduced from Losos (2009), with the kind permission of the author and publisher.

Having discussed how to detect competition in the fossil record and how to link micro- and macroevolution, we are now in position to consider the final two questions that we first posed in chapter 1 regarding competitively mediated selection's possible role in evolution. First, why is evolution generally divergent in nature? Second, why is evolution frequently escalatory?

To address the first question, we examine competition's role in initiating adaptive radiation. To address the second question, we explore competition's potential contribution to two widespread trends in macroevolution: the tendency for many evolutionary lineages to increase in both body size and complexity over evolutionary time.

ADAPTIVE RADIATION

Adaptive radiation occurs when the members of a single phylogenetic lineage diversify relatively rapidly into a large number of descendant lineages that occupy a wide variety of ecological niches (Simpson 1953; Schluter 2000; Wilson 2010). Adaptive radiation consists of two key components (Gavrilets and Losos 2009): (1) the rapid proliferation of new species within a clade; and (2) the adaptive diversification of the clade's constituent species into diverse niches. Hallmark examples of adaptive radiation are the explosive diversification of silversword plants (Barrier et al. 1999) and fruit flies (Carson and Kaneshiro 1976) on the Hawaiian archipelago; cichlid fishes in African Rift lakes (Seehausen 2006); *Anolis* lizards on the Greater Antilles (Losos 2009); and Darwin's finches on the Galápagos archipelago (Grant and Grant 2008). Other possible examples include the diversification of mammals after the dinosaurs' demise (Rose 2006; Archibald 2011) and the appearance of nearly all animal phyla in an interval of less than 100 million years during the Cambrian period (542 to 488 million years ago; Marshall 2006).

Most present-day species are thought to have arisen during past adaptive radiations in which many branches have subsequently become extinct (Simpson 1953). Therefore, identifying the underlying mechanisms that produce adaptive radiation has long been a primary goal of evolutionary biology.

In studying the causes of adaptive radiation, a recurring theme is ecological opportunity (Mayr 1942; Simpson 1944, 1953; Lack 1947; Ehrlich and Raven 1964; Schluter 2000; Gillespie 2009; Losos and Mahler 2010; Yoder et al. 2010). Recall from chapter 3 that "ecological opportunity" is typically defined as "a wealth of evolutionarily accessible resources little used by competing taxa" (Schluter 2000, p. 69). As noted in chapter 3, however, ecological opportunity encompasses more than just enhanced access to *resources*. Specifically, populations may also encounter ecological opportunity through enhanced access to reproductive-trait space. For example, signaling space may become available that is not used by other species (we describe one such example below).

Studies of both living and extinct forms have revealed that lineages are most likely to undergo adaptive radiation when presented with ecological opportunity (Schluter 2000; Gillespie 2009; Losos and Mahler 2010; Yoder et al. 2010; see also previous section).

Ecological opportunity is crucial for adaptive radiation for the same reason it is essential for divergent character displacement: in the absence of ecological opportunity, new adaptive traits (and new species) are less likely to succeed evolutionarily (see chapters 3, 5, and 6). Simpson (1944) first identified three sources of ecological opportunity, two of which are extrinsic factors—dispersal to a new environment and the extinction of antagonists—and one of which is an intrinsic factor—the acquisition of a key innovation that makes new resources available for exploitation (Gillespie 2009; Yoder et al. 2010).

Regarding extrinsic factors, a lineage may find itself in the presence of ecological opportunity when it colonizes a new habitat containing few competitors and thereby encounters a wide range of resources, reproductive-trait space, or both. For example, adaptive radiation often occurs following the colonization of a newly formed island, where ecological opportunity is almost always present (Losos and Ricklefs 2009). A lineage may also find itself in the presence of ecological opportunity in the aftermath of a mass extinction, if many of its competitors have gone extinct (Erwin 2001). Indeed, the fossil record is replete with examples in which the extinction or reduction of one group of organisms is followed or accompanied by the proliferation and diversification of an ecologically similar group (for example, Figure 9.7).

Alternatively, a lineage may experience ecological opportunity because of an intrinsic factor—the acquisition of a key innovation. A "key innovation" is a trait, or a suite of traits, that allow an organism to exploit a new niche or utilize an existing one in a novel manner and thereby outcompete its competitors (Heard and Hauser 1995; Hunter 1998). Essentially, organisms create their own ecological opportunity through the evolution of such a key innovation. After the appearance of this key innovation, the lineage diversifies into a wide variety of new niches—that is, it undergoes adaptive radiation.

For example, the radiation of terrestrial plants from aquatic ancestors in the Silurian period (430 million years ago) may have been triggered by the appearance of two key morphological features that protect against desiccation—a waxy cuticle and the stomata (Taylor et al. 2009). The individuals in which these features initially evolved were presumably selectively favored because they were able to avail themselves of the abundant ecological opportunity in the terrestrial environment and thereby escape both ecological and reproductive competition in the aquatic environment. Once on land, however, these early colonists diversified rapidly into numerous niches, presumably as an adaptive response to escape competition from each other (see chapter 5). Similarly, cichlids are a group of fishes that evolved a unique morphological adaptation: they possess a second set of jaws in their throat (Liem 1974). These pharyngeal jaws freed up their other set of jaws to assume diverse functions, from crushing snail shells to plucking scales off of other fish (see chapter 5). By enabling individuals to lessen competition for food by shifting to underutilized resources, the evolution of these pharyngeal jaws may have contributed to this group's remarkable—and repeated—bouts of adaptive radiation (Galis and Metz 1998; for example, there are over 500 species in Africa's Lake Victoria alone; Seehausen 2006).

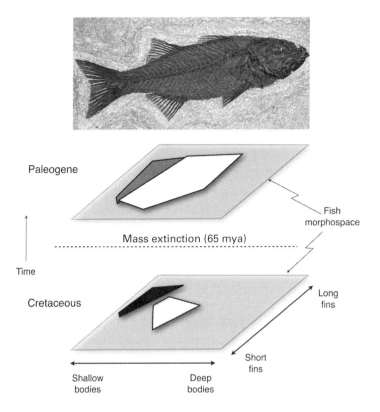

FIGURE 9.7.
The adaptive radiation of the spiny-finned fishes (Acanthomorpha) in the Paleogene period may have been prompted by the disappearance of their competitors at the end of the Cretaceous period, 65 million years ago (mya). A representative acanthomorph, *Mioplosus labracoides*—a voracious predator from the Green River Shale of Wyoming, United States (50 mya)—is pictured at top. Variation in body shape ("morphospace") among different species of acanthomorphs (white polygons) was much smaller during the Cretaceous than during the ensuing Paleogene. During the Cretaceous, no spiny-fins occupied the region of morphospace depicted by the black polygon, which describes the shapes of elongate, fast-swimming non-acanthomorph fishes. When these non-acanthomorph lineages became extinct at the end of the Cretaceous, the acanthomorphs expanded their morphospace (as depicted by the larger white polygon). Several lineages of acanthomorphs even evolved into the region of morphospace formerly occupied by non-acanthomorphs (now depicted by the gray polygon) in the Paleogene. The lineages of acanthomorphs that replaced these extinct elongate, fast-swimming non-acanthomorphs include some of the modern ocean's top predators, such as billfish, barracudas, and tuna. Redrawn from Alfaro and Santini (2010) based on Friedman (2010). Photo of fossil *Mioplosus* by D. Pfennig.

Not all key innovations are strictly morphological features (Hunter 1998). Key innovations could also include the appearance of new behaviors, such as innovative means of communicating (see below); novel ecological interactions, such as herbivory (Ehrlich and Raven 1964; Mitter et al. 1988); new ways of altering the physical environment, such as "ecosystem engineering" (*sensu* Jones et al. 1994; Erwin 2008) or "niche construction" (*sensu* Odling-Smee et al. 2003); or novel means of generating phenotypes, such as through developmental or phenotypic plasticity (West-Eberhard 2003; Pfennig et al. 2010; see chapters 4 and 5).

Additionally, key innovations are not restricted to features that enhance resource competition; they can also improve opportunities for successful *reproduction*. A possible example comes from the adaptive radiation of mormyrid fishes from Africa (commonly known as "elephant fish" because of their proboscis-like snouts). This radiation was not accompanied by the appearance of diverse strategies for acquiring resources; closely related species differ only slightly in traits associated with resource use (Arnegard et al. 2010). Instead, the radiation was accompanied by rapid diversification in a novel communication channel for sexual signaling: a high-frequency electrical channel. This group may have undergone an adaptive radiation once they began to use this new signaling channel, which opened up formerly empty reproductive trait space (mormyroid fishes are unique among all African animals in using a high-frequency electrical channel to communicate [Hopkins 1986]). Indeed, this adaptive radiation was accompanied by an evolutionary change in a region of the fish brain that is devoted to the analysis of communication signals (Carlson et al. 2011).

In this way, neural and behavioral innovations may drive the diversification of signals and thereby possibly promote adaptive radiation. Indeed, compared with traits involved in the acquisition of resources, those involved in successful reproduction may often evolve faster during adaptive radiation (Gonzalez-Voyer and Kolm 2011). Much greater attention should be given to the potential role that competitively mediated selection plays in promoting adaptive radiation through the evolution of novel reproductive traits, especially new sexual signaling systems.

In short, whether they involve features that enhance access to resources or reproduction, key innovations allow organisms to exploit new niche space and thereby outcompete their competitors. Consequently, key innovations enable a group of organisms to radiate into a diverse array of novel niches.

An increase in ecological opportunity is a *condition* that allows an adaptive radiation to proceed—it should not be viewed as the actual *trigger*. The trigger is often competitively mediated selection: many (perhaps most) adaptive radiations were likely set in motion by competitively mediated selection (Lack 1945, 1947; Simpson 1953; Brown and Wilson 1956; Schluter 2000; Mayr 2001; Grant and Grant 2008; Losos and Mahler 2010; Wilson 2010). Indeed, recall from chapter 2 that competition is ubiquitous, and that it tends to favor those individuals that differ from their competitors in how they acquire resources or successful reproduction. By favoring individuals that can capitalize on available ecological opportunity by expressing novel (and initially rare) resource-use or reproductive traits, competitively mediated selection fuels adaptive radiation (Simpson 1953; Schluter 2000; Losos and Mahler 2010). Although other agents of selection—such as predation (Nosil and Crespi 2006; Langerhans et al. 2004, 2007) or abrupt climate change (Erwin 2009)—may similarly cause lineages to take advantage of ecological opportunity and diversify, there are relatively few purported cases (Losos and Mahler 2010). By contrast, there are numerous adaptive radiations in which competition is thought to have played a key role (for example, see Benkman 2003; Calsbeek and Cox 2010; reviewed in Simpson 1953; Orr and Smith 1998; Schluter 2000; Losos and Mahler 2010; Wilson 2010). Moreover, Yoder and Nuismer (2010) have shown theoretically that competitive interactions are more likely than predator-prey interactions to promote diversification.

In principal, stochastic processes (for example, random genetic drift) could also explain how an ancestral lineage could evolve to take advantage of available ecological opportunity by moving into new niches and thereby diversifying (McShea and Brandon 2010). However, random processes cannot explain why lineages diverge more than expected by chance; why these groups evolve adaptive, specialized traits for occupying different niches; and why such adaptation involves species proliferation at the same time that members of a clade adapt to diverse ecological niches. Such diversity is more likely promoted by selection, and that source of selection is often (albeit not always) competition.

Interspecific competition is not the only form of competition that can instigate adaptive radiation; *intra*specific competition (and the resulting intraspecific character displacement; see chapter 5) may also be an important contributor. Recall from chapters 5 and 8 that intraspecific character displacement may promote speciation, which is a key component of adaptive radiation. Additionally, studies of natural populations have revealed that adaptive radiation's other key component—increased phenotypic diversity—is often present in populations that experience both ecological opportunity *and* high levels of intraspecific competition (for example, Nosil and Reimchen 2005; Parent and Crespi 2009; Martin and Pfennig 2010a; see also chapter 5). Indeed, experiments have demonstrated that, when ecological opportunity is available, intraspecific competition can cause populations to undergo rapid, adaptive diversification into new niches (for example, Rainey and Travisano 1998; Bolnick 2001; Martin and Pfennig 2010a; Bono et al., in press). Thus, the available data suggest that intraspecific competition can be decisive in spurring adaptive radiation.

In the next two sections, we examine in greater detail the possible contribution of competitively mediated selection to each of adaptive radiation's two hallmark components: the proliferation of new species and divergent evolution.

SPECIES PROLIFERATION

We begin by considering the contribution of character displacement to the production of numerous new species. When discussing character displacement, we have generally

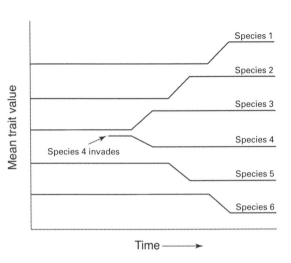

FIGURE 9.8.
A successive series of character displacement and speciation events may unfold when, in a community filled with many similar species, one species pair undergoes character displacement. This may trigger divergence and the evolution of reproductive isolation between numerous other community members.

focused on interactions between pairs of species. Further, in chapter 8, we explored how character displacement could both finalize and initiate *single* speciation events. However, we still need to explain how character displacement can promote the rapid *proliferation* of species in order to explain its role in adaptive radiation. Here we consider two ways in which character displacement can trigger the rapid production of numerous new species.

The first way in which character displacement can trigger the rapid proliferation of numerous new species is through the process of community-wide character displacement (*sensu* Dayan and Simberloff 2005). Most species belong to an ecological community in which they compete with *multiple* species. Thus, when character displacement initially occurs between a species pair (for example, when a new species invades a resident species' habitat), it is likely to also affect other species within the same community. Indeed, an initial instance of ecological or reproductive character displacement between a pair of species may precipitate a succession of character displacement events between numerous interacting species within the same community (Figure 9.8).

This process has consequences for speciation and adaptive radiation. We showed in chapter 8 how character displacement could play an important role in facilitating speciation. Therefore, a series of character displacement events may also drive numerous, rapid speciation events. In particular, each member of the community undergoing this community-wide character displacement could become reproductively isolated from any conspecific populations that are in allopatry (see chapter 8), thereby promoting numerous speciation events. The resulting pattern would be one in which the ancestral clade (that is, the clade that includes both allopatric and sympatric populations of each species) becomes more species rich as well as more diverse in ecological and reproductive traits over time.

A second situation in which character displacement can promote a rapid proliferation of species transpires when a given species co-occurs with different heterospecifics in different parts of its geographic range. As noted in chapter 8, if a species interacts with different heterospecifics across different populations, local adaptation of reproductive or resource-use traits in response to these interactions may generate reproductive isolation among these separate conspecific populations, thereby creating "speciation cascades" (*sensu* Pfennig and Ryan 2006; see also Howard 1993; Lemmon 2009). Thus, multiple speciation events—and possibly even adaptive radiations—may arise as a by-product of character displacement.

DIVERGENT EVOLUTION

We now consider character displacement's contribution to the second of adaptive radiation's two key components: the differentiation of species into numerous niches. Throughout this book, we have described how character displacement can promote divergence between competitors (see especially chapters 1 and 2). Indeed, character displacement can even lead to the evolution of entirely new resource-use or reproductive traits (see chapters 4 and 5). When competitively mediated selection fuels adaptive radiation, character displacement minimizes competitive interactions by driving competitors into diverse, and sometimes novel, ecological or reproductive niches—and maintaining them there (Figure 9.9). Moreover, as we described in the previous section, an initial instance of character displacement between a pair of species may precipitate an entire series of character displacement events between numerous interacting species within the same community (Figure 9.8), thereby causing a small group of ancestral species to diversify rapidly into a large number of descendant species that occupy a wide variety of ecological niches. In other words, an initial instance of character displacement between a pair of species may precipitate an adaptive radiation.

Yet, as first noted by Darwin (1859 [2009]), competitively mediated selection may play an even more fundamental role in evolution beyond driving divergence during adaptive radiation. Specifically, such selection may explain why even closely related species typically differ from each other phenotypically. In other words, as Darwin (1859 [2009]) first realized, competitively mediated selection could account for the divergent nature of evolution *generally*.

In chapter 1, we described how Darwin (1859 [2009]) maintained that competitively mediated divergent selection could explain why evolution has produced a distinctive tree-like typology. Recall that Darwin's imagery of evolution (unique at that time) was one in which ancestors and descendants split and thereafter continually diverged from one another (Figure 1.2B). Darwin (1859 [2009]) specifically suggested that this pattern of divergent evolution reflected the tendency for the strength of competition to increase with increasing taxonomic similarity between competitors. Darwin (1859 [2009], p. 110) maintained that competition is the most severe between the most closely related individuals, because these individuals should generally be the most similar to each other in resource use (see chapter 1). This claim was central to Darwin's arguments for explaining why evolution, on the grand scale, is largely divergent.

Over the years, Darwin's claim concerning the role of competitively mediated selec-

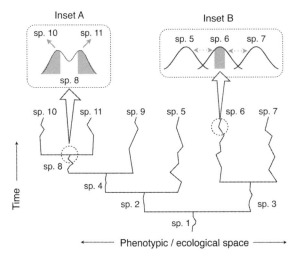

FIGURE 9.9.
How competitively mediated selection promotes divergent trait evolution and adaptive radiation. When ecological opportunity is available, competitively mediated disruptive selection can act within species to favor divergent ecomorphs, which may ultimately become distinct species (depicted in inset A, where selection favors extreme traits [shaded portion] in species 8, which become new species 10 and 11). Competitively mediated selection may subsequently minimize competitive interactions between species by enhancing phenotypic and ecological differences between them. However, when ecological opportunity becomes reduced, competitively mediated selection may favor reduced trait variance—that is, enhanced specialization. As shown in inset B, this selection favors average trait values within each species (again, selection favors the shaded portion), such that species remain distinct in ecological/phenotypic space even though overall trait disparity within the clade no longer increases.

tion in explaining the divergent nature of evolution has either met with skepticism or been misunderstood (see discussion in chapter 1). We therefore briefly review the empirical support for Darwin's claim before returning to the role of divergent selection in adaptive radiation.

In chapter 2, we presented evidence to suggest that the strength of competitively mediated divergent selection increases the more similar two individuals are to each other ecologically and phenotypically (see Figures 2.5, 2.6). A recent study—which (like Gause's [1934] experiments) used bactivorous protist species (*Paramecium*) in a multigenerational experiment—confirmed that the strength of competitively mediated divergent selection also increases with *phylogenetic* similarity between competitors (Violle et al. 2011). In this experiment, both the frequency and tempo of competitive exclusion increased with increasing phylogenetic relatedness between competing species (Figure 9.10), indicating that phylogenetic similarity can serve as a proxy of ecological similarity (Violle et al. 2011).

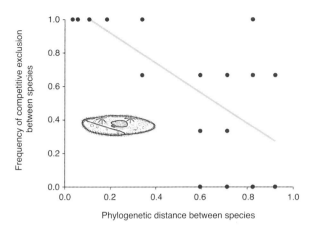

FIGURE 9.10.
Experimental evidence that the likelihood of competitive exclusion occurring between any two species depends, in part, on their phylogenetic distance (see also Fig. 2.5). In this experiment, the researchers examined the outcome of competition between species pairs of all possible pairwise combinations of 10 species of freshwater bacterivorous ciliated protists. After eight weeks, either competitive exclusion or stable coexistence was observed (the frequency of competitive exclusion was estimated as the fraction of times each species pair exhibited competitive exclusion out of three replicates for that pairwise species interaction treatment). In this experiment, phylogenetic distances were calculated as the pairwise branch lengths of the phylogeny. Redrawn from Violle et al. (2011).

Therefore, empirical data support Darwin's (1859 [2009]) claim that the intensity of competitively mediated divergent selection increases as two competitors become more similar to each other ecologically, phenotypically, and phylogenetically. Of course, evolution is not always divergent. Indeed, adaptation to similar selective pressures can promote *convergent* evolution between unrelated taxa. Examples include the similar body forms of bony fishes, sharks, ichthyosaurs, and dolphins (Carroll 1997, pp. 266–275), and the similar water-saving adaptations of plants from the cactus family (Cactaceae) and the euphorbia family (Euphorbiaceae) from different arid regions of the world (Ingrouille and Eddie 2006, pp. 301–303). Moreover, competitively mediated selection is not the only process that could conceivably explain the divergent nature of evolution (Schluter 2000). Nevertheless, character displacement appears to play a significant role in divergent evolution, as we have described throughout this text.

In returning to the role of divergent selection in adaptive radiation, we are not suggesting that character displacement will cause lineages to continue to diverge *pheno-*

typically over time. Indeed, although character displacement is expected to generate divergence, it is not expected to generate ever increasing phenotypic diversity (Doebeli 1996). For example, all adaptive radiations eventually come to an end, meaning that, over time, the rates at which new species and new phenotypes arise within a clade slow down to normal, "background" rates (Nee et al. 1992; Seehausen 2006; McPeek 2008; Phillimore and Price 2008; Rabosky and Lovette 2008; Linder 2009; Gonzalez-Voyer and Kolm 2011). Moreover, studies from the fossil record of morphological "disparity"— a measure of the diversity of morphological types expressed by different taxa within a clade (Gould 1991; Foote 1997; Wills 2001)—reveal that the greatest rate of increase in disparity often occurs early in a clade's history (reviewed in Wagner 2010). In these clades, this initial rapid increase in disparity typically remains unsurpassed during the entire history of the clade (note, however, that some clades show the opposite pattern of low early disparity followed by high disparity later in the clade's history [Wagner 2010]; further, a comparative analysis of body size and shape in numerous animal clades failed to support the early-burst model of adaptive radiation [Harmon et al. 2010]). Although not a necessary component of character displacement, early increased morphological disparity is consistent with the idea that character displacement promotes *rapid* divergence into new niches.

A pattern of high early disparity has also been reported from an analysis of morphological variation *within* species. Webster (2007) investigated long-term trends in morphological variation within individual species of trilobites over this group's 270-million-year history (from the Early Cambrian period, when they first appeared in the fossil record, through the Late Permian period, when they went extinct). Trilobites are among the most diverse of all fossil groups, they have a rich fossil record, and they are highly variable morphologically. As a group, they show a pattern of high early disparity: depending on the morphological character one examines, maximal disparity among different species of trilobites was achieved in either the Early Cambrian (about 542 to 513 million years ago) or the Ordovician period (about 488 to 444 million years ago). A similar pattern—an early peak in morphological variation—also occurs *within* species. Trilobites from the Cambrian period exhibit significantly more variation within species than younger and/or more derived taxa (Webster 2007). Generally, within-species morphological variation declines gradually from this Cambrian peak to the Late Permian mass extinction (251 million years ago; Webster 2007). Thus an early burst of morphological disparity may be expressed both among and within species.

Because many clades may show an early rapid increase in disparity followed by a slowdown through time, it is useful to ask what factor(s) could explain such a pattern. One possibility is that, when a clade's constituent populations or species initially experience conditions favorable for an adaptive radiation, they expand the range of niches that they each occupy (see the discussion in chapter 5 on intraspecific character displacement). As populations and species expand niche breadth, ecological opportunity becomes exhausted (Schluter 2000; Reznick and Ricklefs 2009; Losos and Mahler

2010) and, as a consequence, the adaptive radiation is expected to grind to a halt (recall from above the importance of ecological opportunity in fostering adaptive radiation). The gradual depletion of ecological opportunity during the course of an adaptive radiation may thereby generate a pattern in which disparity culminates early in a clade's history.

Yet, when ecological opportunity is depleted, competition is expected to intensify if an increasing number of species compete for the same range of resources and reproductive-trait space. In such situations, character displacement is expected to assume an additional, new role in macroevolution. Rather than promoting the evolution of new resource-use or reproductive phenotypes (as it does in the early phase of an adaptive radiation; for example, see inset A in Figure 9.9), it may begin to promote the evolution of increased specialization. During this second phase, individual lineages undergo character displacement that results in reduced niche width as they subdivide the available niche space (see inset B in Figure 9.9; see also Box 1.1 and chapter 6). In other words, rather than promoting shifts to a new trait value, character displacement reduces trait *variation*. Consequently, in the period following an adaptive radiation, as new lineages arise, different lineages should become more specialized and therefore more restrictive in the range of resource-use or reproductive phenotypes that they express, even though the clade as a whole may not change in the range of such phenotypes expressed.

A major clade that appears to have increased in specialization—but not in morphological disparity—is the phylum Arthropoda, which includes crabs, spiders, insects, trilobites, and eurypterids. Arthropods are unsurpassed among all animal phyla in terms of both species diversity and the variety of ecological niches that they occupy (Brusca and Brusca 2003). Much of their success is tied to a key innovation: jointed appendages (Knoll and Carroll 1999). Modifications of these appendages provide arthropods with diverse ways of moving, finding food, and seeking mates (Grimaldi and Engel 2005).

Morphometric analyses (Wills et al. 1994) have revealed that extant arthropods (as a group) show no more morphological disparity than do arthropods from the Cambrian period (that is, 542 to 488 million years ago, when the initial diversification of arthropods occurred; Brusca and Brusca 2003). Yet arthropods have apparently become more specialized over this time period, as evidenced by increased specialization of their jointed appendages. Indeed, recent genera have more diverse appendage morphologies than do their Cambrian counterparts (Wills and Fortney 2000). Furthermore, appendages in Cambrian arthropods are less likely to be differentiated from other appendages on the same body than are appendages in modern forms (Wills and Fortney 2000). Thus, although arthropods as a group have not increased since the Cambrian in diversity of morphological types (that is, in disparity), they have become much more species rich and specialized.

In summary, character displacement potentially explains the divergent nature of adaptive radiations specifically and evolution generally (Figure 9.9). However, phenotypic divergence driven by competition is not expected to continue without end. Character displacement may play a primary role in initiating divergent evolutionary trajectories of interacting lineages that can become magnified later by other selective and nonselective processes (*sensu* McShea and Brandon 2010). Moreover, although character displacement may ultimately cease to promote phenotypic divergence between lineages, it may subsequently assume a new role in macroevolution in which it promotes the evolution of increased ecological specialization.

Having examined character displacement's role in adaptive radiation, we now turn to its possible contribution to two widespread trends in macroevolution: the evolution of increased body size and increased complexity.

EVOLUTIONARY ESCALATION

Above we focused on character displacement's role in mediating *divergent* trait evolution and increased *specialization*. As we first pointed out in Box 1.1, both types of trait evolution are possible manifestations of character displacement. However, character displacement can assume other manifestations. For example, competition may favor the evolution of increased competitive ability (Box 1.1). In other words, rather than shifting to an alternative resource or reproductive trait-space, and rather than reducing the range of resources or reproductive trait-space that it utilizes, a focal species may become better at acquiring its current resource or reproductive trait-space in the face of competition. In this way, character displacement can lead to evolutionary escalation of competitive ability (by "escalation," we follow Vermeij's [1987] use of the term to mean that improvements in any given evolutionary lineage are matched by improvements in that lineage's "enemies"—in this case, in its competitors—thereby fueling an ongoing series of adaptations and counter-adaptations).

To understand how character displacement contributes to escalation, consider that an important consequence of competition is that species often need to keep evolving just to maintain the status quo in relation to their competitors, an idea that Van Valen (1973) termed the "Red Queen effect" (after the Red Queen in Lewis Carroll's *Through the Looking Glass*, who remarks, "it takes all the running you can do, to keep in the same place"). The Red Queen hypothesis asserts that each species has to "run" (that is, evolve) as fast as possible just to stay in the same "place" (that is, maintain evolutionary fitness), because of ongoing adaptation in its rivals (that is, its competitors, predators, prey, parasites, and hosts). This reciprocal evolution ("co-evolution") resulting from antagonistic species interactions has been likened to an arms race (Cott 1940; Dawkins and Krebs 1979). During co-evolutionary arms races, a lineage comes under strong selection for traits that offset improvements made by its opponents, which are in turn under strong selection to become (from the focal lineage's viewpoint) more malevolent over evolutionary time (Dawkins 1986). This process can lead to the evolution of a series of adaptations and counter-adaptations by both parties—that is, escalation.

The focus of research on such escalatory interactions has historically been on predator-prey and parasite-host interactions (for example, see Vermeij 1987; Brodie 1999; Foitzik et al. 2001; Buckling and Rainey 2002b; Decaestecker et al. 2007; Davies 2011). However, competitively mediated selection can also drive escalation (Arthur 1982; Vermeij 1987; Ridley 1996). Indeed, we explained in chapter 6 how asymmetries between interacting species in access to a superior niche following character displacement can explain why competitively mediated selection might often foster escalation in competitive ability, rather than divergence onto alternative resources. In other words, escalation in competitive ability would likely not generate niche shifts. However, as we highlight below, escalatory trait evolution could lead to niche shifts if adaptations and counter-adaptations enable species to ultimately take advantage of underutilized or novel niches. Although escalation has not traditionally been viewed as a manifestation of character displacement, escalatory change in features associated with resource acquisition or reproduction may be a relatively common evolutionary response to competitively mediated selection (Ridley 1996; see also Box 1.1), especially in the context of community-wide character displacement.

Escalatory co-evolution driven by competition may have fueled many large-scale trends in the history of life (Table 9.1). We begin by focusing on one such trend—the widespread tendency of many lineages to increase in body size over evolutionary time. As background, evolutionary scientists have long noted that the maximal body size of many evolutionary lineages has tended to increase over time, a trend so pervasive that it has been dubbed "Cope's rule" (after the person who discovered this pattern, E. D. Cope; see Cope 1896; Benton 2002). A compelling example of this macroevolutionary pattern is shown in Figure 9.1. Similar patterns have been documented in diverse plant, invertebrate, and vertebrate taxa (reviewed in Kingsolver and Pfennig 2004; Payne et al. 2009).

Before delving into whether competitively mediated selection may account for Cope's rule, we need to ask whether Cope's rule represents a "driven" trend or a "passive" trend (*sensu* McShea 1994). With a driven trend, change is more likely to occur in one direction than in the other (McShea 1994). By contrast, a passive trend evolves in only one direction because the other direction is bounded by an impassable limit. Because life started small and could therefore only evolve in one direction in terms of body size—to increase—some have argued that Cope's rule is merely a passive trend (Gould 1997).

However, evolution toward larger body size has been shown to be a driven trend in at least some clades, such as the horse family, Equidae. During the past 50 million years, not only did the maximal and mean body size of horses increase, so did their minimum body size (McShea 1994). Moreover, an analysis of ancestor-descendant pairs showed that size increases were much more frequent than size decreases (McShea 1994).

Furthermore, a general tendency to evolve greater body size can be ascribed to a specific cause—selection. In particular, studies of selection in contemporary organisms have demonstrated that directional selection generally favors larger size in diverse organisms (Kingsolver and Pfennig 2004). Moreover, both natural selection and sexual selection generally favor increased body size (Kingsolver and Pfennig 2004). In other words, bigger is generally fitter, in terms of both survival and reproduction. If we

TABLE 9.1. Macroevolutionary trends that may reflect competitively mediated escalation

Trend	Source
Repeated evolution of multicellularity from unicellular ancestors	Grosberg and Strathmann (2007)
Repeated transition from water to land in plants and animals	Clack (2002); Taylor et al. (2009)
Independent evolution of the tree habit in lycophytes, ferns, gymnosperms, and angiosperms	Niklas (1997); Hickey (2003)
Repeated evolution of flight in invertebrates and vertebrates	Rayner (1990); Grimaldi and Engel (2005)
Tendency of many lineages to increase in body size and complexity over evolutionary time	Bonner (1988); Kingsolver and Pfennig (2004)

assume that similar selective pressures operated in the past, then selection becomes a plausible mechanism to explain the general tendency throughout the history of life for lineages to increase in body size.

Given that Cope's rule is (at least in some lineages) a driven trend, and given that selection generally favors large body size, what are the agents of this selection? A primary agent of selection that favors larger body size is competition (Bonner 1988; Kingsolver and Pfennig 2004). Generally, larger individuals are favored over smaller individuals when competing for either resources (Peters 1983) or mates (Andersson 1994).

Bonner (1988) suggested how competitively mediated selection can promote the evolution of larger body size. His hypothesis regarding selection's role in this process is synonymous with character displacement (Ridley 1996; Bonner also includes drift as a possible mechanism promoting larger size, but a trend driven by drift would not constitute character displacement). Specifically, Bonner maintained that:

> ... in any ecological setting, since all the intermediate sizes are represented at all times ... the only place for expansion, for pioneering, is in the upper end of the size scale. This is a place where totally new species can appear, either by selection or by drift, that will not meet stiff competition. They are pioneers that have avoided competition by finding a new niche that is above all the others. Elephants and giant sequoias are obvious examples. Therefore the very slow progress of maximum size increase over the entire course of evolution ... represents the rate at which new, upper size limit niches have been invaded over the last 3 million years or so. (Bonner 1988, p. 59)

Thus, because smaller organisms are much more abundant than larger organisms in any given ecological community (Peters 1983; Bonner 1988), selection can reduce competition (for resources or mates) by promoting the evolution of larger body size (Bonner 1988). Essentially, when it comes to body size evolution, there is always "room at the top." Put another way, because the largest individuals in most ecological communities have an advantage when competing for limited resources (Peters 1983) or successful reproduction (Andersson 1994), community-wide character displacement may be manifested as a tendency for species to evolve toward increasingly larger body size as a means of reducing competition with other members of the community (see also the discussion above on tiering). Such a race to the top may engender escalation in body size.

If competitively mediated selection generally favors ever-larger body size, then why are many extant taxa not at their maximum size? For instance, the largest land invertebrate—the 2-meter-long *Arthropleura* (possibly a millipede)—lived during the Carboniferous period, 300 million years ago. There are at least four non–mutually exclusive reasons why many extant taxa are not at their maximum body size.

First, mass extinctions often appear to be size selective. Generally, larger individuals are more vulnerable to the deteriorating environmental conditions that typify mass extinctions (LaBarbera 1986; Arnold et al. 1995; Alroy 1999), because, for example, they require more food. Thus, in repeated evolutionary bouts, a clade may increase in average body size during "normal" times (when large size is favored), but then the larger constituents may go extinct during episodes of mass extinction, after which the remaining lineages may undergo increasing body-size evolution once again (Kingsolver and Pfennig 2004).

Second, many extant taxa may not be at their maximum size because selection for decreased development time may often oppose selection for increased size, resulting in an absence of evolutionary change in size (Kingsolver and Pfennig 2004). Larger organisms typically have longer generation times (Bonner 1988), and short generation time can be favored over longer generation time.

Third, a clade may have achieved larger body size earlier in its evolutionary history, but the largest members of this clade may have gone extinct during a mass extinction, leaving only the smaller members of the clade behind. In the intervening recovery period, some other clade may have taken over the race to the top before the surviving members of the original clade could do so.

Finally, other factors likely also play a role in limiting the size of some taxa, including biotic factors such as increasingly effective predators (Endler 1991), as well as abiotic factors, such as changes in oxygen levels in the atmosphere (Ward 2006).

In light of the different factors that can contribute to escalation and the evolution of increased body size, additional research is needed to understand competition's role in body-size evolution over macroevolutionary time. Nevertheless, the existing evidence suggests that competition plays an important role, albeit not a solitary one, in generating large-scale trends in body-size evolution.

Not only might competitively mediated selection propel a trend toward larger organisms, it could also drive another macroevolutionary trend: the tendency for organisms to increase in complexity over macroevolutionary time. Although complexity is difficult to define and therefore measure (reviewed in McShea 1991, 1996), a striking pattern in the history of life is an increase in the maximum level of hierarchical organization (one measure of complexity), in which new types of organisms have emerged from the integrated association of lower-level "entities" (such as individual cells) that were previously capable of independent reproduction (Maynard Smith and Szathmáry 1995; McShea and Changizi 2003). For example, the first multicellular organisms evolved from single-celled organisms, thereby forming a new, more complex type of organism (Grosberg and Strathmann 2007).

Evolutionary biologists have long sought to explain why complexity might increase over evolutionary time, and many researchers have suggested that natural selection might favor complexity (see, for example, Rensch 1960a, b; Waddington 1969; Bonner 1988; for an opposing viewpoint, see McShea and Brandon 2010). According to one hypothesis (Waddington 1969; Bonner 1988), during the course of a lineage's evolution, as niches become fully filled, selection strongly favors those individuals that can invade new niches. Because earlier organisms were smaller and generally also less complex (complexity increases with an organism's body size; Bell and Mooers 1997), species could possibly escape competition by evolving to become more complex. The evolution of multicellularity, for example, made possible a division of labor among different cell types in which some cell lines become specialized on resource acquisition (they might become, for instance, roots in plants or the gut in animals), thereby possibly conferring an advantage when competing for scarce resources (Bonner 1988; Maynard Smith and Szathmáry 1995; Grosberg and Strathmann 2007). Indeed, greater complexity not only enables organisms to compete for resources more effectively, but it can also allow organisms to create their own ecological opportunity through the production of novel, specialized phenotypes (see the previous section on adaptive radiation and chapter 5).

In sum, through the history of life, competitively mediated selection may have generally favored "pioneers" that were both larger and more complex and therefore better able to invade vacant niches. If these ideas are correct, then in promoting evolutionary escalation, character displacement may have acted as a key—although by no means the exclusive—driver of the trends toward both increased body size *and* increased complexity. In other words, throughout the history of life, character displacement may have had far-reaching impacts that account not only for why living things are so diverse, but also for why they tend to be larger and more complex.

MACROEVOLUTION: RED QUEEN OR COURT JESTER?

In the discussion above, we considered how biotic factors—specifically, competitive interactions—could account for both the divergent nature of evolution and the widespread tendency of evolutionary lineages to undergo escalation (and the phenotypic ramifications of that escalation). We do not intend to imply, however, that other biotic factors, such as predation, or abiotic factors, such as climate, are unimportant in explaining macroevolutionary patterns.

Indeed, as noted in the introduction to this chapter, the relative importance of biotic versus abiotic factors in macroevolution is a key unresolved problem in evolutionary biology. Recall that, according to the Red Queen hypothesis, interactions among organisms can promote macroevolutionary change (but see Venditti et al. 2010). However, an alternative hypothesis asserts that abiotic factors—especially climate change, tectonic events, and asteroid impacts—are the principal drivers of macroevolution (Benton 2009). The latter hypothesis has been dubbed the "Court Jester hypothesis" (Barnosky 2001).

Generally, there is a lack of consensus concerning the relative importance in macroevolution of Red Queen effects versus Court Jester effects (Benton 2009). The Red Queen hypothesis has been widely supported by studies from contemporary populations, especially in regard to the co-evolution of predators and their prey (for example, see Brodie 1999) and parasites and their hosts (for example, see Foitzik et al. 2001; Decaestecker et al. 2007; Davies 2011). Moreover, the Red Queen hypothesis may explain certain macroevolutionary patterns, such as the tendency for evolutionary lineages to diverge from one another in ecologically relevant traits and to undergo escalatory trait evolution (including, possibly, the evolution of increased body size and complexity; see above).

Nevertheless, the possibility remains that abiotic factors may be more important than biotic factors in explaining macroevolutionary patterns (Benton 2009, 2010; Hunt 2010). Clearly, there are abiotic factors—such as extinction associated with asteroid impacts (Alvarez et al. 1980; Schulte et al. 2010) and abrupt climate change (Erwin 2009)—that can leave major imprints on the history of life. Moreover, although biotic factors and abiotic factors are sometimes treated as mutually exclusive drivers of macroevolution, they likely often act in concert. For example, as described above, an evolutionary lineage may evolve larger body size over time because of competitive co-evolution between and within species. As a result of evolving larger size, however, these lineages may be more prone to going extinct following, for example, climate change. Note that, in this particular example, neither biotic nor abiotic factors would completely explain macroevolutionary change in body size in any given lineage. Only by taking both factors into account would we have a complete picture of the causes of body size evolution. Similarly, mass extinction owing to an asteroid impact may provide the ecological opportunity that enables competitively mediated adaptive radiation to occur.

In sum, both biotic and abiotic factors likely play crucial roles in driving macroevolution (Benton 2009), and a biotic factor that may be especially important in fueling macroevolution is competition.

SUMMARY

The relationship between micro- and macroevolution has long been the focus of debate in evolutionary biology. Much of this controversy centers on whether the same processes

> **BOX 9.1.** Suggestions for Future Research
>
> - Identify additional examples, if any, of character displacement and competitive exclusion from the fossil record.
> - Identify additional cases in which an adaptive radiation unfolded following the appearance of a key innovation that improved opportunities for successful reproduction (for example, see Arnegard et al. 2010).
> - Gather additional empirical data to determine whether the intensity of competitively mediated divergent selection increases the more similar two species are ecologically, phenotypically, and/or phylogenetically.
> - Use experimental evolution studies to ascertain the relative importance of competition versus other biotic and abiotic agents of selection in promoting adaptive radiation.
> - Determine whether competitively mediated selection (whether arising from competition for resources or reproduction), as opposed to other selective agents (for example, predation), generally favors larger and more complex organisms.
> - Identify how biotic and abiotic factors contribute jointly (or independently) to macroevolutionary patterns.

cause evolution at both scales. Although many evolutionary biologists have asserted that most macroevolutionary phenomena reflect microevolutionary processes acting over extended periods of evolutionary time, this connection between micro- and macroevolution is often difficult to demonstrate. In this chapter, we examined whether certain macroevolutionary phenomena—such as adaptive radiation, the divergent nature of evolution, and the widespread tendency for lineages to increase in size and complexity—stem ultimately from competitively mediated selection acting on individual organisms. Although character displacement is not alone in contributing to macroevolution, the study of character displacement potentially helps unite microevolution and macroevolution. We list some key challenges for future research in Box 9.1.

FURTHER READING

Arnegard, M. E., P. B. McIntyre, L. J. Harmon, M. L. Zelditch, W. G. R. Crampton, J. K. Davis, J. P. Sullivan et al. 2010. Sexual signal evolution outpaces ecological divergence during electric fish species radiation. *American Naturalist* 176:335–356. This paper describes how competition for available reproductive-trait space fosters adaptive radiation.

Bonner, J. T. 1988. *The evolution of complexity by means of natural selection*. Princeton University Press: Princeton, NJ. This highly readable book describes how competition may promote an increase in size and complexity over time.

Gillespie, R. 2009. Adaptive radiation. Pp. 143–154. In: *The Princeton guide to ecology*. S. A. Levin (ed). Princeton University Press: Princeton, NJ. This book chapter presents a comprehensive overview of the causes of adaptive radiation.

Losos, J. B., and D. L. Mahler. 2010. Adaptive radiation: The interaction of ecological opportunity, adaptation, and speciation. Pp. 381–420. In: *Evolution since Darwin*. M. A. Bell, D. J. Futuyma, W. F. Eanes, and J. S. Levinton (eds). Sinauer: Sunderland, MA. This book chapter focuses on the importance of ecological opportunity and competition in promoting adaptive radiation.

Wagner, P. J. 2010. Paleontological perspectives on morphological evolution. Pp. 451–478. In: *Evolution since Darwin*. M. A. Bell, D. J. Futuyma, W. F. Eanes, and J. S. Levinton (eds). Sinauer: Sunderland, MA. This book chapter describes how disparity is measured in the fossil record, and it also reviews the patterns that have emerged from such analyses.

10

MAJOR THEMES AND UNSOLVED PROBLEMS

At the outset of this book, we emphasized that a longstanding problem in both evolutionary biology and ecology is to explain why there are so many different kinds of living things and why even closely related organisms tend to differ from one another phenotypically (for example, see Figure 1.1). We described how Darwin (1859 [2009]) addressed this problem by advancing his principle of divergence of character. According to this principle, the origin of species and the evolution of phenotypic differences between them stem ultimately from divergent selection acting to minimize competitive interactions between initially similar groups of individuals. We also noted that, despite the significance of this principle for Darwin's thinking, it is often misunderstood and under-appreciated today.

The central aim of this book has been to explore the role of competition in generating and maintaining biodiversity. In particular, we have examined how competitively mediated selection can promote trait evolution through the evolutionary process of character displacement (Brown and Wilson 1956; Grant 1972). As we have seen, character displacement is well supported theoretically and empirically, and it provides a compelling explanation for why living things diversify.

Our goal was not merely to establish that competition promotes trait evolution, however. More importantly, we sought to investigate character displacement's causes and its consequences. Our premise was that, by understanding these causes and consequences, we could gain crucial insights into such fundamental issues in evolutionary biology and ecology as how new traits and new species arise, why species diversify, how they coexist, and why they occur where they do.

In this chapter, we begin by reviewing the major themes that we emphasized throughout the book. In re-examining these themes, we describe how an understanding of the causes and consequences of character displacement can indeed shed light onto the origins, abundance, and distribution of biodiversity. However, as we have emphasized throughout this book, much remains to be learned about character displacement. Therefore, in the second half of the chapter, we discuss some unanswered questions that promise to be profitable avenues for future research.

MAJOR THEMES OF THE BOOK

What have we learned about competition's role as an organizing force in nature since Darwin (1859 [2009]) first proposed his principle of divergence of character 150 years ago? Here we underscore seven themes that we have emphasized in this book.

CHARACTER DISPLACEMENT IS A PROCESS, NOT A PATTERN

An issue that historically has hampered progress in studying character displacement is how to define it. In chapter 1, we advocated defining character displacement as an evolutionary *process*; that is, as trait evolution that arises as an adaptive response to resource competition or deleterious reproductive interactions between species (*sensu* Grant 1972). Although some authors continue to define character displacement as a geographical pattern of trait variation—specifically, one in which species show exaggerated divergence in resource-use or reproductive traits in sympatry—doing so fosters confusion. This is because a pattern of exaggerated divergence in sympatry can come about through processes other than competitively mediated selection (see chapter 1). Furthermore, defining character displacement as the *specific* pattern of exaggerated divergence in sympatry ignores the fact that competitively mediated selection can produce patterns of trait evolution other than exaggerated divergence in sympatry (see Box 1.1 and also below).

Confusion over what is—and what is not—character displacement can be ameliorated by defining character displacement as a process. Moreover, by defining *both* ecological and reproductive character displacement as processes (rather than defining the former as a process and the latter as a pattern, as is sometimes done; see chapter 1) enables researchers to focus more effectively on how these two forms of character displacement interact. In short, to understand the causes and consequences of diversity, it is important to clarify the agents that foster this diversity. Focusing on process, rather than pattern, enables researchers to identify these agents.

CHARACTER DISPLACEMENT CAN PRODUCE DIFFERENT FORMS OF TRAIT EVOLUTION

Although competitively mediated selection often causes sympatric species to diverge from one another (see chapter 1), it can lead to forms of trait evolution other than *diver-*

gent trait evolution. For instance, competitively mediated selection might promote the evolution of specialization (Box 1.1 Figure 1C). Specifically, in situations where ecological opportunity is absent, species may evolve to utilize a more restrictive range of resources or reproductive-trait space as an adaptive response to interspecific competition. Alternatively, competitively mediated selection might favor the evolution of increased competitive ability (Box 1.1 Figure 1C). In particular, in situations where losing a competitive interaction is especially costly (as discussed in chapter 6), selection might favor a focal species to become better at acquiring its current resource (for example) in the face of competition rather than shift to an alternative resource. Finally, in certain situations, competitively mediated selection can actually lead to character *convergence* between interacting species (Box 1.1).

Recognizing that competitively mediated selection can generate alternative forms of trait evolution is important because these various forms can have different ramifications. For instance, when a lineage evolves increased competitive ability, it exerts selection on interacting lineages to also become better competitors, which imposes additional selection on the original lineage, and so on. This process can fuel evolutionary escalation; that is, it can result in an arms race between competitors (see below and chapter 9). Broadening our view of character displacement to include phenomena other than divergent trait evolution in sympatry will help clarify competition's role in promoting diversity.

ECOLOGICAL AND REPRODUCTIVE CHARACTER DISPLACEMENT INTERACT

Although Darwin (1859 [2009]) held that selection that lessens competition for resources drives character displacement (a process that he dubbed "divergence of character"), abundant evidence now indicates that selection can act similarly to reduce reproductive interactions. Indeed, as we have stressed throughout the book, selection stemming from reproductive competition can produce trait evolution rivaling that generated by resource competition.

Perhaps more importantly, these two forms of character displacement probably often interact for the simple reason that organisms that are similar enough to engage in competition for resources are also likely to impede each other's ability to reproduce successfully. In this way, ecological character displacement and reproductive character displacement will often become intertwined. Once one form occurs, it can often lead to changes (for example, in phenotypes, habitat use, or both) that generate the variation on which selection can act to promote the alternative form of character displacement. Consequently, one form of character displacement might facilitate the other form (chapter 3). Understanding the interplay of these two processes is important because reproductive character displacement may represent the missing link between Darwin's divergence of character and the origin of species (Pfennig and Pfennig 2010).

PHENOTYPIC PLASTICITY CAN MEDIATE CHARACTER DISPLACEMENT

Phenotypic plasticity might play a key role in mediating character displacement. Indeed, when facing competition, the individuals of many species facultatively express traits that lessen the fitness costs of resource competition or deleterious reproductive interactions (chapters 2 and 4). For example, an individual might facultatively express an alternative resource-use phenotype that is distinct from that produced by a competitor, thereby minimizing resource competition (see Table 2.1).

Such competitively mediated plasticity has not traditionally been viewed as a mechanism of character displacement, primarily because many researchers have held that phenotypic plasticity is a nongenetic response that is incapable of driving adaptive evolution. Yet heritable variation typically exists within natural populations in the tendency to respond facultatively to a competitor, and such variation can serve as the target of competitively mediated selection that diverges between interacting species. Thus, phenotypic plasticity is capable of driving adaptive evolution. Additionally, competitively mediated plasticity can satisfy the widely accepted criteria for demonstrating character displacement (see Table 1.2 for criteria). In short, environmentally induced shifts in resource-use or reproductive traits can constitute character displacement (Box 2.2).

Moreover, character displacement might begin as an initial phase in which trait divergence is environmentally triggered. Only in a later phase might trait divergence become genetically fixed in different populations and species (Pfennig and Pfennig 2012). This plasticity-first hypothesis for the evolution of character displacement has increasing empirical support, and it may be a more common evolutionary route to character displacement than is generally assumed (see chapter 4).

CHARACTER DISPLACEMENT PROMOTES DIVERSIFICATION AT MULTIPLE LEVELS

Character displacement potentially promotes diversification at multiple levels of biological organization, including at the level of individual organisms, populations, species, and even entire clades. At the individual level, variation in the costs or benefits of avoiding resource competition or reproductive interactions between species can lead to individual variation in the expression of traits that minimize such interactions (chapter 3). Moreover, intraspecific character displacement generates increased phenotypic variation within populations. Additionally, intraspecific character displacement can favor the evolution of discrete, alternative phenotypes within populations (ecomorphs), including phenotypes that are entirely novel (chapter 5). At the population level, character displacement drives divergence between populations that are in sympatry with a heterospecific competitor and allopatric populations. Moreover, because character displacement may transpire differently among sympatric populations, different sympatric populations may also diverge (chapters 3 and 8). At the species level, character displacement promotes divergence between species that compete with each other. Furthermore, character displacement may both initiate and finalize the process of speciation (chapter 8). Finally, at the level of clades, character displacement potentially plays a key role in triggering adaptive radiation (chapter 9).

The diversifying nature of character displacement is therefore not limited to simply exaggerating existing differences between species. At any of the above levels of biological organization, the expression of character displacement can explain diversity. If some groups are more likely to undergo character displacement than others, then such variation in character displacement can ultimately explain why some groups are more diverse than others.

COMPETITIVELY MEDIATED SELECTION AND SEXUAL SELECTION INTERACT

Sexual selection can be a potent driver of diversification both within and between species (Andersson 1994). Given that both competitively mediated selection and sexual selection can affect traits associated with reproduction, understanding how these two forms of selection interact is essential for explaining diversity's origins.

As described in chapter 7, competitively mediated selection can impact sexual selection in at least two important ways. First, both ecological and reproductive character displacement can alter sexual signals, mate preferences, and the habitat in which these traits are expressed. Character displacement thereby alters the phenotypic targets of sexual selection as well as the selective context in which these traits are both expressed and perceived. Second, competitively mediated selection can have important effects on the underlying fitness consequences of sexual selection. For example, sexual selection theory generally predicts that females should prefer exaggerated traits because these indicate male quality. However, if heterospecifics possess elaborate traits, selection that minimizes reproductive interactions between species should promote the evolution of preferences for less-exaggerated signals that are most dissimilar from those heterospecifics. Yet, by adopting such preferences, females might forego information about a prospective conspecific mate's quality (Pfennig 2000b). Character displacement can therefore explain population and species diversity in sexual signaling, complexity in reproductive traits, and (in some cases) variation in the underlying fitness effects of mate choice and sexual signaling.

Not only can character displacement impact sexual selection, but the converse also holds. In particular, sexual selection can either facilitate or impede character displacement, depending on whether or not traits that minimize heterospecific interactions also enhance or decrease an individual's mating success. Nevertheless, when sexual selection and character displacement reinforce each other, diversity in reproductive traits and the potential for speciation can be greatly enhanced (Figure 7.4; chapter 8).

CHARACTER DISPLACEMENT HAS MACROEVOLUTIONARY IMPLICATIONS

Evolutionary scientists have long debated the possible connection between microevolution (that is, evolutionary processes acting within species) and macroevolution (that is, major evolutionary change above the species level, including long-term evolutionary trends). The study of character displacement potentially helps unite microevolution and macroevolution. Indeed, many macroevolutionary phenomena can potentially be explained in terms of selection acting to minimize costly competitive interactions among *individual* organisms (both heterospecifics and conspecifics).

For example, because character displacement is thought to promote both the proliferation of species and the differentiation of species into numerous ecological niches—the two key components of adaptive radiation—character displacement has long been viewed as underlying adaptive radiation. Character displacement may also play a crucial role in the macroevolutionary trends toward increased body size and increased complexity. In particular, because larger and more complex individuals and species are often favored during competition, competitively mediated selection might have acted as an important driver of both increased body size and increased complexity over the course of the history of life (see chapter 9). In this way, life's complexity may owe its existence, at least in part, to competitively mediated selection.

SOME UNSOLVED PROBLEMS

Throughout this book, we have emphasized open questions regarding character displacement. Indeed, in many sections, we have suggested what *might* or *could* occur, because so little is known in some areas. Here we highlight seven major areas that promise to be profitable for future research.

First, we need to understand more about whether the two major forms of character displacement—ecological character displacement and reproductive character displacement—impact each other, and if they do, how they affect each other. In chapter 3, we described how one form of character displacement could influence the likelihood of the other form transpiring. Specifically, we hypothesized that, once one form of character displacement has initially occurred, it can promote changes in either phenotypes or habitat that subsequently facilitate the alternative form. Of the limited research that has been conducted on this topic, most has concentrated on how ecological character displacement might jump-start reproductive character displacement. By contrast, we know relatively little of how reproductive character displacement might also facilitate ecological character displacement (but see Konuma and Chiba 2007). Moreover, rather than one form of character displacement facilitating the other, ecological or reproductive character displacement might preclude the alternative form of character displacement. Generally, more theoretical and empirical research is needed to determine how ecological or reproductive character displacement preclude or facilitate each other—or, indeed, whether they actually interact at all. Such research is crucial, because, as noted in chapter 3, understanding the conditions under which character displacement is more likely to proceed could explain ecological and evolutionary patterns of diversity.

A second, related area that requires attention is to determine how character displacement (whether reproductive or ecological) and sexual selection interact. Unfortunately, researchers generally continue to study character displacement and sexual selection separately, as Darwin did. Although these barriers are breaking down, we still lack a fundamental understanding of: (1) how interactions between species alter mate choice, male competition, and sexual signaling *within* each such species; (2) how such interactions alter the fitness consequences of mate choice and mate attraction; (3) the degree to which sexual selection promotes—or inhibits—divergence in response to competitive interactions between species; and (4) how character displacement and sexual selection interact to promote reproductive isolation between populations within species. By addressing these issues, we will gain greater insight into how and why sexual selection varies both between and within species (for further discussion, see Ryan 1998; Ptacek 2000; Boughman 2002). We will also better understand sexual selection's role in the origins and maintenance of trait and species diversity.

A third major area requiring attention is the proximate mechanisms mediating character displacement. Two questions, in particular, stand out. First, what is the source of the phenotypic variation on which selection acts during character displacement? Second, how do different sources of phenotypic variation affect the speed and manner in which character displacement unfolds? Although character displacement is assumed to reflect allelic differences between populations and species, it may alternatively arise through competitively mediated phenotypic plasticity. Indeed, as we described in chapter 4, character displacement might pass through an initial phase in which species differences arise through phenotypic plasticity before transitioning to a later phase in which these differences become fixed and subsequently expressed constitutively. Although this evolutionary scenario has some support (see chapter 4), additional tests are needed.

A fourth major area that promises profitable avenues for future research concerns the role of intraspecific character displacement in fostering species diversity. In chapter 5, we described how character displacement acting within species could promote the evolution of resource or mating polymorphism in which discrete, alternative morphs differing in resource use or reproductive tactics coexist within the same population. We also noted that many evolutionary biologists have long contended that such polymorphisms might represent a critical, early stage in the formation of new species. Indeed, we described (chapter 8) how intraspecific character displacement might initiate and facilitate the process of speciation. However, we also described (chapter 5) how the presence of a resource polymorphism might lessen a species' risk of extinction. Further research is needed to determine whether resource or mating polymorphism, generated by intraspecific character displacement, does indeed foster species diversity by increasing the speciation rate, decreasing the extinction rate, or both.

A fifth major area of future research centers on the explicit identification of character displacement's effects on ecological processes. In particular, as described in this book, the study of character displacement has traditionally emphasized how competitively mediated selection promotes trait evolution. Although character displacement is expected to promote species coexistence, relatively few studies make direct connections between the evolution of traits that arise as a consequence of competition and the effects of these traits on population growth, dynamics, or persistence. Indeed, this is especially true for reproductive traits, the evolution of which would be expected to impact population reproductive rates and growth directly (Gröning and Hochkirch 2008).

Making such connections is critical because they would allow researchers to distinguish the degree to which communities and species richness are structured by character displacement rather than by species sorting. Perhaps more importantly, however, establishing direct links between character displacement and its ecological impacts could explain variation in community structure, distributions of populations across space and time, and even the expression of individual species' ranges. In particular, because both the strength of competitively mediated selection and the ability to respond to such selection varies for different populations and species (chapter 3), the outcome of competitive interactions—that is, the outcome of character displacement as opposed to competitive or reproductive exclusion—also necessarily varies spatially and temporally. Identifying the means by which this variation generates differences in community structure, patterns of species richness, and the distribution of species in space and time represents a major frontier at the intersection of evolution and ecology.

A sixth major area requiring more research is that of character displacement's role in macroevolution. Character displacement has long been viewed as playing a key role in macroevolution because of its effects on promoting adaptive radiation. However, character displacement's importance in macroevolution may go beyond its role in instigating adaptive radiation. Indeed, macroevolution encompasses a broad array of phenomena in which character displacement might play a key role; these phenomena range from the origin of novel, complex traits to long-term evolutionary trends. Regarding the former, we noted in chapter 5 that phenotypic novelties associated with resource use or reproduction (which frequently are the main features that differ between different species) might often arise as a direct consequence of competitively mediated selection. However, these ideas require more rigorous tests to establish the degree to which they are correct.

Regarding character displacement's possible role in long-term evolutionary trends, among the most pronounced such trends in the history of life are the widespread tendencies for many evolutionary lineages to increase in both body size and complexity over time (Bonner 1988; McShea 1996; Payne et al. 2009). In chapter 9, we described how such macroevolutionary trends might stem from competitively mediated selection acting on *individual* organisms. In particular, amplifying on an idea first proposed by

Bonner (1988), we hypothesized that—as a means of escaping competition—selection often favors those individuals that can outcompete others or invade new niches. Because earlier organisms were generally smaller and simpler, later-evolving species might have been able to escape competition by evolving to become larger and more complex. Thus, throughout the history of life, competitively mediated selection might have favored larger and more complex "pioneers" that were equipped to invade these niches.

Whether character displacement can act as a driver of both increased body size and increased complexity over the course of long periods of evolutionary time has not been tested empirically, however. Indeed, macroevolutionary trends are difficult, if not impossible, to study experimentally. Nevertheless, there are ways of working around this problem. For example, the increasing use of experiments with replicated populations of rapidly evolving microbes has provided key insights into the role of ecological interactions in driving long-term macroevolutionary patterns (reviewed in Kassen 2009; for specific examples of such experiments, see Rainey and Travisano 1998; Tyerman et al. 2008; Bono et al. in press). Additionally, greater attention should be paid to the possibility that competition can be detected in the fossil record (see chapter 9). By using these and other approaches, evolutionary scientists can make substantial headway in understanding the role that microevolutionary processes—such as competitively mediated selection—may have in generating macroevolutionary patterns.

Finally—and perhaps most critically—more information is needed on competitively mediated selection's contribution to diversification relative to other possible selective agents, such as predation (Endler 1991) and climate change (Erwin 2009). Although competition has been regarded since Darwin's time as a key agent of evolutionary diversification (see chapter 1)—and abundant evidence exists to suggest that competitively mediated selection can indeed promote diversification (see chapter 9)—much less is known about the possible contributions of other selective agents.

For example, the most spectacular of all adaptive radiations is the "Cambrian explosion," a unique episode in Earth's history when nearly all animal phyla first appear in the fossil record, starting about 542 million years ago and ending less than 100 million years later (Gould 1989). The causes of this remarkable radiation have been much debated (reviewed in Marshall 2006). Some researchers have advocated for abiotic factors as the primary cause, noting that this radiation came at a time when the Earth was undergoing dramatic physical changes. For instance, the ocean's chemistry was changing drastically, as levels of oxygen in both the seawater and atmosphere were rising rapidly (Canfield et al. 2007). Heightened levels of oxygen would have made two characteristics of large, complex animals possible: higher activity levels and larger body size (Knoll and Carroll 1999). Other researchers, however, have maintained that antagonistic species interactions—specifically, predation—drove this radiation. According to this line of reasoning, predation favored diverse adaptations in prey as well as diverse counter-adaptations in predators, such as larger body size (Peters 1983), the ability to see (with the emergence of eyes; Parker 2003), and the ability to produce body armor and spines (with the appearance of biomineralization; Towe 1970). By contrast, competition is rarely considered to be a cause of the Cambrian explosion. Were other agents, such as ocean chemistry and predation, more crucial than competition in instigating the Cambrian explosion, or was competition critical in promoting this radiation, as it appears to have been in more recent adaptive radiations (see chapter 9)? Answering this question could help explain animal life's earliest diversification.

Additionally, we know surprisingly little about the ways that different selective agents interact in shaping ecological and evolutionary processes (MacColl 2011). Yet both theoretical and empirical studies suggest that different selective agents can indeed interact, sometimes in counterintuitive ways. For instance, the interaction between biotic and abiotic agents may have been crucial in driving macroevolution. We described one such possible example in chapter 9, when hypothesizing that competitive co-evolution might have promoted larger body size, but that larger size might have subsequently rendered these lineages more prone to going extinct when the Earth underwent dramatic physical changes.

In short, not only do we need to identify the selective agents that promote diversification, but we also need to learn more about how various selective agents interact. Such an understanding could come from meta-analyses of existing studies as well as new experiments that manipulate different selective agents and measure the effects of selection on rates and extent of diversification of adaptive traits. Regardless of the approach used, understanding the interplay of competition with other sources of selection—both from biotic and abiotic factors—will lend greater insight into the evolutionary and ecological origins of diversity.

SUMMARY

Character displacement is well supported theoretically and empirically, and it plays a key, and often decisive, role in generating and maintaining biodiversity. Yet a number of important issues regarding the causes and consequences of character displacement remain unresolved, and additional research on this topic promises to have far-reaching ramifications. Indeed, because character displacement is central to the origins, abundance, and distribution of biodiversity, understanding its causes and consequences can shed light on some of the most fundamental issues in evolutionary biology and ecology, including how new species arise, diversify, and coexist.

REFERENCES

Abrahamson, W. G., M. D. Eubanks, C. P. Blair, and A. V. Whipple. 2001. Gall flies, inquilines, and goldenrods: A model for host-race formation and sympatric speciation. *American Zoologist* 41:928–938.
Abrams, P. A. 1987. Alternative models of character displacement and niche shift. I. Adaptive shifts in resource use when there is competition for nutritionally nonsubstitutable resources. *Evolution* 41:651–661.
———. 2000. Character shifts of prey species that share predators. *American Naturalist* 156:S45–S61.
Abzhanov, A., W. P. Kuo, C. Hartman, B. R. Grant, P. R. Grant, and C. J. Tabin. 2006. The calmodulin pathway and evolution of elongate beak morphology in Darwin's finches. *Nature* 442:563–567.
Abzhanov, A., M. Protas, B. R. Grant, P. R. Grant, and C. J. Tabin. 2004. Bmp4 and morphological variation of beaks in Darwin's finches. *Science* 305:1462–1465.
Adachi-Hagimori, T., K. Miura, and Y. Abe. 2011. Gene flow between sexual and asexual strains of parasitic wasps: A possible case of sympatric speciation caused by a parthenogenesis-inducing bacterium. *Journal of Evolutionary Biology* 24:1254–1262.
Adams, C. E., and F. A. Huntingford. 2004. Incipient speciation driven by phenotypic plasticity? Evidence from sympatric populations of Arctic charr. *Biological Journal of the Linnean Society* 81:611–618.
Adams, D. C. 2004. Character displacement via aggressive interference in Appalachian salamanders. *Ecology* 85:2664–2670.
———. 2010. Parallel evolution of character displacement driven by competitive selection in terrestrial salamanders. *BMC Evolutionary Biology* 10:1–10.

Agrawal, A. A. 2001. Phenotypic plasticity in the interactions and evolution of species. *Science* 294:321–326.

Aguirre, W. E., K. E. Ellis, M. Kusenda, and M. A. Bell. 2008. Phenotypic variation and sexual dimorphism in anadromous threespine stickleback: Implications for postglacial adaptive radiation. *Biological Journal of the Linnean Society* 95:465–478.

Alexandrou, M. A., C. Oliveira, M. Maillard, R. A. R. McGill, J. Newton, S. Creer, and M. I. Taylor. 2011. Competition and phylogeny determine community structure in Müllerian co-mimics. *Nature* 469:84–88.

Alfaro, M., and F. Santini. 2010. A flourishing of fish forms. *Nature* 464:840–842.

Allen, R. M., Y. M. Buckley, and D. J. Marshall. 2008. Offspring size plasticity in response to intraspecific competition: An adaptive maternal effect across life-history stages. *American Naturalist* 171:225–237.

Alroy, J. 1998. Cope's rule and the dynamics of body mass evolution in North American fossil mammals. *Science* 280:731–734.

———. 1999. Putting North America's end-Pleistocene megafaunal extinction into context: Large-scale analyses of spatial patterns, extinction rates, and size distributions. Pages 105–143 in R. D. E. MacPhee, ed. *Extinctions in near time: Causes, contexts, and consequences.* Kluwer Academic / Plenum Publishers: New York.

Alroy, J., M. Aberhan, D. J. Bottjer, M. Foote, F. T. Fursich, P. J. Harries, A. J. W. Hendy et al. 2008. Phanerozoic trends in the global diversity of marine invertebrates. *Science* 321:97–100.

Alvarez, L. W., W. Alvarez, F. Asaro, and H. V. Michel. 1980. Extra-terrestrial cause for the Cretaceous-Tertiary extinction. *Science* 208:1095–1108.

Amarasekare, P. 2009. Competition and coexistence in animal communities. Pages 196–201 in S. A. Levin, ed. *The Princeton guide to ecology.* Princeton University Press: Princeton, NJ.

Anderson, C. N., and G. F. Grether. 2010. Character displacement in the fighting colours of *Hetaerina* damselflies. *Proceedings of the Royal Society B: Biological Sciences* 277:3669–3675.

Andersson, M. 1994. *Sexual selection.* Princeton University Press: Princeton, NJ.

Andrewartha, H. G., and L. C. Birch. 1954. *The distribution and abundance of animals.* University of Chicago Press: Chicago, IL.

Archibald, J. D. 2011. *Extinction and radiation: How the fall of dinosaurs led to the rise of mammals.* Johns Hopkins University Press: Baltimore, MD.

Armbruster, W. S., M. E. Edwards, and E. M. Debevec. 1994. Floral character displacement generates assemblage structure of western Australian triggerplants (*Stylidium*). *Ecology* 75:315–329.

Arnegard, M. E., P. B. McIntyre, L. J. Harmon, M. L. Zelditch, W. G. R. Crampton, J. K. Davis, J. P. Sullivan et al. 2010. Sexual signal evolution outpaces ecological divergence during electric fish species radiation. *American Naturalist* 176:335–356.

Arnold, A. J., D. C. Kelly, and W. C. Parker. 1995. Causality and Cope's rule: Evidence from planktonic foraminifera. *Journal of Paleontology* 69:203–210.

Arnold, M. L. 1997. *Natural hybridization and evolution.* Oxford University Press: Oxford, UK.

Arnqvist, G., and L. Rowe. 2005. *Sexual conflict*. Princeton University Press: Princeton, NJ.

Arthur, W. 1982. The evolutionary consequences of interspecific competition. *Advances in Ecological Research* 12:127–187.

Ashton, P. S., T. J. Givnish, and S. Appanah. 1988. Staggered flowering in the Dipterocarpaceae: New insights into floral induction and the evolution of mast fruiting in the aseasonal tropics. *American Naturalist* 132:44–66.

Aubret, F., and R. Shine. 2009. Genetic assimilation and the postcolonization erosion of phenotypic plasticity in island Tiger Snakes. *Current Biology* 19:1932–1936.

Auld, J. R., A. A. Agrawal, and R. A. Relyea. 2010. Re-evaluating the costs and limits of adaptive phenotypic plasticity. *Proceedings of the Royal Society B: Biological Sciences* 277:503–511.

Ausich, W. I. 1980. A model for niche differentiation in Lower Mississippian crinoid communities. *Journal of Paleontology* 54:273–288.

Ausich, W. I., and D. J. Bottjer. 1982. Tiering in suspension-feeding communities on soft substrata throughout the Phanerozoic. *Science* 216:173–174.

———. 2003. Sessile invertebrates. Pages 384–386 in D. E. G. Briggs, and P. R. Crowther, eds. *Paleobiology II*. Blackwell: Malden, MA.

Backstrom, N., J. Lindell, Y. Zhang, E. Palkopoulou, A. Qvarnstrom, G. P. Sætre, and H. Ellegren. 2010. A high-density scan of the Z chromosome in *Ficedula* flycatchers reveals candidate loci for diversifying selection. *Evolution* 64:3461–3475.

Badyaev, A. V., G. E. Hill, M. L. Beck, A. A. Dervan, R. A. Duckworth, K. J. McGraw, P. M. Nolan et al. 2002. Sex-biased hatching order and adaptive population divergence in a passerine bird. *Science* 295:316–318.

Baldwin, J. M. 1896. A new factor in evolution. *American Naturalist* 30:441–451.

———. 1902. *Development and evolution*. Macmillan: New York.

Bambach, R. K. 1985. Classes and adaptive variety: The ecology of diversification in marine faunas through the Phanerozoic. Pages 191–253 in J. W. Valentine, ed. *Phanerozoic diversity patterns: Profiles in macroevolution*. Princeton University Press: Princeton, NJ.

Bank, C., J. Hermisson, and M. Kirkpatrick. 2011. Can reinforcement complete speciation? *Evolution* 66:229–239.

Barlow, N., ed. 1959. *The autobiography of Charles Darwin 1809-1882, with original omissions restored*. Harcourt, Brace and Company: New York.

Barluenga, M., and A. Meyer. 2004. The Midas cichlid species complex: Incipient sympatric speciation in Nicaraguan cichlid fishes? *Molecular Ecology* 13:2061–2076.

———. 2010. Phylogeography, colonization and population history of the Midas cichlid species complex (*Amphilophus* spp.) in the Nicaraguan crater lakes. *BMC Evolutionary Biology* 10.

Barluenga, M., K. N. Stölting, W. Salzburger, M. Muschick, and A. Meyer. 2006. Sympatric speciation in Nicaraguan crater lake cichlid fish. *Nature* 439:719–723.

Barnosky, A. D. 2001. Distinguishing the effects of the Red Queen and the Court Jester on Miocene mammal evolution in the Northern Rocky Mountains. *Journal of Vertebrate Paleontology* 21:172–185.

Barnwell, C. V., and M. A. F. Noor. 2008. Failure to replicate two mate preference QTLs across multiple strains of *Drosophila pseudoobscura*. *Journal of Heredity* 99:653–656.

Barrett, R. D. H., and G. Bell. 2006. The dynamics of diversification in evolving *Pseudomonas* populations. *Evolution* 60:484–490.

Barrett, R. D. H., and D. Schluter. 2008. Adaptation from standing genetic variation. *Trends in Ecology and Evolution* 23:38–44.

Barrier, M., B. G. Baldwin, R. H. Robichaux, and M. D. Purugganan. 1999. Interspecific hybrid ancestry of a plant adaptive radiation: Allopolyploidy of the Hawaiian Silversword Alliance (Asteraceae) inferred from floral homeotic gene duplications. *Molecular Biology and Evolution* 16:1105–1113.

Barton, N. H. 2001. Adaptation at the edge of a species' range. Pages 365–392 in J. Silvertown, and J. Antonovics, eds. *Integrating ecology and evolution in a spatial context*. Blackwell: New York.

Barton, N. H., and G. M. Hewitt. 1985. Analysis of hybrid zones. *Annual Review of Ecology and Systematics* 16:113–148.

———. 1989. Adaptation, speciation and hybrid zones. *Nature* 341:497–503.

Basolo, A. L. 1990. Female preference predates the evolution of the sword in swordtail fish. *Science* 250:808–810.

Bateman, A. J. 1948. Intra-sexual selection in *Drosophila*. *Heredity* 2:349–368.

Bateson, W. 1909. Heredity and variation in modern lights. Pages 85–101 in A. C. Seward, ed. *Darwin and modern science*. Cambridge University Press: Cambridge, UK.

Behm, J. E., A. R. Ives, and J. W. Boughman. 2010. Breakdown in postmating isolation and the collapse of a species pair through hybridization. *American Naturalist* 175:11–26.

Bell, G., and A. Gonzalez. 2011. Adaptation and evolutionary rescue in metapopulations experiencing environmental deterioration. *Science* 332:1327–1330.

Bell, G., and A. O. Mooers. 1997. Size and complexity among multicellular organisms. *Biological Journal of the Linnean Society* 60:345–363.

Benkman, C. W. 1988. A 3:1 ratio of mandible crossing direction in White-winged Crossbills. *The Auk* 105:578–579.

———. 1996. Are the ratios of bill crossing morphs in crossbills the result of frequency-dependent selection? *Evolutionary Ecology* 10:119–126.

———. 2003. Divergent selection drives the adaptive radiation of crossbills. *Evolution* 57:1176–1181.

Benton, M. J. 1983. Large-scale replacements in the history of life. *Nature* 302:16–17.

———. 1987. Progress and competition in macroevolution. *Biological Reviews of the Cambridge Philosophical Society* 62:305–338.

———. 1996. On the nonprevalence of competitive replacement in the evolution of tetrapods. Pages 185–210 in D. Jablonski, D. H. Erwin, and J. Lipps, eds. *Evolutionary paleobiology*. University of Chicago Press: Chicago, IL.

———. 2002. Cope's rule. Pages 209–210 in M. Pagel, ed. *Encyclopedia of evolution*. Oxford University Press: Oxford, UK.

———. 2009. The red queen and the court jester: Species diversity and the role of biotic and abiotic factors through time. *Science* 323:728–732.

———. 2010. New take on the Red Queen. *Nature* 463:306–307.

Berlocher, S. H., and J. L. Feder. 2002. Sympatric speciation in phytophagous insects: Moving beyond controversy? *Annual Review of Entomology* 47:773–815.

Bernays, E. 1986. Diet-induced head allometry among foliage-chewing insects and its importance for granivores. *Science* 231:495–497.

Biesmeijer, J. C., J. A. P. Richter, M. A. J. P. Smeets, and M. J. Sommeijer. 1999. Niche differentiation in nectar-collecting stingless bees: The influence of morphology, floral choice and interference competition. *Ecological Entomology* 24:380–388.

Birkhead, T. R. 2000. *Promiscuity: An evolutionary history of sperm competition.* Harvard University Press: Cambridge, MA.

Blair, W. F. 1974. Character displacement in frogs. *American Zoologist* 14:1119–1125.

Blondel, J., P. C. Dias, P. Perret, M. Maistre, and M. M. Lambrechts. 1999. Selection-based biodiversity at a small spatial scale in a low-dispersing insular bird. *Science* 285:1399–1402.

Boag, P. T., and P. R. Grant. 1984. The classic case of character release: Darwin's finches (*Geospiza*) on Isla Daphne Major, Galápagos. *Biological Journal of the Linnean Society* 22:243–287.

Bolnick, D. I. 2001. Intraspecific competition favours niche width expansion in *Drosophila melanogaster*. *Nature* 410:463–466.

———. 2004. Can intraspecific competition drive disruptive selection? An experimental test in natural populations of sticklebacks. *Evolution* 58:608–618.

Bolnick, D. I., P. Amarasekare, M. S. Araujo, R. Burger, J. M. Levine, M. Novak, V. H. W. Rudolf et al. 2011. Why intraspecific trait variation matters in community ecology. *Trends in Ecology and Evolution* 26:183–192.

Bolnick, D. I., and M. Doebeli. 2003. Sexual dimorphism and adaptive speciation: Two sides of the same ecological coin. *Evolution* 57:2433–2449.

Bolnick, D. I., and B. M. Fitzpatrick. 2007. Sympatric speciation: Models and empirical evidence. *Annual Review of Ecology, Evolution and Systematics* 38:459–487.

Bolnick, D. I., and O. L. Lau. 2008. Predictable patterns of disruptive selection in stickleback in postglacial lakes. *American Naturalist* 172:1–11.

Bolnick, D. I., R. Svanbäck, L. H. Yang, J. M. Davis, C. D. Hulsey, and M. L. Forister. 2003. The ecology of individuals: Incidence and implications of individual specialization. *American Naturalist* 161:1–28.

Bonner, J. T. 1988. *The evolution of complexity by means of natural selection.* Princeton University Press: Princeton, NJ.

Bono, L. M., C. L. Gensel, D. W. Pfennig, and C. L. Burch. In press. Competition and the origins of novelty: Experimental evolution of niche-width expansion in a virus. *Biology Letters.*

Borgia, G. 2006. Preexisting male traits are important in the evolution of elaborated male sexual display. *Advances in the Study of Behavior* 36:249–303.

Boughman, J. W. 2001. Divergent sexual selection enhances reproductive isolation in sticklebacks. *Nature* 411:944–948.

———. 2002. How sensory drive can promote speciation. *Trends in Ecology and Evolution* 17:571–577.

———. 2007. Condition-dependent expression of red colour differs between stickleback species. *Journal of Evolutionary Biology* 20:1577–1590.

Boughman, J. W., H. D. Rundle, and D. Schluter. 2005. Parallel evolution of sexual isolation in sticklebacks. *Evolution* 59:361–373.

Boul, K. E., W. C. Funk, C. R. Darst, D. C. Cannatella, and M. J. Ryan. 2007. Sexual selection drives speciation in an Amazonian frog. *Proceedings of the Royal Society B: Biological Sciences* 274:399–406.

Bradbury, J. W., and S. L. Vehrencamp. 1998. *Principles of animal communication.* Sinauer: Sunderland, MA.

Bragg, A. N. 1965. *Gnomes of the night: The spadefoot toads.* University of Pennsylvania Press: Philadelphia, PA.

Bridle, J. R., and T. H. Vines. 2007. Limits to evolution at range margins: When and why does adaptation fail? *Trends in Ecology and Evolution* 22:140–147.

Britch, S. C., M. L. Cain, and D. J. Howard. 2001. Spatio-temporal dynamics of the *Allonemobius fasciatus-A. socius* mosaic hybrid zone: A 14-year perspective. *Molecular Ecology* 10:627–638.

Brodie, E. D. 1999. Predator-prey arms races. *BioScience* 49:557–568.

Brower, A. V. Z. 1996. Parallel race formation and the evolution of mimicry in *Heliconius* butterflies: A phylogenetic hypothesis from mitochondrial DNA sequences. *Evolution* 50:195–221.

Brown, J. H., and D. W. Davidson. 1977. Competition between seed-eating rodents and ants in desert ecosystems. *Science* 196:880–882.

Brown, J. S., and T. L. Vincent. 1992. Organization of predator-prey communities as an evolutionary game. *Evolution* 46:1269–1283.

Brown, W. L., and E. O. Wilson. 1956. Character displacement. *Systematic Zoology* 5:49–64.

Browne, J. 1995. *Charles Darwin: Voyaging.* Princeton University Press: Princeton, NJ.

Bruno, J. F., J. J. Stachowicz, and M. E. Bertness. 2003. Inclusion of facilitation into ecological theory. *Trends in Ecology and Evolution* 18:119–125.

Brusca, R. C., and G. J. Brusca. 2003. *Invertebrates.* Sinauer: Sunderland, MA.

Buckling, A., and P. B. Rainey. 2002a. The role of parasites in sympatric and allopatric host diversification. *Nature* 420:496–499.

———. 2002b. Antagonistic coevolution between a bacterium and a bacteriophage. *Proceedings of the Royal Society B: Biological Sciences* 269:931–936.

Burns, J. H., and S. Y. Strauss. 2011. More closely related species are more ecologically similar in an experimental test. *Proceedings of the National Academy of Sciences of the United States of America* 108:5302–5307.

Bush, G. L., and R. K. Butlin. 2004. Sympatric speciation in insects. Pages 229–248 in U. Dieckmann, M. Doebeli, J. A. J. Metz, and D. Tautz, eds. *Adaptive speciation.* Cambridge University Press: Cambridge, UK.

Butler, J., J. M. Jia, and J. Dyer. 1997. Simulation techniques for the sensitivity analysis of multi-criteria decision models. *European Journal of Operational Research* 103:531–546.

Butler, M. A., S. A. Sawyer, and J. B. Losos. 2007. Sexual dimorphism and adaptive radiation in *Anolis* lizards. *Nature* 447:202–205.

Butlin, R. K. 1987. Speciation by reinforcement. *Trends in Ecology and Evolution* 2:8–13.

———. 2005. Recombination and speciation. *Molecular Ecology* 14:2621–2635.

Butlin, R. K., J. Galindo, and J. W. Grahame. 2008. Sympatric, parapatric or allopatric: The most important way to classify speciation? *Philosophical Transactions of the Royal Society B: Biological Sciences* 363:2997–3007.

Butlin, R. K., and M. G. Ritchie. 1994. Behaviour and speciation. Pages 43–79 in P. J. B. Slater and T. R. Halliday, eds. *Behaviour and speciation*. Cambridge University Press: Cambridge, UK.

Byers, J. E. 2002. Impact of non-indigenous species on natives enhanced by anthropogenic alteration of selection regimes. *Oikos* 97:449–458.

Cahill, J. F., G. G. McNickle, J. J. Haag, E. G. Lamb, S. M. Nyanumba, and C. C. St. Clair. 2010. Plants integrate information about nutrients and neighbors. *Science* 328:1657.

Cain, M. L., W. D. Bowman, and S. D. Hacker. 2008. *Ecology*. Sinauer: Sunderland, MA.

Calsbeek, R. 2009. Experimental evidence that competition and habitat use shape the individual fitness surface. *Journal of Evolutionary Biology* 22:97–108.

Calsbeek, R., and R. M. Cox. 2010. Experimentally assessing the relative importance of predation and competition as agents of selection. *Nature* 465:613–616.

Calsbeek, R., and T. B. Smith. 2008. Experimentally replicated disruptive selection on performance traits in a Caribbean lizard. *Evolution* 62:478–484.

Calsbeek, R., T. B. Smith, and C. Bardeleben. 2007. Intraspecific variation in *Anolis sagrei* mirrors the adaptive radiation of Greater Antillean anoles. *Biological Journal of the Linnean Society* 90:189–199.

Candolin, U. 2003. The use of multiple cues in mate choice. *Biological Reviews* 78:575–595.

Canfield, D. E., S. W. Poulton, and G. M. Narbonne. 2007. Late-Neoproterozoic deep-ocean oxygenation and the rise of animal life. *Science* 315:92–95.

Cardoso, G. C., and T. D. Price. 2010. Community convergence in bird song. *Evolutionary Ecology* 24:447–461.

Carlson, B. A., S. M. Hasan, M. Hollmann, D. B. Miller, L. J. Harmon, and M. E. Arnegard. 2011. Brain evolution triggers increased diversification of electric fishes. *Science* 332:583–586.

Carranza, A., O. Defeo, and M. Arim. 2011. Taxonomic relatedness and spatial structure of a shelf benthic gastropod assemblage. *Diversity and Distributions* 17:25–34.

Carroll, R. L. 1997. *Patterns and processes of vertebrate evolution*. Cambridge University Press: Cambridge, UK.

Carroll, S. B. 2005. *Endless forms most beautiful*. W. W. Norton: New York.

Carroll, S. B., J. K. Grenier, and S. D. Waertherbee. 2001. *From DNA to diversity: Molecular genetics and the evolution of animal design*. Blackwell: Malden, MA.

Carroll, S. P., H. Dingle, and S. P. Klassen. 1997. Genetic differentiation of fitness-associated traits among rapidly evolving populations of the soapberry bug. *Evolution* 51:1182–1188.

Carroll, S. P., S. P. Klassen, and H. Dingle. 1998. Rapidly evolving adaptations to host ecology and nutrition in the soapberry bug. *Evolution and Ecology* 12:955–968.

Carroll, S. P., M. Marler, R. Winchell, and H. Dingle. 2003. Evolution of cryptic flight morph and life history differences during host race radiation in the soapberry bug, *Jadera haematoloma* Herrich-Schaeffer (Hemiptera: Rhopalidae). *Annals of the Entomological Society of America* 96:135–143.

Carson, H. L., and K. Y. Kaneshiro. 1976. *Drosophila* of Hawaii: Systematics and ecological genetics. *Annual Review of Ecology and Systematics* 7:311–345.

Carson, H. L., and A. R. Templeton. 1984. Genetic revolutions in relation to speciation phenomena: The founding of new populations. *Annual Review of Ecology and Systematics* 15:97–131.

Caruso, C. M. 2000. Competition for pollination influences selection on floral traits of *Ipomopsis aggregata*. *Evolution* 54:1546–1557.

Caruso, C. M., and M. Alfaro. 2000. Interspecific pollen transfer as a mechanism of competition: Effect of *Castilleja linariaefolia* pollen on seed set of *Ipomopsis aggregata*. *Canadian Journal of Botany* 78:600–606.

Case, T. J., R. D. Holt, M. A. McPeek, and T. H. Keitt. 2005. The community context of species' borders: Ecological and evolutionary perspectives. *Oikos* 108:28–46.

Case, T. J., and R. Sidell. 1983. Pattern and chance in the structure of model and natural communities. *Evolution* 37:832–849.

Case, T. J., and M. L. Taper. 2000. Interspecific competition, environmental gradients, gene flow, and the coevolution of species' borders. *American Naturalist* 155:583–605.

Caswell, H. 1976. Community structure: A neutral model analysis. *Ecological Monographs* 46:327–354.

Catchpole, C. K., and P. J. B. Slater. 2008. *Bird song: Biological themes and variations*. Cambridge University Press: Cambridge, UK.

Cavender-Bares, J., D. D. Ackerly, D. A. Baum, and F. A. Bazzaz. 2004. Phylogenetic overdispersion in Floridian oak communities. *American Naturalist* 163:823–843.

Chamberlain, N. L., R. I. Hill, D. D. Kapan, L. E. Gilbert, and M. R. Kronforst. 2009. Polymorphic butterfly reveals the missing link in ecological speciation. *Science* 326:847–850.

Charlesworth, B., D. Charlesworth, and N. H. Barton. 2003. The effects of genetic and geographic structure on neutral variation. *Annual Review of Ecology, Evolution and Systematics* 34:99–125.

Charlesworth, B., M. Nordborg, and D. Charlesworth. 1997. The effects of local selection, balanced polymorphism and background selection on equilibrium patterns of genetic diversity in subdivided populations. *Genetical Research* 70:155–174.

Chase, J. M., and M. A. Leibold. 2003. *Ecological niches: Linking classical and contemporary approaches*. University of Chicago Press: Chicago, IL.

Chave, J. 2009. Competition, neutrality, and community organization. Pages 264–273 in S. A. Levin, ed. *The Princeton guide to ecology*. Princeton University Press: Princeton, NJ.

Chek, A. A., J. P. Bogart, and S. C. Lougheed. 2003. Mating signal partitioning in multi-species assemblages: A null model test using frogs. *Ecology Letters* 6:235–247.

Chesson, P. 2000. Mechanisms of maintenance of species diversity. *Annual Review of Ecology and Systematics* 31:343–366.

Chevin, L. M., R. Lande, and G. M. Mace. 2010. Adaptation, plasticity, and extinction in a changing environment: Towards a predictive theory. *PLoS Biology* 8:e1000357.

Chew, F. S., and R. K. Robbins. 1984. Egg-laying in butterflies. Pages 65–79 in R. I. Vane-Wright, and P. R. Ackery, eds. *The biology of butterflies*. Academic Press: London.

Chunco, A. J., J. S. McKinnon, and M. R. Servedio. 2007. Microhabitat variation and sexual selection can maintain male color polymorphisms. *Evolution* 61:2504–2515.

Clack, J. 2002. *Gaining ground*. University of Indiana Press: Bloomington, IN.

Clark, J. S. 2010. Individuals and the variation needed for high species diversity in forest trees. *Science* 327:1129–1132.

Clarke, B. C. 1975. The contribution of ecological genetics to evolutionary theory: Detecting the direct effects of natural selection on particularly polymorphic loci. *Genetics (Supplement)* 79:101–113.

Clutton-Brock, T. H. 1991. *The evolution of parental care.* Princeton University Press: Princeton, NJ.

Cody, M. L. 1969. Convergent characteristics in sympatric species: A possible relation to interspecific competition and aggression. *Condor* 71:222–239.

———. 1973. Character convergence. *Annual Review of Ecology and Systematics* 4:189–211.

Collins, J. P., and J. E. Cheek. 1983. Effect of food and density on development of typical and cannibalistic salamander larvae in *Ambystoma tigrinum nebulosum*. *American Zoologist* 23:77–84.

Connell, J. H. 1961a. The effects of competition, predation by *Thais lapillus*, and other factors on natural populations of the barnacle *Balanus balanoides*. *Ecological Monographs* 31:61–104.

———. 1961b. The influence of interspecific competition and other factors on the distribution of the barnacle *Chthamalus stellatus*. *Ecology* 42:710–723.

———. 1980. Diversity and the coevolution of competitors, or the ghost of competition past. *Oikos* 35:131–138.

———. 1983. On the prevalence and relative importance of interspecific competition: Evidence from field experiments. *American Naturalist* 122:661–696.

Cooley, J. R. 2007. Decoding asymmetries in reproductive character displacement. *Proceedings of the Academy of Natural Sciences of Philadelphia* 156:89–96.

Cooper, P. 1988. Paleoecology: Paleoecosystems, paleocommunities. *Geoscience Canada* 15:199–208.

Cope, E. D. 1896. *The primary factors of organic evolution.* Open Court Publishing: Chicago, IL.

Corl, A., A. R. Davis, S. R. Kuchta, and B. Sinervo. 2010. Selective loss of polymorphic mating types is associated with rapid phenotypic evolution during morphic speciation. *Proceedings of the National Academy of Sciences of the United States of America* 107:4254–4259.

Costa, G. C., L. J. Vitt, E. R. Pianka, D. O. Mesquita, and G. R. Colli. 2008. Optimal foraging constrains macroecological patterns: Body size and dietary niche breadth in lizards. *Global Ecology and Biogeography* 17:670–677.

Cott, H. B. 1940. *Adaptive coloration in animals.* Methuen: London.

Coyne, J. A., and H. A. Orr. 1989. Patterns of speciation in *Drosophila*. *Evolution* 43:362–381.

———. 2004. *Speciation.* Sinauer: Sunderland, MA.

Craig, T. P., J. K. Itami, W. G. Abrahamson, and J. D. Horner. 1993. Behavioral evidence for host-race formation in *Eurosta solidaginis*. *Evolution* 47:1696–1710.

Crespi, B. J. 1988. Adaptation, compromise and constraint: The development, morphometrics, and behavioral basis of a fighter-flier polymorphism in male *Hoplothrips karnyi* (Insecta: Thysanoptera). *Behavioral Ecology and Sociobiology* 23:93–104.

Crozier, R. H. 1974. Niche shape and genetic aspects of character displacement. *American Zoologist* 14:1151–1157.

Cucherousset, J., A. Acou, S. Blachet, J. A. Britton, W. R. C. Beaumont, and R. E. Gozlan.

2011. Fitness consequences of individual specialization in resource use and trophic morphology in European eels. *Oecologia* 167:75–84.

Dambroski, H. R., and J. L. Feder. 2007. Host plant and latitude-related diapause variation in *Rhagoletis pomonella*: A test for multifaceted life history adaptation on different stages of diapause development. *Journal of Evolutionary Biology* 20:2101–2112.

Dame, E. A., and K. Petren. 2006. Behavioural mechanisms of invasion and displacement in Pacific island geckos (*Hemidactylus*). *Animal Behaviour* 71:1165–1173.

Danforth, B. N., and J. L. Neff. 1992. Male polymorphism and polyethism in *Perdita texana* (Hymenoptera, Andrenidae). *Annals of the Entomological Society of America* 85:616–626.

Darwin, C. 1859 (2009). *The annotated origin: A facsimile of the first edition of* On the origin of species. J. T. Costa, annotator. Belknap Press: Cambridge, MA.

———. 1871. *The descent of man, and selection in relation to sex.* John Murray: London.

Davidson, A. M., M. Jennions, and A. B. Nicotra. 2011. Do invasive species show higher phenotypic plasticity than native species and, if so, is it adaptive? A meta-analysis. *Ecology Letters* 14:419–431.

Davies, N. B. 2011. Cuckoo adaptations: Trickery and tuning. *Journal of Zoology* 284:1-14.

Davies, N. B., J. R. Krebs, and S. A. West. 2012. *An introduction to behavioural ecology* (4th edn.). Wiley-Blackwell: Oxford, UK.

Dawkins, R. 1986. *The blind watchmaker.* Norton: New York.

Dawkins, R., and J. R. Krebs. 1979. Arms races between and within species. *Proceedings of the Royal Society B: Biological Sciences* 205:489–511.

Day, T., J. Pritchard, and D. Schluter. 1994. A comparison of two sticklebacks. *Evolution* 48:1723–1734.

Dayan, T., and D. Simberloff. 1994. Character displacement and morphological variation among British and Irish mustelids. *Ecology* 75:1063–1073.

———. 2005. Ecological and community-wide character displacement: The next generation. *Ecology Letters* 8:875–894.

Dayan, T., D. S. Simberloff, E. Tchhernov, and Y. Yom-Tov. 1990. Feline canines: Community-wide character displacement among the small cats of Israel. *American Naturalist* 136:39–60.

Decaestecker, E., S. Gaba, J. A. M. Raeymaekers, R. Stoks, L. Van Kerckhoven, D. Ebert, and L. De Meester. 2007. Host-parasite "Red Queen" dynamics archived in pond sediment. *Nature* 450:870–873.

Deitloff, J., J. O. Church, D. C. Adams, and R. G. Jaeger. 2009. Interspecific agonistic behaviors in a salamander community: Implications for alpha-selection. *Herpetologica* 65:174–182.

Denno, R. F., M. S. McClure, and J. R. Ott. 1995. Interspecific interactions in phytophagous insects: Competition reexamined and resurrected. *Annual Review of Entomology* 40:297–331.

DeWitt, T. J., A. Sih, and D. S. Wilson. 1998. Costs and limits of phenotypic plasticity. *Trends in Ecology and Evolution* 13:1–8.

Dhondt, A. A. 2012. *Interspecific competition in birds.* Oxford University Press: Oxford, UK.

Diamond, J., S. L. Pimm, M. E. Gilpin, and M. LeCroy. 1989. Rapid evolution of character displacement in Myzomelid honeyeaters. *American Naturalist* 134:675–708.

Dieckmann, U., and M. Doebeli. 1999. On the origin of species by sympatric speciation. *Nature* 400:354–357.

Dieckmann, U., M. Doebeli, J. A. J. Metz, and D. Tautz, eds. 2004. *Adaptive speciation*. Cambridge University Press: Cambridge, UK.

Dijkstra, P. D., C. Hemelrijk, O. Seehausen, and T. G. G. Groothuis. 2008. Color polymorphism and intrasexual competition in assemblages of cichlid fish. *Behavioral Ecology* 20:138–144.

Dobzhansky, T. 1934. Studies on hybrid sterility. I. Spermatogenesis in pure and hybrid *Drosophila pseudoobscura*. *Zeitschrift für Zellforschung und Mikroskopische Anatomie* 21:169–221.

———. 1937. *Genetics and the origin of species*. Columbia University Press: New York.

———. 1940. Speciation as a stage in evolutionary divergence. *American Naturalist* 74:312–321.

———. 1951. *Genetics and the origin of species*. Columbia University Press: New York.

Doebeli, M. 1996. An explicit genetic model for ecological character displacement. *Ecology* 77:510–520.

———. 2011. *Adaptive diversification*. Princeton University Press: Princeton, NJ.

Doebeli, M., and U. Dieckmann. 2000. Evolutionary branching and sympatric speciation caused by different types of ecological interactions. *American Naturalist* 156:S77–S101.

Donohue, K., and J. Schmitt. 1998. Maternal environmental effects in plants: Adaptive plasticity? Pages 137–158 in T. A. Mousseau, and C. W. Fox, eds. *Maternal effects as adaptations*. Oxford University Press: Oxford, UK.

Dorchin, N., E. R. Scott, C. E. Clarkin, M. P. Luongo, S. Jordan, and W. G. Abrahamson. 2009. Behavioural, ecological and genetic evidence confirm the occurrence of host-associated differentiation in goldenrod gall-midges. *Journal of Evolutionary Biology* 22:729–739.

Dornelas, M., S. R. Connolly, and T. P. Hughes. 2006. Coral reef diversity refutes the neutral theory of biodiversity. *Nature* 440:80–82.

Drés, M., and J. Mallet. 2002. Host races in plant-feeding insects and their importance in sympatric speciation. *Philosophical Transactions of the Royal Society B: Biological Sciences* 357:471–492.

Drewry, G. E., and A. S. Rand. 1983. Characteristics of an acoustic community: Puerto Rican frogs of the genus *Eleutherodactylus*. *Copeia* 1983:941–953.

Duffy, S., C. L. Burch, and P. E. Turner. 2007. Evolution of host specificity drives reproductive isolation among RNA viruses. *Evolution* 61:2614–2622.

Durrett, R., and J. Schweinsberg. 2004. Approximating selective sweeps. *Theoretical Population Biology* 66:129–138.

Dybzinski, R., and D. Tilman. 2009. Competition and coexistence in plant communities. Pages 186–195 in S. A. Levin, ed. *The Princeton guide to ecology*. Princeton University Press: Princeton, NJ.

Eberhard, W. G. 1982. Beetle horn dimorphism: Making the best of a bad lot. *American Naturalist* 119:420–426.

———. 1996. *Female control: Sexual selection by cryptic female choice*. Princeton University Press: Princeton, NJ.

Edelaar, P., E. Postma, P. Knops, and R. Phillips. 2005. No support for a genetic basis of mandible crossing direction in crossbills (*Loxia* spp.). *Auk* 122:1123–1129.

Ehrlich, P. R., and P. H. Raven. 1964. Butterflies and plants: A study of coevolution. *Evolution* 18:586–608.

Elias, M., Z. Gompert, C. D. Jiggins, and K. Willmott. 2008. Mutualistic interactions drive ecological niche convergence in a diverse butterfly community. *PLoS Biology* 6:e300.

Elmer, K. R., T. K. Lehtonen, A. F. Kautt, C. Harrod, and A. Meyer. 2010. Rapid sympatric ecological differentiation of crater lake cichlid fishes within historic times. *BMC Biology* 8:60.

Elmer, K. R., T. K. Lehtonen, and A. Meyer. 2009. Color assortative mating contributes to sympatric divergence of Neotropical cichlid fish. *Evolution* 63:2750–2757.

Elton, C. S. 1927 (1943). *Animal ecology*. Sidgwick and Jackson: London.

Emelianov, I., M. Drés, W. Baltensweiler, and J. Mallet. 2001. Host-induced assortative mating in host races of the larch budmoth. *Evolution* 55:2002–2010.

Emelianov, I., J. Mallet, and W. Baltensweiler. 1995. Genetic differentiation in *Zeiraphera diniana* (Lepidoptera, Tortricidae, the Larch Budmoth): Polymorphism, host races or sibling species. *Heredity* 75:416–424.

Emelianov, I., F. Simpson, P. Narang, and J. Mallet. 2003. Host choice promotes reproductive isolation between host races of the larch budmoth *Zeiraphera diniana*. *Journal of Evolutionary Biology* 16:208–218.

Emlen, D. J. 1997. Alternative reproductive tactics and male dimorphism in the horned beetle *Onthophagus acuminatus*. *Behavioral Ecology and Sociobiology* 41:335–341.

———. 2000. Integrating development with evolution: A case study with beetle horns. *BioScience* 50:403–418.

———. 2008. The evolution of animal weapons. *Annual Review of Ecology, Systematics, and Evolution* 39:387–413.

Emlen, D. J., L. C. Lavine, and B. Ewen-Campen. 2007. On the origin and evolutionary diversification of beetle horns. *Proceedings of the National Academy of Sciences of the United States of America* 104:8661–8668.

Emlen, S. T. 1976. Lek organization and mating strategies in bullfrog. *Behavioral Ecology and Sociobiology* 1:283–313.

Emlen, S. T., and L. W. Oring. 1977. Ecology, sexual selection, and the evolution of mating systems. *Science* 197:215–223.

Endler, J. A. 1977. *Geographic variation, speciation, and clines*. Princeton University Press: Princeton, NJ.

———. 1986. *Natural selection in the wild*. Princeton University Press: Princeton, NJ.

———. 1991. Interactions between predators and prey. Pages 169–196 in J. R. Krebs, and N. B. Davies, eds. *Behavioural ecology: An evolutionary approach*. London, Blackwell.

———. 1992. Signals, signal conditions, and the direction of evolution. *American Naturalist* 139:S125–S153.

Endler, J. A., and A. L. Basolo. 1998. Sensory ecology, receiver biases and sexual selection. *Trends in Ecology and Evolution* 13:415–420.

Erwin, D. H. 2000. Macroevolution is more than repeated rounds of microevolution. *Evolution and Development* 2:78–84.

———. 2001. Lessons from the past: Biotic recoveries from mass extinctions. *Proceedings of the National Academy of Sciences of the United States of America* 98:5399–5401.

———. 2008. Macroevolution of ecosystem engineering, niche construction and diversity. *Trends in Ecology and Evolution* 23:304–310.

———. 2009. Climate as a driver of evolutionary change. *Current Biology* 19:R575–R583.

Esselstyn, J. A., S. P. Maher, and R. M. Brown. 2011. Species interactions during diversification and community assembly in an island radiation of shrews. *PLoS One* 6:e21885.

Estrada, C., and C. D. Jiggins. 2008. Interspecific sexual attraction because of convergence in warning colouration: Is there a conflict between natural and sexual selection in mimetic species? *Journal of Evolutionary Biology* 21:749–760.

Etges, W. J., C. C. de Oliveira, E. Gragg, D. Ortíz-Barrientos, M. A. F. Noor, and M. G. Ritchie. 2007. Genetics of incipient speciation in *Drosophila mojavensis*. I. Male courtship song, mating success, and genotype x environment interactions. *Evolution* 61:1106–1119.

Etges, W. J., C. C. de Oliveira, M. G. Ritchie, and M. A. F. Noor. 2009. Genetics of incipient speciation in *Drosophila mojavensis*: II. Host plants and mating status influence cuticular hydrocarbon QTL expression and G x E interactions. *Evolution* 63:1712–1730.

Ewald, P. W. 1994. *The evolution of infectious disease.* Oxford University Press: Oxford, UK.

Farrell, D. B. 1998. "Inordinate fondness" explained: Why are there so many beetles? *Science* 281:553–557.

Fawcett, T. W., and R. A. Johnstone. 2003. Optimal assessment of multiple cues. *Proceedings of the Royal Society B: Biological Sciences* 270:1637–1643.

Feder, J. L., C. A. Chilcote, and G. L. Bush. 1989. Are the apple maggot, *Rhagoletis pomonella*, and blueberry maggot, *Rhagoletis mendax*, distinct species? Implications for sympatric speciation. *Entomological Experiments and Applications* 51:113–123.

Felsenstein, J. 1981. Skepticism toward Santa Rosalia, or why are there so few kinds of animals? *Evolution* 35:124–138.

Fenchel, T. 1975. Character displacement and coexistence in mud snails (Hydrobiidae). *Oecologia* 20:19–32.

Ferveur, J. F. 2005. Cuticular hydrocarbons: Their evolution and roles in *Drosophila* pheromonal communication. *Behavior Genetics* 35:279–295.

Filchak, K. E., J. B. Roethele, and J. L. Feder. 2000. Natural selection and sympatric divergence in the apple maggot *Rhagoletis pomonella*. *Nature* 407:739–742.

Fisher, R. A. 1915. The evolution of sexual preference. *Eugenics Review* 7:184–192.

———. 1930 (1999). *The genetical theory of natural selection. A complete variorum edition.* Oxford University Press: Oxford, UK.

Foitzik, S., C. J. De Heer, D. N. Hunjan, and J. M. Herbers. 2001. Coevolution in host-parasite systems: Behavioural strategies of slave-making ants and their hosts. *Proceedings of the Royal Society B: Biological Sciences* 268:1139–1146.

Foote, M. 1997. The evolution of morphological diversity. *Annual Review of Ecology and Systematics* 28:129–152.

———. 2000. Origination and extinction components of diversity: General problems. Pages 74–102 in D. E. Erwin, and S. L. Wing, eds. *Deep time: Paleobiology's perspective*, Supplement to *Paleobiology*.

Fordyce, J. A. 2006. The evolutionary consequences of ecological interactions mediated through phenotypic plasticity. *Journal of Experimental Biology* 209:2377–2383.

Fordyce, J. A., and C. C. Nice. 2004. Geographic variation in clutch size and a realized benefit of aggregative feeding. *Evolution* 58:447–450.

Fortney, R. A., and R. M. Owens. 1999. Feeding habits in trilobites. *Palaeontology* 42:429–465.

Frankino, W. A., and D. W. Pfennig. 2001. Condition-dependent expression of trophic polyphenism: Effects of individual size and competitive ability. *Evolutionary Ecology Research* 3:939–951.

Frantz, A., V. Calcagno, L. Mieuzet, M. Plantegenest, and J. C. Simon. 2009. Complex trait differentiation between host-populations of the pea aphid *Acyrthosiphon pisum* (Harris): Implications for the evolution of ecological specialisation. *Biological Journal of the Linnean Society* 97:718–727.

Friedman, M. 2010. Explosive morphological diversification of spiny-finned teleost fishes in the aftermath of the end-Cretaceous extinction. *Proceedings of the Royal Society B: Biological Sciences* 277:1675–1683.

Friesen, M. L., G. Saxer, M. Travisano, and M. Doebeli. 2004. Experimental evidence for sympatric ecological diversification due to frequency-dependent competition in *Escherichia coli*. *Evolution* 58:245–260.

Fuller, R. C. 2008. Genetic incompatibilities in killifish and the role of environment. *Evolution* 62:3056–3068.

Funk, D. J. 1998. Isolating a role for natural selection in speciation: Host adaptation and sexual isolation in *Neochlamisus bebbianae* leaf beetles. *Evolution* 52:1744–1759.

Funk, D. J., P. Nosil, and B. Etges. 2006. Ecological divergence exhibits consistently positive associations with reproductive isolation across disparate taxa. *Proceedings of the National Academy of Sciences of the United States of America* 103:3209–3213.

Futuyma, D. J. 2009. *Evolution*. Sinauer: Sunderland, MA.

Galindo, B. E., V. D. Vacquier, and W. J. Swanson. 2003. Positive selection in the egg receptor for abalone sperm lysin. *Proceedings of the National Academy of Sciences of the United States of America* 100:4639–4643.

Galis, F., and J. A. J. Metz. 1998. Why are there so many cichlid species? *Trends in Ecology and Evolution* 13:1–2.

Galloway, L. F., and J. R. Etterson. 2007. Transgenerational plasticity is adaptive in the wild. *Science* 318:1134–1136.

Gause, G. F. 1934. *The struggle for existence*. Williams and Wilkins: Baltimore.

Gavrilets, S. 2004. *Fitness landscapes and the origin of species*: Monographs in Population Biology, v. 41. Princeton University Press: Princeton, NJ.

Gavrilets, S., and J. B. Losos. 2009. Adaptive radiation: Contrasting theory with data. *Science* 732:732–737.

Gerhardt, H. C. 1991. Female mate choice in treefrogs: Static and dynamic acoustic criteria. *Animal Behaviour* 42:615–635.

———. 1994. Reproductive character displacement of female mate choice in the gray treefrog, *Hyla chrysoscelis*. *Animal Behaviour* 47:959–969.

Gerhardt, H. C., and F. Huber. 2002. *Acoustic communication in insects and anurans: Common problems and diverse solutions.* University of Chicago Press: Chicago, IL.

Gersani, M., J. S. Brown, E. E. O'Brien, G. M. Maina, and Z. Abramsky. 2001. Tragedy of the commons as a result of root competition. *Journal of Ecology* 89:660–669.

Ghalambor, C. K., J. K. McKay, S. Carroll, and D. N. Reznick. 2007. Adaptive versus non-adaptive phenotypic plasticity and the potential for contemporary adaptation to new environments. *Functional Ecology* 21:394–407.

Gibson, G., and I. Dworkin. 2004. Uncovering cryptic genetic variation. *Nature Reviews Genetics* 5:1199–1212.

Gil, D., E. Bulmer, P. Celis, and I. Lopez-Rull. 2008. Adaptive developmental plasticity in growing nestlings: Sibling competition induces differential gape growth. *Proceedings of the Royal Society B: Biological Sciences* 275:549–554.

Gilbert, F. 2005. The evolution of imperfect mimicry in hoverflies. Pages 231–288 in M. D. E. Fellowes, G. J. Holloway, and J. Rolff, eds. *Insect evolutionary ecology.* CABI: Wallingford, UK.

Gilbert, J. J. 1973. Induction and ecological significance of gigantism in the rotifer *Asplancha sieboldi. Science* 181:63–66.

Gilbert, S. F., and D. Epel. 2009. *Ecological developmental biology: Integrating epigenetics, medicine, and evolution.* Sinauer: Sunderland, MA.

Gillespie, R. 2009. Adaptive radiation. Pages 143–154 in S. A. Levin, ed. *The Princeton guide to ecology.* Princeton University Press: Princeton, NJ.

Goldberg, D. E., and A. M. Barton. 1992. Patterns and consequences of interspecific competition in natural communities: A review of field experiments with plants. *American Naturalist* 139:771–801.

Goldberg, E. E., and R. Lande. 2006. Ecological and reproductive character displacement on an environmental gradient. *Evolution* 60:1344–1357.

Gonzalez-Voyer, A., and N. Kolm. 2011. Rates of phenotypic evolution of ecological characters and sexual traits during the Tanganyikan cichlid adaptive radiation. *Journal of Evolutionary Biology* 24:2378–2388.

Gorissen, L., M. Gorissen, and M. Eens. 2006. Heterospecific song matching in two closely related songbirds (*Parus major* and *P. caeruleus*): Great tits match blue tits but not vice versa. *Behavioral Ecology and Sociobiology* 60:260–269.

Gould, S. J. 1989. *Wonderful life: The Burgess Shale and the nature of history.* Norton: New York.

———. 1991. The disparity of the Burgess Shale arthropod fauna and the limits of cladistic analysis: Why we must strive to quantify morphospace. *Paleobiology* 17:411–423.

———. 1997. Cope's rule as psychological artefact. *Nature* 385:199–200.

———. 2002. *The structure of evolutionary theory.* Harvard University Press: Cambridge, MA.

Gould, S. J., and C. B. Calloway. 1980. Clams and brachiopods: Ships that pass in the night? *Paleobiology* 6:383–396.

Grant, B. R., and P. R. Grant. 2010. Songs of Darwin's finches diverge when a new species enters the community. *Proceedings of the National Academy of Sciences of the United States of America* 107:20156–20163.

Grant, P. R. 1972. Convergent and divergent character displacement. *Biological Journal of the Linnean Society* 4:39–68.

———. 1986. *Ecology and evolution of Darwin's finches*. Princeton University Press: Princeton, NJ.

Grant, P. R., and B. R. Grant. 2006. Evolution of character displacement in Darwin's finches. *Science* 313:224–226.

———. 2008. *How and why species multiply: The radiation of Darwin's finches*. Princeton University Press: Princeton, NJ.

Grant, V. 1971. *Plant speciation*. Columbia University Press: New York.

Gray, S. M., L. M. Dill, F. Y. Tantu, E. R. Loew, F. Herder, and J. S. McKinnon. 2008. Environment-contingent sexual selection in a colour polymorphic fish. *Proceedings of the Royal Society B: Biological Sciences* 275:1785–1791.

Gray, S. M., and J. S. McKinnon. 2007. Linking color polymorphism maintenance and speciation. *Trends in Ecology and Evolution* 22:71–79.

Gray, S. M., and B. W. Robinson. 2002. Experimental evidence that competition between stickleback species favours adaptive character divergence. *Ecology Letters* 5:264–272.

Gray, S. M., B. W. Robinson, and K. J. Parsons. 2005. Testing alternative explanations of character shifts against ecological character displacement in brook sticklebacks (*Culaea inconstans*) that coexist with ninespine sticklebacks (*Pungitius pungitius*). *Oecologia* 146:25–35.

Greene, E. 1989. A diet-induced developmental polymorphism in a caterpillar. *Science* 243:643–646.

Greenfield, M. D., and T. E. Shelly. 1985. Alternative mating strategies in a desert grasshopper: Evidence of density dependence. *Animal Behaviour* 33:1192–1210.

Grether, G. F., J. Hudon, and D. F. Millie. 1999. Carotenoid limitation of sexual coloration along an environmental gradient in guppies. *Proceedings of the Royal Society B: Biological Sciences* 266:1317–1322.

Grether, G. F., N. Losin, C. N. Anderson, and K. Okamoto. 2009. The role of interspecific interference competition in character displacement and the evolution of competitor recognition. *Biological Reviews* 84: 617–635.

Grimaldi, D., and M. S. Engel. 2005. *Evolution of the insects*. Cambridge University Press: Cambridge, UK.

Grinnell, J. 1917. The niche relationships of the California thrasher. *Auk* 21:364–382.

———. 1924. Geography and evolution. *Ecology* 5:225–229.

———. 1928 (1943). *Joseph Grinnell's philosophy of nature*. University of California Press: Berkeley, CA.

Gröning, J., and A. Hochkirch. 2008. Reproductive interference between animal species. *Quarterly Review of Biology* 83: 257–282.

Gröning, J., N. Lucke, A. Finger, and A. Hochkirch. 2007. Reproductive interference in two ground-hopper species: Testing hypotheses of coexistence in the field. *Oikos* 116:1449–1460.

Grosberg, R. K., and R. R. Strathmann. 2007. The evolution of multicellularity: A minor major transition? *Annual Review of Ecology Evolution and Systematics* 38:621–654.

Gross, K. 2008. Positive interactions among competitors can produce species-rich communities. *Ecology Letters* 11:929–936.

Gross, M. R. 1982. Sneakers, satellites and parentals: Polymorphic mating strategies in North American sunfishes. *Zeitschrift für Tierpsychologie* 60:1–26.

———. 1985. Disruptive selection for alternative life histories in salmon. *Nature* 313:47–48.

———. 1996. Alternative reproductive strategies and tactics: Diversity within sexes. *Trends in Ecology and Evolution* 11:92–98.

Grover, J. P. 1997. *Resource competition*. Chapman and Hall: London.

Gupta, A. P., and R. C. Lewontin. 1982. A study of reaction norms in natural populations of *Drosophila pseudoobscura*. *Evolution* 36:934–948.

Gurevitch, J., L. L. Morrow, A. Wallace, and J. S. Walsh. 1992. A meta-analysis of competition in field experiments. *American Naturalist* 140:539–572.

Haavie, J., T. Borge, S. Bureš, L. Z. Garamszegi, H. M. Lampe, J. Moreno, A. Qvarnström et al. 2004. Flycatcher song in allopatry and sympatry: Convergence, divergence and reinforcement. *Journal of Evolutionary Biology* 17:227–237.

Hairston, N. G. 1987. *Community ecology and salamander guilds*. Cambridge University Press: Cambridge, UK.

Haldane, J. B. S. 1932 (1993). *The causes of evolution*. Cornell University Press: Ithaca, NY.

Hall, B. K. 1999. *Evolutionary developmental biology*. Kluwer Academic Publishers: Dordrecht, The Netherlands.

Hamilton, A. J., Y. Basset, K. K. Benke, P. S. Grimbacher, S. E. Miller, V. Novotny, G. A. Samuelson et al. 2010. Quantifying uncertainty in estimation of tropical arthropod species richness. *American Naturalist* 176:90–95.

Hamilton, W. D. 1979. Wingless and fighting males in fig wasps and other insects. Pages 167–220 in M. S. Blum, and N. A. Blum, eds. *Sexual selection and reproductive competition in insects*. Academic Press: London.

Hamilton, W. D., and M. Zuk. 1982. Heritable true fitness and bright birds: A role for parasites. *Science* 218:384–387.

Hankison, S. J., and M. R. Morris. 2003. Avoiding a compromise between sexual selection and species recognition: Female swordtail fish assess multiple species-specific cues. *Behavioral Ecology* 14:282–287.

Hansen, T. F., W. S. Armbruster, and L. Antonsen. 2000. Comparative analysis of character displacement and spatial adaptations as illustrated by the evolution of *Dalechampia* blossoms. *American Naturalist* 156:S17–S34.

Hardin, G. 1960. The competitive exclusion principle. *Science* 131:1292–1297.

Härdling, R., and H. Kokko. 2005. The evolution of prudent choice. *Evolutionary Ecology Research* 7:697–715.

Harmon, L. J., J. B. Losos, T. J. Davies, R. G. Gillespie, J. L. Gittleman, W. Bryan Jennings, K. H. Kozak et al. 2010. Early bursts of body size and shape evolution are rare in comparative data. *Evolution* 64:2385–2396.

Harmon, L. J., B. Matthews, S. D. Roches, J. M. Chase, J. B. Shurin, and D. Schluter. 2009. Evolutionary diversification in stickleback affects ecosystem functioning. *Nature* 458:1167–1170.

Harper, G. R., and D. W. Pfennig. 2008. Selection overrides gene flow to break down maladaptive mimicry. *Nature* 451:1103–1106.

Harper, J. L. 1977. *Population biology of plants*. Academic Press: London, UK.

Harris, R. N. 1987. Density-dependent paedomorphosis in the salamander *Notophthalmus viridescens dorsalis*. *Ecology* 68:705–712.

Harrison, R. G. 1993. Hybrid zones and the evolutionary process. Oxford University Press: Oxford, UK.

———. 1998. Linking evolutionary pattern and process. Pages 493–505 in D. J. Howard, and S. H. Berlocher, eds. *Endless forms: Species and speciation*. Oxford University Press: Oxford, UK.

———. 2010. Understanding the origin of species: Where have we been? Where are we going? Pages 319–346 in M. A. Bell, D. J. Futuyma, W. F. Eanes, and J. S. Levinton, eds. *Evolution since Darwin: The first 150 years*. Sinauer: Sunderland, MA.

Hart, M. W., and R. R. Strathmann. 1994. Functional consequences of phenotypic plasticity in echinoid larvae. *Biological Bulletin* 186:291–299.

Hatfield, T. 1997. Genetic divergence in adaptive characters between sympatric species of sticklebacks. *American Naturalist* 149:1009–1029.

Hatfield, T., and D. Schluter. 1996. A test for sexual selection on hybrids of two sympatric sticklebacks. *Evolution* 50:2429–2434.

———. 1999. Ecological speciation in sticklebacks: Environment-dependent hybrid fitness. *Evolution* 53:866–873.

Heard, S. B., and D. L. Hauser. 1995. Key evolutionary innovations and their ecological mechanisms. *Historical Biology* 10:151–173.

Hebets, E. A., and D. R. Papaj. 2005. Complex signal function: Developing a framework of testable hypotheses. *Behavioral Ecology and Sociobiology* 57:197–214.

Heinrich, B. 1976. Resource partitioning among some eusocial insects: Bumblebees. *Ecology* 57:874–889.

Hendry, A. P. 2009. Ecological speciation! Or lack thereof? *Canadian Journal of Fisheries and Aquatic Sciences* 66:1383–1398.

Hendry, A. P., P. R. Grant, B. R. Grant, H. A. Ford, M. J. Brewer, and J. Podos. 2006. Possible human impacts on adaptive radiation: Beak size bimodality in Darwin's finches. *Proceedings of the Royal Society B: Biological Sciences* 273:1887–1894.

Hendry, A. P., S. K. Huber, L. F. De Leon, A. Herrel, and J. Podos. 2009. Disruptive selection in a bimodal population of Darwin's finches. *Proceedings of the Royal Society B: Biological Sciences* 276:753–759.

Hendry, A. P., J. K. Wenburg, P. Bentzen, E. C. Volk, and T. P. Quinn. 2000. Rapid evolution of reproductive isolation in the wild: Evidence from introduced salmon. *Science* 290:516–518.

Henry, K. S., and J. R. Lucas. 2008. Coevolution of auditory sensitivity and temporal resolution with acoustic signal space in three songbirds. *Animal Behaviour* 76:1659–1671.

Hermoyian, C. S., L. R. Leighton, and P. Kaplan. 2002. Testing the role of competition in fossil communities using limiting similarity. *Geology* 30:15–18.

Herrel, A., K. Huyghe, B. Vanhooydonck, T. Backeljau, K. Breugelmans, I. Grbac, R. Van Damme et al. 2008. Rapid large-scale evolutionary divergence in morphology and performance associated with exploitation of a different dietary resource. *Proceedings of the National Academy of Sciences of the United States of America* 105:4792–4795.

Hess, H., W. I. Ausich, C. E. Brett, and M. J. Simms. 1999. *Fossil crinoids*. Oxford University Press: Oxford, UK.

Hickey, L. J. 2003. *The forest primeval: The geological history of wood and petrified forests*. Peabody Museum: New Haven, CT.

Higashi, M., G. Takimoto, and N. Yamamura. 1999. Sympatric speciation by sexual selection. *Nature* 402:523–526.

Higgie, M., and M. W. Blows. 2007. Are traits that experience reinforcement also under sexual selection? *American Naturalist* 170:409–420.

———. 2008. The evolution of reproductive character displacement conflicts with how sexual selection operates within a species. *Evolution* 62:1192–1203.

Higgie, M., S. Chenoweth, and M. W. Blows. 2000. Natural selection and the reinforcement of mate recognition. *Science* 290:519–521.

Höbel, G., and H. C. Gerhardt. 2003. Reproductive character displacement in the acoustic communication system of green tree frogs (*Hyla cinerea*). *Evolution* 57:894–904.

Hochkirch, A., J. Gröning, and A. Bucker. 2007. Sympatry with the devil: Reproductive interference could hamper species coexistence. *Journal of Animal Ecology* 76:633–642.

Hoekstra, H. E., and J. A. Coyne. 2007. The locus of evolution: Evo devo and the genetics of adaptation. *Evolution* 61:995–1016.

Hoffmann, A. A., and L. H. Rieseberg. 2008. Revisiting the impact of inversions in evolution: From population genetic markers to drivers of adaptive shifts and speciation? *Annual Review of Ecology Evolution and Systematics* 39:21–42.

Hoffman, E. A., and D. W. Pfennig. 1999. Proximate causes of cannibalistic polyphenism in larval tiger salamanders. *Ecology* 80:1076–1080.

Holmes, R. T., and F. A. Pitelka. 1968. Food overlap among coexisting Sandpipers on northern Alaskan tundra. *Systematic Zoology* 17:305–318.

Holt, R. D. 1977. Predation, apparent competition and the structure of prey communities. *Theoretical Population Biology* 12:197–229.

Holt, R. D., and G. A. Polis. 1997. A theoretical framework for intraguild predation. *American Naturalist* 149:745–764.

Hopkins, C. D. 1986. Behavior of Mormyridae. Pages 527–576 in T. H. Bullock, and W. Heiligenberg, eds. *Electroreception*. Wiley: New York.

Hopkins, R., D. A. Levin, and M. D. Rausher. 2012. Molecular signatures of selection on reproductive character displacement of flower color in *Phlox drummondii*. *Evolution* 66:469–485.

Hopkins, R., and M. D. Rausher. 2011. Identification of two genes causing reinforcement in the Texas wildflower *Phlox drummondii*. *Nature* 469:411–414.

Hori, M. 1993. Frequency-dependent natural selection in the handedness of scale-eating cichlid fish. *Science* 260:216–219.

Hoskin, C. J., M. Higgie, K. R. McDonald, and C. Moritz. 2005. Reinforcement drives rapid allopatric speciation. *Nature* 437:1353–1356.

Houde, A. E. 1988. Genetic difference in female choice between two guppy populations. *Animal Behaviour* 36:510–516.

———. 1997. *Sex, color, and mate choice in guppies.* Princeton University Press: Princeton, NJ.

Howard, D. J. 1993. Reinforcement: Origin, dynamics, and fate of an evolutionary hypothesis. Pages 46–69 in R. G. Harrison, ed. *Hybrid zones and the evolutionary process.* Oxford University Press: Oxford, UK.

———. 1999. Conspecific sperm and pollen precedence and speciation. *Annual Review of Ecology and Systematics* 30:109–132.

Hubbell, S. P. 2001. *The unified theory of biodiversity and biogeography.* Princeton University Press: Princeton, NJ.

Huber, S. K., L. F. De Leon, A. P. Hendry, E. Bermingham, and J. Podos. 2007. Reproductive isolation of sympatric morphs in a population of Darwin's finches. *Proceedings of the Royal Society B: Biological Sciences* 274:1709–1714.

Huber, S. K., and J. Podos. 2006. Beak morphology and song features covary in a population of Darwin's finches (*Geospiza fortis*). *Biological Journal of the Linnean Society* 88:489–498.

Hugall, A. F., and D. Stuart-Fox. 2012. Accelerated speciation in colour-polymorphic birds. *Nature* 485:631–634.

Hulsey, C. D., D. A. Hendrickson, and F. J. G. de Leon. 2005. Trophic morphology, feeding performance and prey use in the polymorphic fish *Herichthys minckleyi*. *Evolutionary Ecology Research* 7:303–324.

Hulsey, C. D., J. Marks, D. A. Hendrickson, C. A. Williamson, A. E. Cohen, and M. J. Stephens. 2006. Feeding specialization in *Herichthys minckleyi*: A trophically polymorphic fish. *Journal of Fish Biology* 68:1399–1410.

Hunt, D. E., L. A. David, D. Gevers, S. P. Preheim, E. J. Alm, and M. F. Polz. 2008. Resource partitioning and sympatric differentiation among closely related bacterioplankton. *Science* 320:1081–1085.

Hunt, G. 2010. Evolution in fossil lineages: Paleontology and *The Origin of Species*. *American Naturalist* 176:S61–S76.

Hunter, J. P. 1998. Key innovations and the ecology of macroevolution. *Trends in Ecology and Evolution* 13:31–36.

Hutchinson, G. E. 1957. Concluding remarks. *Cold Springs Harbor Symposium in Quantitative Biology* 22:415–427.

———. 1959. Homage to Santa Rosalia or why are there so many kinds of animals? *American Naturalist* 93:145–159.

———. 1961. The paradox of the plankton. *American Naturalist* 95:137–145.

Ingrouille, M. J., and B. Eddie. 2006. *Plants: Diversity and evolution.* Cambridge University Press: Cambridge, UK.

Jablonski, D. 1986. Background and mass extinctions: The alternation of macroevolutionary regimes. *Science* 231:129–133.

———. 2008. Biotic interactions and macroevolution: Extensions and mismatches across scales and levels. *Evolution* 62:715–739.

Jaenike, J., K. A. Dyer, C. Cornish, and M. S. Minhas. 2006. Asymmetrical reinforcement and *Wolbachia* infection in *Drosophila*. *PLoS Biology* 4:1852–1862.

Jennions, M. D., and M. Petrie. 1997. Variation in mate choice and mating preferences:

A review of causes and consequences. *Biological Reviews of the Cambridge Philosophical Society* 72:283–327.

Jerison, H. J. 1973. *Evolution of the brain and intelligence.* Academic Press: New York.

Jiggins, C. D., R. E. Naisbit, R. L. Coe, and J. Mallet. 2001. Reproductive isolation caused by colour pattern mimicry. *Nature* 411:302–305.

Jones, A. G., E. M. Adams, and S. J. Arnold. 2002a. Topping off: A mechanism of first-male sperm precedence in a vertebrate. *Proceedings of the National Academy of Sciences of the United States of America* 99:2078–2081.

Jones, A. G., J. R. Arguello, and S. J. Arnold. 2002b. Validation of Bateman's principles: A genetic study of sexual selection and mating patterns in the rough-skinned newt. *Proceedings of the Royal Society B: Biological Sciences* 269:2533–2539.

Jones, A. G., G. Rosenqvist, A. Berglund, S. J. Arnold, and J. C. Avise. 2000. The Bateman gradient and the cause of sexual selection in a sex-role-reversed pipefish. *Proceedings of the Royal Society B: Biological Sciences* 267:677–680.

Jones, A. G., G. Rosenqvist, A. Berglund, and J. C. Avise. 2005. The measurement of sexual selection using Bateman's principles: An experimental test in the sex-role-reversed pipefish *Syngnathus typhle*. *Integrative and Comparative Biology* 45:874–884.

Jones, C. D. 1998. The genetic basis of *Drosophila sechellia*'s resistance to a host plant toxin. *Genetics* 149:1899–1908.

Jones, C. G., J. H. Lawton, and M. Shachak. 1994. Organisms as ecosystem engineers. *Oikos* 69:373–386.

Jones, J. M. 1973. Effects of thirty years hybridization on the toads *Bufo americanus* and *Bufo woodhousii fowleri* at Bloomington, Indiana. *Evolution* 27:435–448.

Jukema, J., and T. Piersma. 2006. Permanent female mimics in a lekking shorebird. *Biology Letters* 2:161–164.

Kamada, N., Y.-G. Kim, H. P. Sham, B. A. Vallance, J. L. Puente, E. C. Martens, and G. Núñez. 2012. Regulated virulence controls the ability of a pathogen to compete with the gut microbiota. *Science* 336:1325–1329.

Kaplan, I., and R. F. Denno. 2007. Interspecific interactions in phytophagous insects revisited: A quantitative assessment of competition theory. *Ecology Letters* 10: 977–994.

Karlsson, K., F. Eroukhmanoff, R. Hardling, and E. I. Svensson. 2010. Parallel divergence in mate guarding behaviour following colonization of a novel habitat. *Journal of Evolutionary Biology* 23:2540–2549.

Kassen, R. 2009. Toward a general theory of adaptive radiation: Insights from microbial experimental evolution. *Annals of the New York Academy of Sciences: The Year in Evolutionary Biology* 1168:3–22.

Kawano, K. 2002. Character displacement in giant rhinoceros beetles. *American Naturalist* 159:255–271.

———. 2003. Character displacement in stag beetles (Coleoptera: Lucanidae). *Annals of the Entomological Society of America* 96:503–511.

Keddy, P. A. 2001. *Competition* (2nd edn.). Kluwer Academic Publishers: Dordrecht, The Netherlands.

Kellogg, D. E. 1975. Character displacement in radiolarian genus, *Eucyrtidium*. *Evolution* 29:736–749.

Kelly, J. K., and M. A. F. Noor. 1996. Speciation by reinforcement: A model derived from studies of *Drosophila*. *Genetics* 143:1485–1497.

King, M. C., and A. C. Wilson. 1975. Evolution at two levels in humans and chimpanzees. *Science* 188:107–116.

Kingsolver, J. G., H. E. Hoekstra, J. M. Hoekstra, D. Berrigan, N. Vignieri, C. E. Hill, A. Hoang et al. 2001. The strength of phenotypic selection in natural populations. *American Naturalist* 157:245–261.

Kingsolver, J. G., and D. W. Pfennig. 2004. Individual-level selection as a cause of Cope's rule of phyletic size increase. *Evolution* 58:1608–1612.

———. 2007. Patterns and power of phenotypic selection in nature. *BioScience* 57:561–572.

Kingsolver, J. G., G. J. Ragland, and J. G. Schlichta. 2004. Quantitative genetics of continuous reaction norms: Thermal sensitivity of caterpillar growth rates. *Evolution* 58:1521–1529.

Kirkpatrick, M. 1982. Sexual selection and the evolution of female choice. *Evolution* 36:1–12.

Kirkpatrick, M., and N. H. Barton. 1997. Evolution of a species' range. *American Naturalist* 150:1–23.

Kirschel, A. N. G., H. Slabbekoorn, D. T. Blumstein, R. E. Cohen, S. R. de Kort, W. Buermann, and T. B. Smith. 2011. Testing alternative hypotheses for evolutionary diversification in an African songbird: Rainforest refugia versus ecological gradients. *Evolution* 65:3162–3174.

Kishi, S., T. Nishida, and Y. Tsubaki. 2009. Reproductive interference determines persistence and exclusion in species interactions. *Journal of Animal Ecology* 78:1043–1049.

Klopfer, P. H. 1962. *Behavioral aspects of ecology*. Princeton University Press: Princeton, NJ.

Kluge, J., and M. Kessler. 2011. Phylogenetic diversity, trait diversity and niches: Species assembly of ferns along a tropical elevational gradient. *Journal of Biogeography* 38:394–405.

Knoll, A. H., and S. B. Carroll. 1999. Early animal evolution: Emerging views from comparative biology and geology. *Science* 284:2129–2137.

Kokko, H., and R. Brooks. 2003. Sexy to die for? Sexual selection and the risk of extinction. *Annales Zoologici Fennici* 40:207–219.

Kolbe, J. J., and J. B. Losos. 2005. Hind-limb length plasticity in *Anolis carolinensis*. *Journal of Herpetology* 39:674–678.

Kondrashov, A. S., and F. A. Kondrashov. 1999. Interactions among quantitative traits in the course of sympatric speciation. *Nature* 400:351–354.

Konuma, J., and S. Chiba. 2007. Ecological character displacement caused by reproductive interference. *Journal of Theoretical Biology* 247:354–364.

Kooyman, R., M. Rossetto, W. Cornwell, and M. Westoby. 2011. Phylogenetic tests of community assembly across regional to continental scales in tropical and subtropical rain forests. *Global Ecology and Biogeography* 20:707–716.

Kopp, M., and R. Tollrian. 2003. Trophic size polyphenism in *Lembadion bullinum*: Costs and benefits of an inducible offense. *Ecology* 84:641–651.

Kozak, G. M., M. L. Head, and J. W. Boughman. 2011. Sexual imprinting on ecologically divergent traits leads to sexual isolation in sticklebacks. *Proceedings of the Royal Society B: Biological Sciences* 278:2604–2810.

Kozak, K. H., and J. J. Wiens. 2006. Does niche conservatism promote speciation? A case study in North American salamanders. *Evolution* 60:2604–2621.

Krebs, J. R., and N. B. Davies. 1993. *An introduction to behavioural ecology* (3rd ed.). Blackwell: Malden, MA.

Kronforst, M. R., L. G. Young, D. D. Kapan, C. McNeely, R. J. O'Neill, and L. E. Gilbert. 2006. Linkage of butterfly mate preference and wing color preference cue at the genomic location of wingless. *Proceedings of the National Academy of Sciences of the United States of America* 103:6575–6580.

Krupa, J. J. 1989. Alternative mating tactics in the Great Plains Toad. *Animal Behaviour* 37:1035–1043.

Kuno, E. 1992. Competitive exclusion through reproductive interference. *Researches on Population Ecology* 34:275–284.

Kurdziel, J. P., and L. L. Knowles. 2002. The mechanisms of morph determination in the amphipod *Jassa*: Implications for the evolution of alternative male phenotypes. *Proceedings of the Royal Society B: Biological Sciences* 269:1749–1754.

LaBarbera, M. 1986. The evolution and ecology of body size. Pages 69–98 in D. M. Raup, and D. Jablonski, eds. *Patterns and processes in the history of life*. Springer-Verlag: Berlin.

Lack, D. 1945. The Galápagos finches (Geospizinae): A study in variation. *Occasional Papers of the California Academy of Sciences* 21:1–159.

———. 1947. *Darwin's finches*. Cambridge University Press: Cambridge, UK.

———. 1954. *The natural regulation of animal numbers*. Carendon Press: Oxford, UK.

Lahti, D. C., N. A. Johnson, B. C. Ajie, S. P. Otto, A. P. Hendry, D. T. Blumstein, R. G. Coss et al. 2009. Relaxed selection in the wild. *Trends in Ecology and Evolution* 24:487–496.

Lande, R. 1981. Models of speciation by sexual selection on polygenic traits. *Proceedings of the National Academy of Sciences of the United States of America* 78:3721–3725.

———. 1982. Rapid origin of sexual isolation and character divergence in a cline. *Evolution* 36:213–223.

———. 2009. Adaptation to an extraordinary environment by evolution of phenotypic plasticity and genetic assimilation. *Journal of Evolutionary Biology* 22:1435–1446.

Langerhans, R. B., M. E. Gifford, and E. O. Joseph. 2007. Ecological speciation in *Gambusia* fishes. *Evolution* 61:2056–2074.

Langerhans, R. B., C. A. Layman, M. Shokrollahi, and T. J. DeWitt. 2004. Predator-driven phenotypic diversification in *Gambusia affinis*. *Evolution* 58:2305–2318.

Lank, D. B., C. M. Smith, O. Hanotte, T. Burke, and F. Cooke. 1995. Genetic polymorphism for alternative mating behaviour in lekking male ruff *Philomachus pugnax*. *Nature* 378:59–62.

Leal, M., and L. J. Fleishman. 2002. Evidence for habitat partitioning based on adaptation to environmental light in a pair of sympatric lizard species. *Proceedings of the Royal Society B: Biological Sciences* 269:351–359.

Leal, M., and J. B. Losos. 2010. Communication and speciation. *Nature* 467:159–160.

Ledón-Rettig, C. C., and D. W. Pfennig. 2011. Emerging model systems in eco-evo-devo: The environmentally responsive spadefoot toad. *Evolution and Development* 13:391–400.

Ledón-Rettig, C. C., D. W. Pfennig, and E. J. Crespi. 2010. Diet and hormone manipulations reveal cryptic genetic variation: Implications for the evolution of novel feeding strategies. *Proceedings of the Royal Society B: Biological Sciences* 277:3569–3578.

Ledón-Rettig, C. C., D. W. Pfennig, and N. Nascone-Yoder. 2008. Ancestral variation and the potential for genetic accommodation in larval amphibians: Implications for the evolution of novel feeding strategies. *Evolution and Development* 10:316–325.

Lee, H. J., S. Pittlik, J. C. Jones, W. Salzburger, M. Barluenga, and A. Meyer. 2010. Genetic support for random mating between left and right-mouth morphs in the dimorphic scale-eating cichlid fish *Perissodus microlepis* from Lake Tanganyika. *Journal of Fish Biology* 76:1940–1957.

Lehmann, G. U. C., K. G. Heller, and A. W. Lehmann. 2001. Male bushcrickets favoured by parasitoid flies when acoustically more attractive for conspecific females (Orthoptera: Phanopteridae / Diptera: Tachinidae). *Entomologia Generalis* 25:135–140.

Leichty, A. R., D. W. Pfennig, C. D. Jones, and K. S. Pfennig. 2012. Relaxed genetic constraint is ancestral to the evolution of phenotypic plasticity. *Integrative and Comparative Biology* 52:16–30.

Lemmon, E. M. 2009. Diversification of conspecific signals in sympatry: Geographic overlap drives multidimensional reproductive character displacement in frogs. *Evolution* 63:1155–1170.

Levene, H. 1953. Genetic equilibrium when more than one ecological niche is available. *American Naturalist* 87:331–333.

Levin, D. A. 1985. Reproductive character displacement in *Phlox*. *Evolution* 39:1275–1281.

Levin, S. A. (ed.) 2009. *The Princeton guide to ecology*. Princeton University Press: Princeton, NJ.

Levine, J. M., and J. HilleRisLambers. 2009. The importance of niches for the maintenance of species diversity. *Nature* 461: 254–257.

Levins, R. 1968. *Evolution in changing environments*. Princeton University Press: Princeton, NJ.

Levinton, J. S. 2001. *Genetics, paleontology, and macroevolution*. Cambridge University Press: Cambridge, UK.

Lewin, R. 1983. Santa Rosalia was a goat. *Science* 221:636–639.

Liem, K. F. 1974. Evolutionary strategies and morphological innovations: Cichlid pharyngeal jaws. *Systematic Zoology* 22:425–441.

Liem, K. F., and L. S. Kaufman. 1984. Intraspecific macroevolution: Functional biology of the polymorphic cichlid species *Cichlasoma minckleyi*. Pages 203–215 in A. A. Echelle, and I. Kornfield, eds. *Evolution of fish species flocks*. University of Maine Press: Orono, ME.

Linder, H. P. 2009. Plant species radiations: Where, when, why? *Philosophical Transactions of the Royal Society B: Biological Sciences* 363:3092–3105.

Liou, L. W., and T. D. Price. 1994. Speciation by reinforcement of premating isolation. *Evolution* 48:1451–1459.

Littlejohn, M. J. 1959. Call differentiation in a complex of seven species of *Crinia* (Anura, Leptodactylidae). *Evolution* 13:452–468.

Loeb, M. L. G., J. P. Collins, and T. J. Maret. 1994. The role of prey in controlling expression of a trophic polymorphism in *Ambystoma tigrinum nebulosum*. *Functional Ecology* 8:151–158.

Lorch, P. D., S. Proulx, L. Rowe, and T. Day. 2003. Condition-dependent sexual selection can accelerate adaptation. *Evolutionary Ecology Research* 5:867–881.

Losos, J. B. 1990. A phylogenetic analysis of character displacement. *Evolution* 44:558–569.

———. 1992. The evolution of convergent structure in Caribbean *Anolis* communities. *Systematic Biology* 41:403–420.

———. 2000. Ecological character displacement and the study of adaptation. *Proceedings of the National Academy of Sciences of the United States of America* 97:5693–5695.

———. 2008. Phylogenetic niche conservatism, phylogenetic signal and the relationship between phylogenetic relatedness and ecological similarity among species. *Ecology Letters* 11:995–1007.

———. 2009. *Lizards in an evolutionary tree: Ecology and adaptive radiation of Anoles*. University of California Press: Berkeley, CA.

Losos, J. B., D. A. Creer, D. Glossip, R. Goellner, A. Hampton, G. Roberts, N. Haskell et al. 2000. Evolutionary implications of phenotypic plasticity in the hindlimb of the lizard *Anolis sagrei*. *Evolution* 54:301–305.

Losos, J. B., T. R. Jackman, A. Larson, K. de Queiroz, and L. Rodriguez-Schettino. 1998. Contingency and determinism in replicated adaptive radiations of island lizards. *Science* 279:2115–2118.

Losos, J. B., and D. L. Mahler. 2010. Adaptive radiation: The interaction of ecological opportunity, adaptation, and speciation. Pages 381–420 in M. A. Bell, D. J. Futuyma, W. F. Eanes, and J. S. Levinton, eds. *Evolution since Darwin: The first 150 years*. Sinauer: Sunderland, MA.

Losos, J. B., and R. E. Ricklefs. 2009. Adaptation and diversification on islands. *Nature* 457:830–836.

Lotka, A. J. 1932. The growth of mixed populations: Two species competing for a common food suppy. *Journal of the Washington Academy of Sciences* 22:461–469.

Lowry, D. B., R. C. Rockwood, and J. H. Willis. 2008. Ecological reproductive isolation of coast and inland races of *Mimulus guttatus*. *Evolution* 62:2196–2214.

Lubchenco, J. 1978. Plant species diversity in a marine intertidal community: Importance of herbivore food preference and algal competitive abilities. *American Naturalist* 112:23–39.

Lupia, R., S. Lidgard, and P. R. Crane. 1999. Comparing palynological abundance and diversity: Implications for biotic replacement during the Cretaceous angiosperm radiation. *Paleobiology* 25:305–340.

Luther, D. 2009. The influence of the acoustic community on songs of birds in a neotropical rain forest. *Behavioral Ecology* 20:864–871.

Lyson, T. R., and N. R. Longrich. 2011. Spatial niche partitioning in dinosaurs from the latest Cretaceous (Maastrichtian) of North America. *Proceedings of the Royal Society B: Biological Sciences* 278:1158–1164.

Maan, M. E., and O. Seehausen. 2011. Ecology, sexual selection and speciation. *Ecology Letters* 14:591–602.

MacArthur, R. H. 1958. Population ecology of some warblers of northeastern coniferous forests. *Ecology* 39:599–619.

———. 1969. Species packing, and what interspecies competition minimizes. *Proceedings of the National Academy of Sciences of the United States of America* 64:1369–1371.

———. 1972. *Geographical ecology: Patterns in the distribution of species*. Harper and Row: New York.

MacArthur, R. H., and R. Levins. 1964. Competition, habitat selection, and character displacement in a patchy environment. *Proceedings of the National Academy of Sciences of the United States of America* 51:1207–1210.

———. 1967. The limiting similarity, convergence and divergence of coexisting species. *American Naturalist* 101:377–385.

MacArthur, R. H., and J. W. MacArthur. 1961. On bird species diversity. *Ecology* 42:594–598.

MacArthur, R. H., and E. O. Wilson. 1967. *The theory of island biogeography.* Princeton University Press: Princeton, NJ.

MacColl, A. D. C. 2011. The ecological causes of evolution. *Trends in Ecology and Evolution* 26:514–522.

MacFadden, B. J. 1992. *Fossil horses: Systematics, paleobiology, and evolution of the family Equidae.* Cambridge University Press: Cambridge, UK.

Macnair, M. R., and M. Gardner. 1998. The evolution of edaphic endemics. Pages 157–171 in D. J. Howard, and S. H. Berlocher, eds. *Endless forms: Species and speciation.* Oxford University Press: Oxford, UK.

Mahall, B. E., and R. M. Callaway. 1991. Root communication among desert shrubs. *Proceedings of the National Academy of Sciences of the United States of America* 88:874–876.

Maherali, H., and J. Klironomos. 2007. Influence of phylogeny on fungal community assembly and ecosystem functioning. *Science* 316:1746–1748.

Mallet, J. 2008. Hybridization, ecological races and the nature of species: Empirical evidence for the ease of speciation. *Philosophical Transactions of the Royal Society B: Biological Sciences* 363:2971–2986.

Mallet, J., and L. E. Gilbert. 1995. Why are there so many mimicry rings? Correlations between habitat, behavior and mimicry in *Heliconius* butterflies. *Biological Journal of the Linnean Society* 55:159–180.

Malthus, T. R. 1797 (1990). *An essay on the principle of population.* Cambridge University Press: Cambridge, UK.

Mani, G. S., and B. C. Clarke. 1990. Mutational order: A major stochastic process in evolution. *Proceedings of the Royal Society B: Biological Sciences* 240:29–37.

Maret, T. J., and J. P. Collins. 1997. Ecological origin of morphological diversity: A study of alternative trophic phenotypes in larval salamanders. *Evolution* 51:898–905.

The Marie Curie SPECIATION Network. 2012. What do we need to know about speciation? *Trends in Ecology and Evolution* 27:27–39.

Marko, P. B. 2005. An intraspecific comparative analysis of character divergence between sympatric species. *Evolution* 59:554–564.

Marshall, C. R. 2006. Explaining the Cambrian "explosion" of animals. *Annual Review of Earth and Planetary Sciences* 34:355–384.

Martin, R. A., and D. W. Pfennig. 2009. Disruptive selection in natural populations: The roles of ecological specialization and resource competition. *American Naturalist* 174:268–281.

———. 2010a. Field and experimental evidence that competition and ecological opportunity promote resource polymorphism. *Biological Journal of the Linnean Society* 100:73–88.

———. 2010b. Maternal investment influences expression of resource polymorphism in amphibians: Implications for the evolution of novel resource-use phenotypes. *PLoS One* 5:e9117.

———. 2011. Evaluating the targets of selection during character displacement. *Evolution* 65:2946–2958.

———. In press. Widespread disruptive selection in the wild is associated with intense resource competition. *BMC Evolutionary Biology*.

Martin, T. E. 1996. Fitness costs of resource overlap among coexisting bird species. *Nature* 380:338–340.

Masel, J., O. D. King, and H. Maughan. 2007. The loss of adaptive plasticity during long periods of environmental stasis. *American Naturalist* 169:38–46.

Mather, K. 1953. The genetical structure of populations. *Symposium of the Society for Experimental Biology* 2:196–216.

———. 1955. Polymorphism as an outcome of disruptive selection. *Evolution* 9:52–61.

Matocq, M. D., and P. J. Murphy. 2007. Fine-scale phenotypic change across a species transition zone in the genus *Neotoma*: Disentangling independent evolution from phylogenetic history. *Evolution* 61:2544–2557.

Matsuda, R. 1987. *Animal evolution in changing environments with special reference to abnormal metamorphosis*. John Wiley and Sons: New York.

Matute, D. R. 2010. Reinforcement of gametic isolation in *Drosophila*. *PLoS Biology* 8: e1000341.

Matute, D. R., and J. A. Coyne. 2010. Intrinsic reproductive isolation between two sister species of *Drosophila*. *Evolution* 64:903–920.

May, R. M. 1973. *Stability and complexity in model ecosystems*. Princeton University Press: Princeton, NJ.

———. 1974. On the theory of niche overlap. *Theoretical Population Biology* 5:297–332.

May, R. M., and R. H. MacArthur. 1972. Niche overlap as a function of environmental variability. *Proceedings of the National Academy of Sciences of the United States of America* 69:1109–1113.

Maynard Smith, J. 1962. Disruptive selection, polymorphism and sympatric speciation. *Nature* 195:60–62.

———. 1966. Sympatric speciation. *American Naturalist* 104:487–490.

———. 1982. *Evolution and the theory of games*. Cambridge University Press: Cambridge, UK.

Maynard Smith, J., and E. Szathmáry. 1995. *The major transitions in evolution*. W. H. Freeman: San Francisco, CA.

Mayr, E. 1942. *Systematics and the origin of species*. Columbia University Press: New York.

———. 1947. Ecological factors in speciation. *Evolution* 1:263–288.

———. 1963. *Animal species and evolution*. Harvard University Press: Cambridge, MA.

———. 1970. *Populations, species, and evolution*. Harvard University Press: Cambridge, MA.

———. 2001. *What evolution is*. Basic Books: New York.

Mayrose, I., S. H. Zhan, C. J. Rothfels, K. Magnuson-Ford, M. S. Barker, L. H. Rieseberg, and S. P. Otto. 2011. Recently formed polyploid plants diversify at lower rates. *Science* 333:1257.

McGill, B. J., B. A. Mauer, and M. D. Weiser. 2006. Empirical evaluation of neutral theory. *Ecology* 87:1411–1423.

McGraw, E. A., X. H. Ye, B. Foley, S. F. Chenoweth, M. Higgie, E. Hine, and M. W. Blows. 2011. High-dimensional variance partitioning reveals the modular genetic basis of adaptive divergence in gene expression during reproductive character displacement. *Evolution* 65:3126–3137.

McKinney, F. K. 1995. One hundred million years of competitive interactions between bryozoan clades: Asymmetrical but not escalating. *Biological Journal of the Linnean Society* 56:465–481.

McKinnon, J. S., S. Mori, B. K. Blackman, L. David, D. M. Kingsley, L. Jamieson, J. Chou et al. 2004. Evidence of ecology's role in speciation. *Nature* 429:294–298.

McNett, G. D., and R. B. Cocroft. 2008. Host shifts favor vibrational signal divergence in *Enchenopa binotata* treehoppers. *Behavioral Ecology* 19:650–656.

McPeek, M. A. 2008. The ecological dynamics of clade diversification and community assembly. *American Naturalist* 172:E270–E284.

McPeek, M. A., and J. M. Brown. 2007. Clade age and not diversification rate explains species richness among animal taxa. *American Naturalist* 169:E97–E106.

McPhail, J. D. 1984. Ecology and evolution of sympatric sticklebacks (*Gasterosteus*): Morphological and genetic evidence for a species pair in Enos Lake, British Columbia. *Canadian Journal of Zoology* 62:1402–1408.

———. 1992. Ecology and evolution of sympatric sticklebacks (*Gasterosteus*): Evidence for a species pair in Paxton Lake, Texada Island, British Columbia. *Canadian Journal of Zoology* 70:361–369.

McShea, D. W. 1991. Complexity and evolution: What everybody already knows. *Biology and Philosophy* 6:303–324.

———. 1994. Mechanisms of large-scale evolutionary change. *Evolution* 48:1747–1763.

———. 1996. Metazoan complexity and evolution: Is there a trend? *Evolution* 50:477–492.

McShea, D. W., and R. N. Brandon. 2010. *Biology's first law*. University of Chicago Press: Chicago, IL.

McShea, D. W., and M. A. Changizi. 2003. Three puzzles in hierarchical evolution. *Integrative and Comparative Biology* 43:74–81.

Medel, R., C. Botto-Mahan, and M. Kalin-Arroyo. 2003. Pollinator-mediated selection on the nectar guide phenotype in the Andean monkey flower, *Mimulus luteus*. *Ecology* 84:1721–1732.

Meyer, A. 1987. Phenotypic plasticity and heterochrony in *Ciclasoma managuense* (Pisces, Cichlidae) and their implications for speciation in cichlid fishes. *Evolution* 41:1357–1369.

Michimae, H., and M. Wakahara. 2002. A tadpole-induced polyphenism in the salamander *Hynobius retardatus*. *Evolution* 56:2029–2038.

Milligan, B. G. 1985. Evolutionary divergence and character displacement in two phenotypically-variable competing species. *Evolution* 39:1207–1222.

Mitter, C., B. Farrell, and B. Wiegmann. 1988. The phylogenetic study of adaptive zones: Has phytophagy promoted insect diversification? *American Naturalist* 132:107–128.

Moczek, A. P. 2005. The evolution and development of novel traits, or how beetles got their horns. *BioScience* 11:935–951.

Moczek, A. P., and D. J. Emlen. 2000. Male horn dimorphism in the scarab beetle *Onthophagus taurus*: Do alternative reproductive tactics favor alternative phenotypes? *Animal Behaviour* 59:459–466.

Moczek, A. P., J. Hunt, D. J. Emlen, and L. W. Simmons. 2002. Evolution of a developmental threshold in exotic populations of a polyphenic beetle. *Evolutionary Ecology Research* 4:587–601.

Moczek, A. P., S. E. Sultan, S. Foster, C. Ledon-Rettig, I. Dworkin, H. F. Nijhout, E. Abouheif et al. 2011. The role of developmental plasticity in evolutionary innovation. *Proceedings of the Royal Society B: Biological Sciences* 278:2705–2713.

Moore, J. A. 1952. Competition between *Drosophila melanogaster* and *Drosophila simulans* II. The improvement of competitive ability through selection. *Proceedings of the National Academy of Sciences of the United States of America* 38:813–817.

Mora, M., D. P. Tittensor, S. Adl, A. G. B. Simpson, and B. Worm. 2011. How many species are there on earth and in the ocean? *PLoS Biology* 9: e1001127.

Morgan, C. L. 1896. On modification and variation. *Science* 4:733–740.

Morton, E. S. 1975. Ecological sources of selection on avian sounds. *American Naturalist* 109:17–34.

Mousseau, T. A., and C. W. Fox, eds. 1998. *Maternal effects as adaptations.* Oxford University Press: Oxford, UK.

Muchhala, N., Z. Brown, W. S. Armbruster, and M. D. Potts. 2010. Competition drives specialization in pollination systems through costs to male fitness. *American Naturalist* 176:732–743.

Muchhala, N., and M. D. Potts. 2007. Character displacement among bat-pollinated flowers of the genus *Burmeistera*: Analysis of mechanism, process and pattern. *Proceedings of the Royal Society B: Biological Sciences* 274:2731–2737.

Muller, H. J. 1939. Reversibility in evolution considered from the standpoint of genetics. *Biological Reviews of the Cambridge Philosophical Society* 14:261–280.

———. 1940. Bearing of the *Drosophila* work on systematics. Pages 185–268 in J. S. Huxley, ed. *The new systematics.* Carendon Press: Oxford, UK.

———. 1942. Isolating mechanisms, evolution, and temperature. *Biological Symposia* 6:71–125.

Nagamitsu, T., and T. Inoue. 1997. Aggressive foraging of social bees as a mechanism of floral resource partitioning in an Asian tropical rainforest. *Oecologia* 110:432–439.

Nee, S., A. O. Mooers, and P. H. Harvey. 1992. Tempo and mode of evolution revealed from molecular phylogenies. *Proceedings of the National Academy of Sciences of the United States of America* 89:8322–8326.

Nevo, E., G. Gorman, M. Soule, S. Y. Yang, R. Clover, and V. Jovanovic. 1972. Competitive exclusion between insular *Lacerta* species (Sauria, Lacertidae). *Oecologia* 10:183–190.

Nichols, D. 1959. Changes in the chalk heart-urchin *Micraster* interpretted in relation to living forms. *Philosophical Transactions of the Royal Society B: Biological Sciences* 242:347–437.

Nielsen, R. 2005. Molecular signatures of natural selection. *Annual Review of Genetics* 39:197–218.

Niklas, K. J. 1997. *The evolutionary biology of plants.* University of Chicago Press: Chicago, IL.

Niklas, K. J., B. H. Tiffney, and A. H. Knoll. 1983. Patterns of vascular land plant diversification. *Nature* 303:614–616.

Nobel, P. S. 1997. Root distribution and seasonal production in the northwestern Sonoran desert for a C3 subshrub, a C4 bunchgrass, and a CAM leaf succulent. *American Journal of Botany* 84:949–955.

Noor, M. A. 1995. Speciation driven by natural selection in *Drosophila*. *Nature* 375:674–675.

Noor, M. A. F., and S. M. Bennett. 2009. Islands of speciation or mirages in the desert? Examining the role of restricted recombination in maintaining species. *Heredity* 103:439–444.

Noor, M. A. F., K. L. Grams, L. A. Bertucci, and J. Reiland. 2001. Chromosomal inversions and the reproductive isolation of species. *Proceedings of the National Academy of Sciences of the United States of America* 98:12084–12088.

Nosil, P. 2007. Divergent host-plant adaptation and reproductive isolation between ecotypes of *Timema cristinae*. *American Naturalist* 169:151–162.

———. 2008. Speciation with gene flow could be common. *Molecular Ecology* 17:2103–2106.

———. 2012. *Ecological speciation*. Oxford University Press: Oxford, UK.

Nosil, P., and B. J. Crespi. 2006. Experimental evidence that predation promotes divergence in adaptive radiation. *Proceedings of the National Academy of Sciences of the United States of America* 103:9090–9095.

Nosil, P., B. J. Crespi, and C. P. Sandoval. 2002. Host-plant adaptation drives the parallel evolution of reproductive isolation. *Nature* 417:440–443.

Nosil, P., L. Harmon, and O. Seehausen. 2009. Ecological explanations for (incomplete) speciation. *Trends in Ecology and Evolution* 24:145–156.

Nosil, P., and T. E. Reimchen. 2005. Ecological opportunity and levels of morphological variance within freshwater stickleback populations. *Biological Journal of the Linnean Society* 86:297–308.

Nosil, P., and H. D. Rundle. 2009. Ecological speciation: Natural selection and the formation of new species. Pages 134–142 in S. A. Levin, ed. *The Princeton guide to ecology*. Princeton University Press: Princeton, NJ.

Nosil, P., and D. Schluter. 2011. The genes underlying the process of speciation. *Trends in Ecology and Evolution* 26:160–167.

Nottebohm, F. 1975. Continental patterns of song variability in *Zonotrichia capensis*: Some possible ecological correlates. *American Naturalist* 109:605–624.

Odling-Smee, F. J., K. N. Laland, and M. W. Feldman. 2003. *Niche construction: The neglected process in evolution*. Princeton University Press: Princeton, NJ.

Odum, E. P. 1959. *Fundamentals of ecology*. Saunders: Philadelphia, PA.

Okuzaki, Y., Y. Takami, and T. Sota. 2010. Resource partitioning or reproductive isolation: The ecological role of body size differences among closely related species in sympatry. *Journal of Animal Ecology* 79:383–392.

Olson, V. A., and I. P. F. Owens. 1998. Costly sexual signals: Are carotenoids rare, risky or required? *Trends in Ecology and Evolution* 13:510–514.

Olsson, M., E. Wapstra, T. Schwartz, T. Madsen, B. Ujvari, and T. Uller. 2011. In hot pursuit: Fluctuating mating system and sexual selection in sand lizards. *Evolution* 65:574–583.

Orr, H. A. 1995. The population genetics of speciation: The evolution of hybrid incompatibilities. *Genetics* 139:1805–1813.

Orr, H. A., and M. Turelli. 2001. The evolution of postzygotic isolation: Accumulating Dobzhansky-Muller incompatibilities. *Evolution* 55:1085–1094.

Orr, M. R., and T. B. Smith. 1998. Ecology and speciation. *Trends in Ecology and Evolution* 13:502–506.

Ortíz-Barrientos, D., B. A. Counterman, and M. A. F. Noor. 2004. The genetics of speciation by reinforcement. *PLoS Biology* 2:2256–2263.

Ortíz-Barrientos, D., and M. A. F. Noor. 2005. Evidence for a one-allele assortative mating locus. *Science* 310:1467–1467.

Otte, D. 1989. Speciation in Hawaiian crickets. Pages 482–526 in D. Otte, and J. A. Endler, eds. *Speciation and its consequences*. Sinauer: Sunderland, MA.

Pacala, S. W., and J. Roughgarden. 1982. Resource partitioning and interspecific competition in two-species insular *Anolis* lizard communities. *Science* 217:444–446.

———. 1985. Population experiments with the *Anolis* lizards of St. Maarten and St. Eustatius. *Ecology* 66:129–141.

Padilla, D. K. 2001. Food and environmental cues trigger an inducible offence. *Evolutionary Ecology Research* 3:15–25.

Paine, R. T. 1966. Food web complexity and species diversity. *American Naturalist* 100:65–75.

———. 1979. Disaster, catastrophe, and local persistence of the sea palm, *Postelsia palmaeformis*. *Science* 205:685–687.

Palumbi, S. R. 1998. Species formation and the evolution of gamete recognition loci. Pages 271–278 in D. J. Howard, and S. H. Berlocher, eds. *Endless forms: Species and speciation*. Oxford University Press: Oxford, UK.

———. 2009. Speciation and the evolution of gamete recognition genes: Pattern and process. *Heredity* 102:66–76.

Papaj, D. R., and R. J. Prokopy. 1989. Ecological and evolutionary aspects of learning in phytophagous insects. *Annual Review of Entomology* 34:315–350.

Parent, C. E., and B. J. Crespi. 2009. Ecological opportunity in adaptive radiation of Galápagos endemic land snails. *American Naturalist* 174:898–905.

Parker, A. 2003. *In the blink of an eye*. Perseus Books: Cambridge, MA.

Parker, G. A. 1970. Sperm competition and its evolutionary consequences in the insects. *Biological Reviews* 45:525–567.

Parker, G. A., and L. Partridge. 1998. Sexual conflict and speciation. *Philosophical Transactions of the Royal Society B: Biological Sciences* 353:261–274.

Passera, L., E. Roncin, B. Kaufmann, and L. Keller. 1996. Increased soldier production in ant colonies exposed to intraspecific competition. *Nature* 379:630–631.

Paterson, H. E. H. 1978. More evidence against speciation by reinforcement. *South African Journal of Science* 74:369–371.

———. 1982. Perspective on speciation by reinforcement. *South African Journal of Science* 78:53–57.

Paull, J. S., R. A. Martin, and D. W. Pfennig. In press. Increased competition as a cost of specialization during the evolution of resource polymorphism. *Biological Journal of the Linnean Society*.

Pavey, S. A., H. Collin, P. Nosil, and S. M. Rogers. 2010. The role of gene expression in ecological speciation. *Annals of the New York Academy of Science: The Year in Evolutionary Biology* 1206:110–129.

Payne, J. L., A. G. Boyer, J. H. Brown, S. Finnegan, M. Kowalewski, R. A. Krause, S. K. Lyons et al. 2009. Two-phase increase in the maximum size of life over 3.5 billion years reflects biological innovation and environmental opportunity. *Proceedings of the National Academy of Sciences of the United States of America* 106:24–27.

Payne, R. B., L. L. Payne, J. L. Woods, and M. D. Sorenson. 2000. Imprinting and the origin of parasite-host associations in brood-parasitic indigobirds, *Vidua chalybeata*. *Animal Behaviour* 59:69–81.

Pearson, D. L., and E. J. Mury. 1979. Character convergence and divergence among tiger beetles (Coleoptera: Cicindelidae). *Ecology* 60:557–566.

Peiman, K. S., and B. W. Robinson. 2007. Heterospecific aggression and adaptive divergence in brook stickleback (*Culaea inconstans*). *Evolution* 61:1327–1338.

———. 2010. Ecology and evolution of resource-related heterospecific aggression. *Quarterly Review of Biology* 85:133–158.

Peters, R. H. 1983. *The ecological implications of body size*. Cambridge University Press: Cambridge, UK.

Pfaender, J., F. W. Miesen, R. K. Hadiaty, and F. Herder. 2011. Adaptive speciation and sexual dimorphism contribute to diversity in form and function in the adaptive radiation of Lake Matano's sympatric roundfin sailfin silversides. *Journal of Evolutionary Biology* 24:2329–2345.

Pfennig, D. W. 1990. The adaptive significance of an environmentally-cued developmental switch in an anuran tadpole. *Oecologia* 85:101–107.

———. 1992. Polyphenism in spadefoot toads as a locally adjusted evolutionarily stable strategy. *Evolution* 46:1408–1420.

———. 2000a. Effect of predator-prey phylogenetic similarity on the fitness consequences of predation: A trade-off between nutrition and disease? *American Naturalist* 155:335–345.

Pfennig, D. W., and J. P. Collins. 1993. Kinship affects morphogenesis in cannibalistic salamanders. *Nature* 362:836–838.

Pfennig, D. W., and W. A. Frankino. 1997. Kin-mediated morphogenesis in facultatively cannibalistic tadpoles. *Evolution* 51:1993–1999.

Pfennig, D. W., and D. W. Kikuchi. 2012. Competition and the evolution of imperfect mimicry. *Current Zoology* 58:607–618.

Pfennig, D. W., and R. A. Martin. 2009. A maternal effect mediates rapid population divergence and character displacement in spadefoot toads. *Evolution* 63:898–909.

———. 2010. Evolution of character displacement in spadefoot toads: Different proximate mechanisms in different species. *Evolution* 64:2331–2341.

Pfennig, D. W., and M. McGee. 2010. Resource polyphenism increases species richness: A test of the hypothesis. *Philosophical Transactions of the Royal Society B: Biological Sciences* 365:577–591.

Pfennig, D. W., and P. J. Murphy. 2000. Character displacement in polyphenic tadpoles. *Evolution* 54:1738–1749.

———. 2002. How fluctuating competition and phenotypic plasticity mediate species divergence. *Evolution* 56:1217–1228.

———. 2003. A test of alternative hypotheses for character divergence between coexisting species. *Ecology* 84:1288–1297.

Pfennig, D. W., and K. S. Pfennig. 2010. Character displacement and the origins of diversity. *American Naturalist* 176:S26–S44.

———. 2012. Development and evolution of character displacement. *Annals of the New York Academy of Sciences: The Year in Evolutionary Biology* 1256:89–107.

Pfennig, D. W., H. K. Reeve, and P. W. Sherman. 1993. Kin recognition and cannibalism in spadefoot toad tadpoles. *Animal Behaviour* 46:87–94.

Pfennig, D. W., and A. M. Rice. 2007. An experimental test of character displacement's role in promoting postmating isolation between conspecific populations in contrasting competitive environments. *Evolution* 61:2433–2443.

Pfennig, D. W., A. M. Rice, and R. A. Martin. 2006. Ecological opportunity and phenotypic plasticity interact to promote character displacement and species coexistence. *Ecology* 87:769–779.

———. 2007. Field and experimental evidence for competition's role in phenotypic divergence. *Evolution* 61:257–271.

Pfennig, D. W., M. A. Wund, E. C. Snell-Rood, T. Cruickshank, C. D. Schlichting, and A. P. Moczek. 2010. Phenotypic plasticity's impacts on diversification and speciation. *Trends in Ecology and Evolution* 25:459–467.

Pfennig, K. S. 1998. The evolution of mate choice and the potential for conflict between species and mate-quality recognition. *Proceedings of the Royal Society B: Biological Sciences* 265:1743–1748.

———. 2000b. Female spadefoot toads compromise on mate quality to ensure conspecific matings. *Behavioral Ecology* 11:220–227.

———. 2003. A test of alternative hypotheses for the evolution of reproductive isolation between spadefoot toads: Support for the reinforcement hypothesis. *Evolution* 57:2842–2851.

———. 2007. Facultative mate choice drives adaptive hybridization. *Science* 318:965–967.

———. 2008. Population differences in condition-dependent sexual selection may promote divergence in non-sexual traits. *Evolutionary Ecology Research* 10:763–773.

Pfennig, K. S., and D. W. Pfennig. 2005. Character displacement as the "best of a bad situation": Fitness trade-offs resulting from selection to minimize resource and mate competition. *Evolution* 59:2200–2208.

———. 2009. Character displacement: Ecological and reproductive responses to a common evolutionary problem. *Quarterly Review of Biology* 84:253–276.

Pfennig, K. S., and M. J. Ryan. 2006. Reproductive character displacement generates reproductive isolation among conspecific populations: An artificial neural network study. *Proceedings of the Royal Society B: Biological Sciences* 273:1361–1368.

———. 2007. Character displacement and the evolution of mate choice: An artificial neural network approach. *Philosophical Transactions of the Royal Society B: Biological Sciences* 362:411–419.

———. 2010. Evolutionary diversification of mating behaviour: Using artificial neural networks to study reproductive character displacement and speciation. Pages 187–214 in C. Tosh, and G. Ruxton, eds. *Modeling perception with artificial neural networks*. Cambridge University Press: Cambridge, UK.

Pfennig, K. S., and M. A. Simovich. 2002. Differential selection to avoid hybridization in two toad species. *Evolution* 56:1840–1848.

Phillimore, A. B., and T. D. Price. 2008. Density-dependent cladogenesis in birds. *PLoS Biology* 6:483–489.

Phillips, P. C. 1996. Waiting for a compensatory mutation: Phase zero of the shifting-balance process. *Genetical Research* 67:271–283.

Pianka, E. R. 1976. Competition and niche theory. Pages 114–141 in R. M. May, ed. *Theoretical ecology: Principles and applications.* Blackwell: London.

———. 2000. *Evolutionary ecology.* Benjamin/Cummings: San Francisco, CA.

Pienaar, J., and J. M. Greeff. 2003. Different male morphs of *Otitesella pseudoserrata* fig wasps have equal fitness but are not determined by different alleles. *Ecology Letters* 6:286–289.

Pigliucci, M. 2010. Phenotypic plasticity. Pages 355–378 in M. Pigliucci, and G. B. Müller, eds. *Evolution: The extended synthesis.* MIT Press: Cambridge, MA.

Pigliucci, M., and C. J. Murren. 2003. Genetic assimilation and a possible evolutionary paradox: Can macroevolution sometimes be so fast as to pass us by? *Evolution* 57:1455–1464.

Pijanowska, J., P. Bernatowicz, and J. Fronk. 2007. Phenotypic plasticity within *Daphnia longispina* complex: Differences between parental and hybrid clones. *Polish Journal of Ecology* 55:761–769.

Pimentel, D., E. H. Feinberg, P. W. Wood, and J. T. Hayes. 1965. Selection, spatial distribution, and the coexistence of competing fly species. *American Naturalist* 99:97–109.

Pitnick, S., and D. J. Hosken. 2010. Postcopulatory sexual selection. Pages 379–399 in D. F. Westneat, and C. W. Fox, eds. *Evolutionary behavioral ecology.* Oxford University Press: Oxford, UK.

Plaistow, S. J., C. T. Lapsley, and T. G. Benton. 2006. Context-dependent intergenerational effects: The interaction between past and present environments and its effect on population dynamics. *American Naturalist* 167:206–215.

Pleasants, J. M. 1980. Competition for bumblebee pollinators in Rocky Mountain plant communities. *Ecology* 61:1446–1459.

Podos, J. 2001. Correlated evolution of morphology and vocal signal structure in Darwin's finches. *Nature* 409:185–188.

Podos, J., and S. Nowicki. 2004. Beaks, adaptation, and vocal evolution in Darwin's finches. *BioScience* 54:501–510.

Podos, J., J. A. Southall, and M. R. Rossi-Santos. 2004. Vocal mechanics in Darwin's finches: Correlation of beak gape and song frequency. *Journal of Experimental Biology* 207:607–619.

Polechová, J., and N. H. Barton. 2005. Speciation through competition: A critical review. *Evolution* 59:1194–1210.

Polis, G. A., C. A. Myers, and R. D. Holt. 1989. The ecology and evolution of intraguild predation: Potential competitors that eat each other. *Annual Review of Ecology and Systematics* 20:297–330.

Pomiankowski, A. 1987. The costs of choice in sexual selection. *Journal of Theoretical Biology* 128:195–218.

Price, T. 1998. Sexual selection and natural selection in bird speciation. *Philosophical Transactions of the Royal Society B: Biological Sciences* 353:251–260.

———. 2008. *Speciation in birds.* Roberts and Company: Greenwood Village, CO.

Price, T., and M. Kirkpatrick. 2009. Evolutionarily stable range limits set by interspecific competition. *Proceedings of the Royal Society B: Biological Sciences* 276:1429–1434.

Price, T. D., A. Qvarnström, and D. E. Irwin. 2003. The role of phenotypic plasticity in driving genetic evolution. *Proceedings of the Royal Society B: Biological Sciences* 270:1433–1440.

Pritchard, J. R., and D. Schluter. 2001. Declining interspecific competition during character displacement: Summoning the ghost of competition past. *Evolutionary Ecology Research* 3:209–220.

Proctor, H. C. 1991. Courtship in the water mite *Neumania papillator*: Males capitalize on female adaptations for predation. *Animal Behaviour* 42:589–598.

Pryke, S. R., and S. Andersson. 2005. Experimental evidence for female choice and energetic costs of male tail elongation in red-collared widowbirds. *Biological Journal of the Linnean Society* 86:35–43.

———. 2008. Female preferences for long tails constrained by species recognition in short-tailed red bishops. *Behavioral Ecology* 19:1116–1121.

Ptacek, M. B. 1992. Calling sites used by male gray treefrogs, *Hyla versicolor* and *Hyla chrysoscelis*, in sympatry and allopatry in Missouri. *Herpetologica* 48:373–382.

———. 2000. The role of mating preferences in shaping interspecific divergence in mating signals in vertebrates. *Behavioural Processes* 51:111–134.

Ptashne, M. 1986. *A genetic switch*. Cell Press: Cambridge, UK.

Rabosky, D. L., and I. J. Lovette. 2008. Density-dependent diversification in North American wood warblers. *Proceedings of the Royal Society B: Biological Sciences* 275:2363–2371.

Radwan, J. 1993. The adaptive significance of male polymorphism in the acarid mite *Caloglyphus berlesei*. *Behavioral Ecology and Sociobiology* 33:201–208.

Raeymaekers, J. A. M., M. Boisjoly, L. Delaire, D. Berner, K. Räsänen, and A. P. Hendry. 2010. Testing for mating isolation between ecotypes: Laboratory experiments with lake, stream and hybrid stickleback. *Journal of Evolutionary Biology* 23:2694–2708.

Raff, R. A., and T. C. Kauffman. 1983. *Genes, embryos, and evolution*. MacMillan: New York.

Rainey, M. M., and G. F. Grether. 2007. Competitive mimicry: Synthesis of a neglected class of mimetic relationships. *Ecology* 88:2440–2448.

Rainey, P. B., A. Buckling, R. Kassen, and M. Travisano. 2000. The emergence and maintenance of diversity: Insights from experimental bacterial populations. *Trends in Ecology and Evolution* 15:243–247.

Rainey, P. B., and M. Travisano. 1998. Adaptive radiation in a heterogeneous environment. *Nature* 394:69–72.

Ramsey, J., H. D. Bradshaw, and D. W. Schemske. 2003. Components of reproductive isolation between the monkeyflowers *Mimulus lewisii* and *M. cardinalis* (Phrymaceae). *Evolution* 57:1520–1534.

Ramsey, J., and D. W. Schemske. 1998. Pathways, mechanisms, and rates of polyploid formation in flowering plants. *Annual Review of Ecology and Systematics* 29:467–501.

Räsänen, K., and L. E. B. Kruuk. 2007. Maternal effects and evolution at ecological timescales. *Functional Ecology* 21:408–421.

Rashed, A., and T. N. Sherratt. 2007. Mimicry in hoverflies (Diptera: Syrphidae): A field test of the competitive mimicry hypothesis. *Behavioral Ecology* 18:337–344.

Ratcliffe, L. M., and P. R. Grant. 1983. Species recognition in Darwin's finches (*Geospiza*, Gould). I. Discrimination by morphological cues. *Animal Behaviour* 31:1139–1153.

Rayner, J. M. V. 1990. Vertebrate flight and the origins of flying vertebrates. Pages 188–217 in K. C. Allen, and D. E. G. Briggs, eds. *Evolution and the fossil record*. Belhaven: London.

Razeto-Barry, P., and K. Maldonado. 2011. Adaptive *cis*-regulatory changes may involve few mutations. *Evolution* 65:3332–3335.

Reifová, R., J. Reif, M. Antczak, and M. W. Nachman. 2011. Ecological character displacement in the face of gene flow: Evidence from two species of nightingales. *BMC Evolutionary Biology* 11:138.

Relyea, R. A. 2002. Costs of phenotypic plasticity. *American Naturalist* 159:272–282.

Rensch, B. 1960a. *Evolution above the species level*. Columbia University Press: New York.

———. 1960b. The laws of evolution. Pages 95–116 in S. Tax, ed. *The evolution of life*. University of Chicago Press: Chicago, IL.

Reznick, D. N., and R. E. Ricklefs. 2009. Darwin's bridge between microevolution and macroevolution. *Nature* 457:837–842.

Rhymer, J. M., and D. Simberloff. 1996. Extinction by hybridization and introgression. *Annual Review of Ecology and Systematics* 27:83–109.

Rice, A. M., A. R. Leichty, and D. W. Pfennig. 2009. Parallel evolution and ecological selection: Replicated character displacement in spadefoot toads. *Proceedings of the Royal Society B: Biological Sciences* 276:4189–4196.

Rice, A. M., and D. W. Pfennig. 2007. Character displacement: In situ evolution of novel phenotypes or sorting of pre-existing variation? *Journal of Evolutionary Biology* 20:448–459.

———. 2008. Analysis of range expansion in two species undergoing character displacement: Why might invaders generally "win" during character displacement? *Journal of Evolutionary Biology* 21:696–704.

———. 2010. Does character displacement initiate speciation? Evidence of reduced gene flow between populations experiencing divergent selection. *Journal of Evolutionary Biology* 23:854–865.

Rice, W. R. 1987. Speciation via habitat specialization: The evolution of reproductive isolation as a correlated character. *Evolutionary Ecology* 1:301–315.

Rice, W. R., and E. E. Hostert. 1993. Laboratory experiments on speciation: What have we learned in 40 years? *Evolution* 47:1637–1653.

Richards, C. M., O. Bossdorf, N. Z. Muth, J. Gurevitch, and M. Pigliucci. 2006. Jack of all trades, master of some? On the role of phenotypic plasticity in plant invasions. *Ecology Letters* 9:981–993.

Ridley, M. 1996. *Evolution*. Blackwell: Cambridge, MA.

———. 2005. *How to read Darwin*. Norton: New York.

Rieseberg, L. H., M. A. Archer, and R. K. Wayne. 1999. Transgressive segregation, adaptation and speciation. *Heredity* 83:363–372.

Ritchie, M. G. 2007. Sexual selection and speciation. *Annual Review of Ecology Evolution and Systematics* 38:79–102.

Robinson, B. W., and R. Dukas. 1999. The influence of phenotypic modifications on evolution: The Baldwin effect and modern perspectives. *Oikos* 85:582–589.

Robinson, B. W., and K. J. Parsons. 2002. Changing times, spaces, and faces: Tests and implications of adaptive morphological plasticity in the fishes of northern postglacial lakes. *Canadian Journal of Fisheries and Aquatic Sciences* 59:1819–1833.

Robinson, B. W., and D. S. Wilson. 1994. Character release and displacement in fish: A neglected literature. *American Naturalist* 144:596–627.

———. 1998. Optimal foraging, specialization, and a solution to Liem's paradox. *The American Naturalist* 151:223–235.

Robinson, B. W., D. S. Wilson, A. S. Margosian, and P. T. Lotito. 1993. Ecological and morphological differentiation of pumpkinseed sunfish in lakes without bluegill sunfish. *Evolutionary Ecology* 7:451–464.

Robinson, B. W., D. S. Wilson, and G. O. Shea. 1996. Trade-offs of ecological specialization: An intraspecific comparison of pumpkinseed sunfish phenotypes. *Ecology* 77:170–178.

Rodriguez, R. L., L. E. Sullivan, and R. B. Cocroft. 2004. Vibrational communication and reproductive isolation in the *Enchenopa binotata* species complex of treehoppers (Hemiptera: Membracidae). *Evolution* 58:571–578.

Rokas, A. 2000. *Wolbachia* as a speciation agent. *Trends in Ecology and Evolution* 15:44–45.

Rose, K. D. 2006. *The beginning of the age of mammals*. The Johns Hopkins University Press: Baltimore, MD.

Rosenthal, G. G., W. E. Wagner, and M. J. Ryan. 2002. Secondary reduction of preference for the sword ornament in the pygmy swordtail *Xiphophorus nigrensis* (Pisces: Poeciliidae). *Animal Behaviour* 63:37–45.

Rosenzweig, M. L. 1978. Competitive speciation. *Biological Journal of the Linnean Society* 10:274–289.

Roughgarden, J. 1972. Evolution of niche width. *American Naturalist* 106:683–718.

———. 1974. Species packing and competition function with illustrations from coral reef fish. *Theoretical Population Biology* 5:163–186.

———. 1976. Resource partitioning among competing species: A coevolutionary approach. *Theoretical Population Biology* 9:388–424.

Rubin, J. A. 1985. Mortality and avoidance of competitive overgrowth in encrusting Bryozoa. *Marine Ecology Progress Series* 23:291–299.

Rueffler, C., T. J. M. Van Dooren, O. Leimar, and P. A. Abrams. 2006. Disruptive selection and then what? *Trends in Ecology and Evolution* 21:238–245.

Rundle, H. D. 2002. A test of ecologically dependent postmating isolation between sympatric sticklebacks. *Evolution* 56:322–329.

Rundle, H. D., L. Nagel, J. W. Boughman, and D. Schluter. 2000. Natural selection and parallel speciation in sympatric sticklebacks. *Science* 287:306–308.

Rundle, H. D., and P. Nosil. 2005. Ecological speciation. *Ecology Letters* 8:336–352.

Rundle, H. D., and D. Schluter. 1998. Reinforcement of stickleback mate preferences: Sympatry breeds contempt. *Evolution* 52:200–208.

———. 2004. Natural selection and ecological speciation in sticklebacks. Pages 192–209 in U. Dieckmann, M. Doebeli, J. A. J. Metz, and D. Tautz, eds. *Adaptive speciation*. Cambridge University Press: Cambridge, UK.

Rundle, H. D., and M. C. Whitlock. 2001. A genetic interpretation of ecologically dependent isolation. *Evolution* 55:198–201.

Ruxton, G. D., T. N. Sherratt, and M. P. Speed. 2004. *Avoiding attack: The evolutionary ecology of crypsis, warning signals and mimicry.* Oxford University Press: Oxford, UK.

Ryals, P. E., H. E. Smith-Somerville, and H. E. Buhse Jr. 2002. Phenotype switching in polymorphic *Tetrahymena*: A single-cell Jekyll and Hyde. *International Review of Cytology* 212:209–238.

Ryan, M. J. 1998. Sexual selection, receiver biases, and the evolution of sex differences. *Science* 281:1999–2003.

Ryan, M. J., and A. S. Rand. 1993. Species recognition and sexual selection as a unitary problem in animal communication. *Evolution* 47:647–657.

Sæther, S. A., G. P. Sætre, T. Borge, C. Wiley, N. Svedin, G. Andersson, T. Veen et al. 2007. Sex chromosome-linked species recognition and evolution of reproductive isolation in flycatchers. *Science* 318:95–97.

Sætre, G. P., T. Moum, S. Bures, M. Kral, M. Adamjan, and J. Moreno. 1997. A sexually selected character displacement in flycatchers reinforces premating isolation. *Nature* 387:589–592.

Sahney, S., M. J. Benton, and P. A. Ferry. 2010. Links between global taxonomic diversity, ecological diversity and the expansion of vertebrates on land. *Biology Letters* 6:544–547.

Sanderson, N. 1989. Can gene flow prevent reinforcement? *Evolution* 43:1223–1235.

Sandoval, C. P. 1994. Differential visual predation on morphs of *Timema cristinae* (Phasmatodeae, Timemidae) and its consequences for host range. *Biological Journal of the Linnean Society* 52:341–356.

Saxer, G., M. Doebeli, and M. Travisano. 2010. The repeatability of adaptive radiation during long-term experimental evolution of *Escherichia coli* in a multiple nutrient environment. *PLoS One* 5.

Schlichting, C. D. 2004. The role of phenotypic plasticity in diversification. Pages 191–200 in T. J. DeWitt, and S. M. Scheiner, eds. *Phenotypic plasticity: Functional and conceptual approaches.* Oxford University Press: Oxford, UK.

———. 2008. Hidden reaction norms, cryptic genetic variation, and evolvability. *Annals of the New York Academy of Science* 1133:187–203.

Schlichting, C. D., and M. Pigliucci. 1998. *Phenotypic evolution: A reaction norm perspective.* Sinauer: Sunderland, MA.

Schluter, D. 1993. Adaptive radiation in sticklebacks: Size, shape, and habitat use efficiency. *Ecology* 74:699–709.

———. 1994. Experimental evidence that competition promotes divergence in adaptive radiation. *Science* 266:798–801.

———. 1996. Adaptive radiation along genetic lines of least resistance. *Evolution* 50:1766–1774.

———. 2000. *The ecology of adaptive radiation.* Oxford University Press: Oxford, UK.

———. 2001. Ecological character displacement. Pages 265–276 in C. W. Fox, D. A. Roff, and D. J. Fairbairn, eds. *Evolutionary ecology: Concepts and case studies.* Oxford University Press, UK.

———. 2002. Character displacement. Pages 149–150 in M. Pagel, ed. *Encyclopedia of evolution.* Oxford University Press, Oxford, UK.

———. 2003. Frequency dependent natural selection during character displacement in sticklebacks. *Evolution* 57:1142–1150.

———. 2009. Evidence for ecological speciation and its alternative. *Science* 323:737–741.

Schluter, D., and J. D. McPhail. 1992. Ecological character displacement and speciation in sticklebacks. *American Naturalist* 140:85–108.

Schluter, D., and L. M. Nagel. 1995. Parallel speciation by natural selection. *American Naturalist* 146:292–301.

Schluter, D., and T. Price. 1993. Honesty, perception and population divergence in sexually selected traits. *Proceedings of the Royal Society B: Biological Sciences* 253:117–122.

Schluter, D., T. D. Price, and P. R. Grant. 1985. Ecological character displacement in Darwin's finches. *Science* 227:1056–1059.

Schmitt, J., S. A. Dudley, and M. Pigliucci. 1999. Manipulative approaches to testing adaptive plasticity: Phytochrome-mediated shade-avoidance responses in plants. *American Naturalist* 154:S43–S54.

Schoener, T. W. 1974. Resource partitioning in ecological communities. *Science* 185:27–39.

———. 1983. Field experiments on interspecific competition. *American Naturalist* 122:240–285.

———. 1986. Resource partitioning. Pages 91–126 in J. Kikkawa, and D. J. Anderson, eds. *Community ecology: Pattern and process*. Blackwell Scientific Publications: Oxford, UK.

———. 2009. Ecological niche. Pages 3–13 in S. A. Levin, ed. *The Princeton guide to ecology*. Princeton University Press: Princeton, NJ.

Schulte, P., L. Alegret, I. Arenillas, J. A. Arz, P. J. Barton, P. R. Bown, T. J. Bralower et al. 2010. The Chicxulub asteroid impact and mass extinction at the Cretaceous-Paleogene boundary. *Science* 327:1214–1218.

Schutz, D., G. Pachler, E. Ripmeester, O. Goffinet, and M. Taborsky. 2010. Reproductive investment of giants and dwarfs: Specialized tactics in a cichlid fish with alternative male morphs. *Functional Ecology* 24:131–140.

Schwander, T., and O. Leimar. 2011. Genes as leaders and followers in evolution. *Trends in Ecology and Evolution* 26:143–151.

Scoville, A. G., and M. E. Pfrender. 2010. Phenotypic plasticity facilitates recurrent rapid adaptation to introduced predators. *Proceedings of the National Academy of Sciences of the United States of America* 107:4260–4263.

Secondi, J., V. Bretagnolle, C. Compagnon, and B. Faivre. 2003. Species-specific song convergence in a moving hybrid zone between two passerines. *Biological Journal of the Linnean Society* 80:507–517.

Seehausen, O. 2004. Hybridization and adaptive radiation. *Trends in Ecology and Evolution* 19:198–207.

———. 2006. African cichlid fish: A model system in adaptive radiation research. *Proceedings of the Royal Society B: Biological Sciences* 273:1987–1998.

Seehausen, O., Y. Terai, I. S. Magalhaes, K. L. Carleton, H. D. J. Mrosso, R. Miyagi, I. van der Sluijs et al. 2008. Speciation through sensory drive in cichlid fish. *Nature* 455:620–623.

Seehausen, O., and J. J. M. van Alphen. 1998. The effect of male coloration on female mate choice in closely related Lake Victoria cichlids (*Haplochromis nyererei* complex). *Behavioral Ecology and Sociobiology* 42:1–8.

Seehausen, O., J. J. M. van Alphen, and F. Witte. 1997. Cichlid fish diversity threatened by eutrophication that curbs sexual selection. *Science* 277:1808–1811.

Seger, J. 1985. Intraspecific resource competition as a cause of sympatric speciation. Pages 43–53 in P. J. Greenwood, P. H. Harvey, and M. Slatkin, eds. *Evolution: Essays in honour of John Maynard Smith*. Cambridge University Press: Cambridge, UK.

Selander, R. K. 1966. Sexual dimorphism and differential niche utilization in birds. *Condor* 68:113–151.

Sepkoski, J. J. 1984. A kinetic-model of Phanerozoic taxonomic diversity. III. Post-Paleozoic families and mass extinction. *Paleobiology* 10:246–267.

———. 1993. Ten years in the library: New data confirm paleontological patterns. *Paleobiology* 19:43–51.

———. 1996. Competition in macroevolution: The double wedge revisited. Pages 211–255 in D. Jablonski, D. E. Erwin, and J. Lipps, eds. *Evolutionary paleobiology*. University of Chicago Press: Chicago, IL.

———. 2003. Competition in evolution. Pages 171–176 in D. E. G. Briggs, and P. R. Crowther, eds. *Paleobiology II*. Blackwell: Malden, MA.

Servedio, M. R. 2008. The role of linkage disequilibrium in the evolution of pre-mating isolation. *Heredity* 102:51–56.

Servedio, M. R., and M. A. F. Noor. 2003. The role of reinforcement in speciation: Theory and data. *Annual Review of Ecology and Systematics* 34: 339–364.

Servedio, M. R., S. A. Sæther, and G.-P. Sætre. 2009. Reinforcement and learning. *Evolutionary Ecology* 23:109–123.

Servedio, M. R., G. S. van Doorn, M. Kopp, A. M. Frame, and P. Nosil. 2011. Magic traits in speciation: "Magic" but not rare? *Trends in Ecology and Evolution* 26:389–397.

Sexton, J. P., P. J. McIntyre, A. L. Angert, and K. J. Rice. 2009. Evolution and ecology of species range limits. *Annual Review of Ecology Evolution and Systematics* 40:415–436.

Shelly, T. E., and M. D. Greenfield. 1989. Satellites and transients: Ecological constraints on alternative mating tactics in male grasshoppers. *Behaviour* 109:200–221.

Shuster, S. M. 1989. The reproductive behavior of α-, β-, and γ-male morphs in *Paracerceis sculpta*. *Evolution* 43:1683–1698.

Shuster, S. M., and M. J. Wade. 1992. Equal mating success among male reproductive strategies in a marine isopod. *Nature* 350:606–661.

———. 2003. *Mating systems and strategies*. Princeton University Press: Princeton, NJ.

Silvertown, J. 2004. Plant coexistence and the niche. *Trends in Ecology and Evolution* 19:605–611.

Simberloff, D. 2004. Community ecology: Is it time to move on? *American Naturalist* 163:787–799.

Simberloff, D., T. Dayan, C. Jones, and G. Ogura. 2000. Character displacement and release in the small Indian Mongoose, *Herpestes javanicus*. *Ecology* 81:2086–2099.

Simberloff, D. S. 1970. Taxonomic diversity of island biotas. *Evolution* 24:23–47.

Simberloff, D. S., and W. Boecklen. 1981. Santa Rosalia reconsidered: Size ratios and competition. *Evolution* 35:1206–1228.

Simmons, L. W. 1994. Courtship role reversal in bush crickets: Another role for parasites? *Behavioral Ecology* 5:259–266.

Simpson, G. G. 1944. *Tempo and mode in evolution.* Columbia University Press: New York.

———. 1953. *The major features of evolution.* Columbia University Press: New York.

Sinervo, B., and C. M. Lively. 1996. The rock-scissors-paper game and the evolution of alternative male strategies. *Nature* 340:240–246.

Sinervo, B., E. Svensson, and T. Comendant. 2000. Density cycles and an offspring quantity and quality game driven by natural selection. *Nature* 406:985–988.

Singer, M. C., and C. S. McBride. 2010. Multitrait, host-associated divergence among sets of butterfly populations: Implications for reproductive isolation and ecological speciation. *Evolution* 64:921–933.

Skúlason, S., S. S. Snorrason, and B. Jónsson. 1999. Sympatric morphs, populations and speciation in freshwater fish with emphasis on arctic charr. Pages 70–92 in A. E. Magurran, and R. M. May, eds. *Evolution of biological diversity.* Oxford University Press: Oxford, UK.

Slabbekoorn, H., and T. B. Smith. 2000. Does bill size polymorphism affect courtship song characteristics in the African finch *Pyrenestes ostrinus*? *Biological Journal of the Linnean Society* 71:737–753.

Slagsvold, T., S. Dale, and A. Kruszewicz. 1995. Predation favors cryptic coloration in breeding male pied flycatchers. *Animal Behaviour* 50:1109–1121.

Slatkin, M. 1979. Frequency- and density-dependent selection on a quantitative character. *Genetics* 93:755–771.

———. 1980. Ecological character displacement. *Ecology* 61:163–177.

Smadja, C., and R. K. Butlin. 2009. On the scent of speciation: The chemosensory system and its role in premating isolation. *Heredity* 102:77–97.

Smith, D. C. 1988. Heritable divergence of *Rhagoletis pomonella* host races by seasonal asynchrony. *Nature* 336:66–67.

Smith, H., and G. C. Whitelam. 1997. The shade avoidance syndrome: Multiple responses mediated by multiple phytochromes. *Plant, Cell, and Environment* 20:840–844.

Smith, L. D., and A. R. Palmer. 1994. Effects of manipulated diet on size and performance of brachyuran crab claws. *Science* 264:710–712.

Smith, R. A., and M. D. Rausher. 2008. Experimental evidence that selection favors character displacement in the ivyleaf morning glory. *American Naturalist* 171:1–9.

Smith, T. B. 1993. Disruptive selection and the genetic basis of bill size polymorphism in the African finch *Pyrenestes*. *Nature* 363:618–620.

Smith, T. B., and S. Skúlason. 1996. Evolutionary significance of resource polymorphisms in fishes, amphibians, and birds. *Annual Review of Ecology and Systematics* 27:111–133.

Snowberg, L. K., and C. W. Benkman. 2007. The role of marker traits in the assortative mating within red crossbills, *Loxia curvirostra* complex. *Journal of Evolutionary Biology* 20:1924–1932.

———. 2009. Mate choice based on a key ecological performance trait. *Journal of Evolutionary Biology* 22:762–769.

Sobel, J. M., G. F. Chen, L. R. Watt, and D. W. Schemske. 2010. The biology of speciation. *Evolution* 64:295–315.

Spencer, H. G., B. H. McArdle, and D. M. Lambert. 1986. A theoretical investigation of speciation by reinforcement. *American Naturalist* 128:241–262.

Stachowicz, J. J. 2001. Mutualism, facilitation, and the structure of ecological communities. *BioScience* 51:235–246.

Stanley, S. M. 1979. *Macroevolution.* W. H. Freeman: San Francisco, CA.

———. 2008. Predation defeats competition on the seafloor. *Paleobiology* 34:1–21.

Steele-Petrovic, M. 1979. The physiological differences between articulate brachiopods and filter-feeding bivalves as a factor in the evolution of marine level bottom communities. *Paleontology* 22:101–134.

Stern, D. L. 2011. *Evolution, development, and the predictable genome.* Roberts and Company: Greenwood Village, CO.

Stiling, P. D. 2012. *Ecology: Global insights and investigation.* McGraw-Hill: New York.

Stinson, K. A., S. A. Campbell, J. R. Powell, B. E. Wolfe, R. M. Callaway, G. C. Thelen, S. G. Hallett et al. 2006. Invasive plant suppresses the growth of native tree seedlings by disrupting belowground mutualisms. *PLoS Biology* 4:e140.

Stone, G. N., P. Willmer, and J. A. Rowe. 1998. Partitioning of pollinators during flowering in an African *Acacia* community. *Ecology* 79:2808–2827.

Stork, N. E. 1993. How many species are there? *Biodiversity and Conservation* 2:215–232.

Strong, D. R., Jr., L. A. Szyska, and D. S. Simberloff. 1979. Tests for community-wide character displacement against null hypotheses. *Evolution* 33:897–913.

Sueur, J. 2002. Cicada acoustic communication: Potential sound partitioning in a multispecies community from Mexico (Hemiptera: Cicadomorpha: Cicadidae). *Biological Journal of the Linnean Society* 75:379–394.

Sullivan, B. K. 1983. Sexual selection in the Great Plains Toad (*Bufo cognatus*). *Behaviour* 84:258–264.

Sulloway, F. J. 1982. Darwin and his finches: The evolution of a legend. *Journal of the History of Biology* 15:1–53.

Sultan, S. E. 2007. Development in context: The timely emergence of eco-devo. *Trends in Ecology and Evolution* 22:575–582.

———. 2011. Evolutionary implications of individual plasticity. Pages 193–203 in S. B. Gissis, and E. Jablonka, eds. *Transformations of Lamarckism: From subtle fluids to molecular biology.* MIT Press: Cambridge, MA.

Sultan, S. E., and F. A. Bazzaz. 1993. Phenotypic plasticity in *Polygonum persicaria*. I. Diversity and uniformity in genotypic norms of reaction to light. *Evolution* 47:1009–1031.

Sultan, S. E., and H. G. Spencer. 2002. Metapopulation structure favors plasticity over local adaptation. *American Naturalist* 160:271–283.

Suzuki, Y., and H. F. Nijhout. 2006. Evolution of a polyphenism by genetic accommodation. *Science* 311:650–652.

Sved, J. A. 1981. A 2-sex polygenic model for the evolution of pre-mating isolation. 1. Deterministic theory for natural populations. *Genetics* 97:197–215.

Svedin, N., C. Wiley, T. Veen, L. Gustafsson, and A. Qvarnström. 2008. Natural and sexual selection against hybrid flycatchers. *Proceedings of the Royal Society B: Biological Sciences* 275:735–744.

Svensson, E. I., F. Eroukhmanoff, K. Karlsson, A. Runemark, and A. Brodin. 2010. A role for learning in population divergence of mate preferences. *Evolution* 64:3101–3113.

Takafuji, A., E. Kuno, and H. Fujimoto. 1997. Reproductive interference and its consequences for the competitive interactions between two closely related *Panonychus* spider mites. *Experimental and Applied Acarology* 21:379–391.

Takahashi, M. K., and M. J. Parris. 2008. Life cycle polyphenism as a factor affecting ecological divergence within *Notophthalmus viridescens*. *Oecologia* 158:23–34.

Taper, M. L., and T. J. Case. 1985. Quantitative genetic models for the coevolution of character displacement. *Ecology* 66:355–371.

———. 1992. Coevolution among competitors. *Oxford Surveys in Evolutionary Biology* 8:63–109.

Taylor, D. L., T. D. Bruns, and S. A. Hodges. 2004. Evidence for mycorrhizal races in a cheating orchid. *Proceedings of the Royal Society B: Biological Sciences* 271:35–43.

Taylor, E. B., J. D. McPhail, and D. Schluter. 1997. History of ecological speciation in sticklebacks: Using experimental and phylogenetic approaches. Pages 511–534 in T. J. Givinish, and K. J. Sytsma, eds. *Molecular evolution and adaptation radiation*. Cambridge University Press: Cambridge, UK.

Taylor, T. N., E. L. Taylor, and M. Krings. 2009. *Paleobotany: The biology and evolution of fossil plants*. Elsevier: New York.

Temeles, E. J., and W. J. Kress. 2003. Adaptation in a plant-hummingbird association. *Science* 300:630–633.

Temeles, E. J., J. S. Miller, and J. L. Rifkin. 2010. Evolution of sexual dimorphism in bill size and shape of hermit hummingbirds (Phaethornithinae): A role for ecological causation. *Philosophical Transactions of the Royal Society B: Biological Sciences* 365:1053–1063.

Temeles, E. J., I. L. Pan, J. L. Brennan, and N. Horwitt. 2000. Evidence for ecological causation of sexual dimorphism in a hummingbird. *Science* 289:441–443.

Thomas, A. T., and P. D. Lane. 1984. Autecology of Silurian trilobites. Pages 55–69 in M. J. Basset, and J. D. Lawson, eds. *Autecology of Silurian Organisms*. Special Papers in Paleontology No. 32. Academic Press: London.

Thomas, F., E. Oget, P. Gente, D. Desmots, and F. Renaud. 1999. Assortative pairing with respect to parasite load in the beetle *Timarcha maritima* (Chrysomelidae). *Journal of Evolutionary Biology* 12:385–390.

Thompson, J. N. 2005. *The geographic mosaic of coevolution*. University of Chicago Press: Chicago, IL.

Thompson, J. N., and O. Pellmyr. 1991. Evolution of oviposition behavior and host preference in Lepidoptera. *Annual Review of Entomology* 36:65–89.

Thornhill, R. 1976. Sexual selection and nuptial feeding behavior in *Bittacus apicalis* (Insecta: Mecoptera). *American Naturalist* 110:529–548.

Tilman, D. 1977. Resource competition between plankton algae: An experimental and theoretical approach. *Ecology* 58:338–348.

Tilman, D., and S. W. Pacala. 1993. The maintenance of species richness in plant communities. Pages 13–25 in R. E. Ricklefs, and D. Schluter, eds. *Species diversity in ecological communities*. University of Chicago Press: Chicago, IL.

Tilmon, K. J. 2008. *The evolutionary biology of herbivorous insects: Specialization, speciation, and radiation*. University of California Press: Berkeley, CA.

Tobias, J. A., J. Aben, R. T. Brumfield, E. P. Derryberry, W. Halfwerk, H. Slabbekoorn, and N. Seddon. 2010. Song divergence by sensory drive in Amazonian birds. *Evolution* 64:2820–2839.

Tomkins, J. L., and G. S. Brown. 2004. Population density drives the local evolution of a threshold dimorphism. *Nature* 431:1099–1103.

Tonnis, B., P. R. Grant, B. R. Grant, and K. Petren. 2005. Habitat selection and ecological speciation in Galápagos warbler finches (*Certhidea olivacea* and *Certhidea fusca*). *Proceedings of the Royal Society B: Biological Sciences* 272:819–826.

Towe, K. M. 1970. Oxygen-collagen priority and the early metazoan fossil record. *Proceedings of the National Academy of Sciences of the United States of America* 65:781–788.

Tregenza, T., and N. Wedell. 2000. Genetic compatibility, mate choice and patterns of parentage: Invited review. *Molecular Ecology* 9:1013–1027.

Trivers, R. L. 1972. Parental investment and sexual selection. Pages 136–179 in B. Campbell, ed. *Sexual selection and the descent of man 1871–1971*. Aldine: Chicago, IL.

Turner, J. R. G. 1981. Adaptation and evolution in *Heliconius*: A defense of neodarwinism. *Annual Review of Ecology and Systematics* 12:99–121.

Tyerman, J. G., M. Bertrand, C. C. Spencer, and M. Doebel. 2008. Experimental demonstration of ecological character displacement. *BMC Evolutionary Biology* 8:34.

Tynkkynen, K., M. J. Rantala, and J. Suhonen. 2004. Interspecific aggression and character displacement in the damselfly *Calopteryx splendens*. *Journal of Evolutionary Biology* 17:759–767.

Vamosi, J. C., T. M. Knight, J. A. Steets, S. J. Mazer, M. Burd, and T. L. Ashman. 2006. Pollination decays in biodiversity hotspots. *Proceedings of the National Academy of Sciences of the United States of America* 103:956–961.

Vamosi, S. M., and D. Schluter. 1999. Sexual selection against hybrids between sympatric stickleback species: Evidence from a field experiment. *Evolution* 53:874–879.

van der Sluijs, I., T. J. M. Van Dooren, K. D. Hofker, J. J. M. van Alphen, R. B. Stelkens, and O. Seehausen. 2008. Female mating preference functions predict sexual selection against hybrids between sibling species of cichlid fish. *Philosophical Transactions of the Royal Society B: Biological Sciences* 363:2871–2877.

van Doorn, G. S., P. Edelaar, and F. J. Weissing. 2009. On the origin of species by natural and sexual selection. *Science* 326:1704–1707.

Van Valen, L. M. 1965. Morphological variation and width of ecological niche. *American Naturalist* 99:377–390.

———. 1973. A new evolutionary law. *Evolutionary Theory* 1:1–30.

Vandermeer, J. H. 1972. Niche theory. *Annual Review of Ecology and Systematics* 3:107–132.

Venditti, C., A. Meade, and M. Pagel. 2010. Phylogenies reveal new interpretation of speciation and the Red Queen. *Nature* 463:349–352.

Vermeij, G. J. 1987. *Evolution and escalation: An ecological history of life*. Princeton University Press: Princeton, NJ.

Via, S., A. C. Bouck, and S. Skillman. 2000. Reproductive isolation between divergent races of pea aphids on two hosts. II. Selection against migrants and hybrids in the parental environments. *Evolution* 54:1626–1637.

Violle, C., D. R. Nemergut, Z. Pu, and L. Jiang. 2011. Phylogenetic limiting similarity and competitive exclusion. *Ecology Letters* 14:782–787.

Volterra, V. 1926. Fluctuations in the abundance of a species considered mathematically. *Nature* 118:558–560.

Waddington, C. H. 1953. Genetic assimilation of an acquired character. *Evolution* 7:118–126.

———. 1957. *The strategy of the genes*. George Allen and Unwin: London.

———. 1969. Paradigm for an evolutionary process. Pages 106–128 in C. H. Waddington, ed. *Towards a theoretical biology*, vol. 2. Aldine: Chicago, IL.

Wade, M. J., and S. J. Arnold. 1980. The intensity of sexual selection in relation to male sexual behaviour, female choice, and sperm precedence. *Animal Behaviour* 28:446–461.

Wagner, P. J. 2010. Paleontological perspectives on morphological evolution. Pages 451–478 in M. A. Bell, D. J. Futuyma, W. F. Eanes, and J. S. Levinton, eds. *Evolution since Darwin: The first 150 years*. Sinauer: Sunderland, MA.

Wagner, W. E. 1996. Convergent song preferences between female field crickets and acoustically orienting parasitoid flies. *Behavioral Ecology* 7:279–285.

Waksman, S. A. 1947. What is an antibiotic or an antibiotic substance? *Mycologia* 39:565–569.

Walls, S. C., S. S. Belanger, and A. R. Blaustein. 1993. Morphological variation in a larval salamander: Dietary induction of plasticity in head shape. *Oecologia* 96:162–168.

Ward, P. D. 2006. *Out of thin air*. Joseph Henry Press: Washington, DC.

Waser, N. M. 1978. Interspecific pollen transfer and competition between co-occurring plant species. *Oecologia* 36:223–236.

———. 1983. Competition for pollination and floral character differences among sympatric plant species: A review of the evidence. Pages 277–293 in C. E. Jones, and R. J. Little, eds. *Handbook of experimental pollination ecology*. Van Nostrand Reinhold: New York.

Webb, C. 2003. A complete classification of Darwinian extinction in ecological interactions. *American Naturalist* 161:181–205.

Webb, C. O. 2000. Exploring the phylogenetic structure of ecological communities: An example for rain forest trees. *American Naturalist* 156:145–155.

Webb, C. O., D. D. Ackerly, M. A. McPeek, and M. J. Donohue. 2002. Phylogenies and community ecology. *Annual Review of Ecology and Systematics* 33:475–505.

Webster, M. 2007. A Cambrian peak in morphological variation within trilobite species. *Science* 317:499–502.

Webster, T. P., and J. M. Burns. 1973. Dewlap color variation and electrophoretically detected sibling species in a Haitian lizard *Anolis brevirostris*. *Evolution* 27:368–377.

Weir, L. K., J. W. A. Grant, and J. A. Hutchings. 2011. The influence of operational sex ratio on the intensity of competition for mates. *American Naturalist* 177:167–176.

Welch, A. M., R. D. Semlitsch, and H. C. Gerhardt. 1998. Call duration as an indicator of genetic quality in male gray tree frogs. *Science* 280:1928–1930.

Wellnhofer, P. 2009. *Archaeopteryx: Icon of evolution*. Verlag Dr. Friedrich Pfeil: Munich.

Werner, E. E. 1977. Species packing and niche complementarity in three sunfishes. *American Naturalist* 111:553–578.

Werner, T. K., and T. W. Sherry. 1987. Behavioral feeding specialization in *Pinaroloxias inornata*, the "Darwin's Finch" of Cocos Island, Costa Rica. *Proceedings of the National Academy of Sciences of the United States of America* 84:5506–5510.

Werren, J. H. 1997. *Wolbachia* and speciation. Pages 245–260 in D. J. Howard, and S. H. Berlocher, eds. *Endless forms: Species and speciation*. Oxford University Press: Oxford, UK.

West, S. A., M. G. Murray, C. A. Machado, A. S. Griffin, and E. A. Herre. 2001. Testing Hamilton's rule with competition between relatives. *Nature* 409:510–513.

West, S. A., I. Pen, and A. S. Griffin. 2002. Cooperation and competition between relatives. *Science* 296:72–75.

West-Eberhard, M. J. 1979. Sexual selection, social competition, and evolution. *Proceedings of the American Philosophical Society* 123:222–234.

———. 1983. Sexual selection, social competition, and speciation. *Quarterly Review of Biology* 58:155–183.

———. 1989. Phenotypic plasticity and the origins of diversity. *Annual Review of Ecology and Systematics* 20:249–278.

———. 1992. Behavior and evolution. Pages 57–75 in P. R. Grant, and H. S. Horn, eds. *Molds, molecules, and metazoa: Growing points in evolutionary biology.* Princeton University Press: Princeton, NJ.

———. 2003. *Developmental plasticity and evolution.* Oxford University Press: Oxford, UK.

———. 2005. Developmental plasticity and the origin of species differences. *Proceedings of the National Academy of Sciences of the United States of America* 102:6543–6549.

Whiteman, H. H. 1994. Evolution of paedomorphosis in salamanders. *Quarterly Review of Biology* 69:205–221.

Whitman, D. W., and A. A. Agrawal. 2009. What is phenotypic plasticity and why is it important? Pages 1–63 in D. W. Whitman, and T. N. Ananthakrishnan, eds. *Phenotypic plasticity of insects.* Science Publishers: Enfield, NH.

Wiens, J. J. 2011. The niche, biogeography and species interactions. *Philosophical Transactions of the Royal Society B: Biological Sciences* 366:2336–2350.

Wiens, J. J., D. D. Ackerly, A. P. Allen, B. L. Anacker, L. B. Buckley, H. V. Cornell, E. I. Damschen et al. 2010. Niche conservatism as an emerging principle in ecology and conservation biology. *Ecology Letters* 13:1310–1324.

Wiley, R. H. 1991. Associations of song properties with habitats for territorial oscine birds of eastern North America. *American Naturalist* 138:973–993.

———. 1994. Errors, exaggeration, and deception in animal communication. Pages 157–189 in L. Real, ed. *Behavioral mechanisms in ecology.* University of Chicago Press: Chicago, IL.

Wiley, R. H., and J. Poston. 1996. Perspective: Indirect mate choice, competition for mates, and coevolution of the sexes. *Evolution* 50:1371–1381.

Wiley, R. H., and D. G. Richards. 1978. Physical constraints on acoustic communication in the atmosphere: Implications for the evolution of animal vocalizations. *Behavioral Ecology and Sociobiology* 3:69–94.

Wilkins, A. S. 2002. *The evolution of developmental pathways.* Sinauer: Sunderland, MA.

Wilkins, J. S. 2009. *Species: A history of the idea.* Species and systematics series. University of California Press: Berkeley, CA.

Williams, C. B. 1964. *Patterns in the balance of nature.* Academic Press: New York.

Williams, E. E. 1972. The origin of faunas. Evolution of lizard congeners in a complex island fauna: A trial analysis. *Evolutionary Biology* 6:47–89.

Williams, G. C. 1966. *Adaptation and natural selection.* Princeton University Press: Princeton, NJ.

Wills, C. 1993. Escape from Stupidworld. *Discover* 14:54–59.

Wills, M. A. 2001. Morphological disparity: A primer. Pages 55–144 in J. M. Adrain, G. D. Edgecombe, and B. S. Lieberman, eds. *Fossils, phylogeny and form: An analytical approach.* Kluwer Academic / Plenum Publishers: New York.

Wills, M. A., D. E. G. Briggs, and R. A. Fortney. 1994. Disparity as an evolutionary index: A comparison of Cambrian and Recent arthropods. *Paleobiology* 20:93–131.

Wills, M. A., and R. A. Fortney. 2000. The shape of life: How much is written in stone? *Bioessays* 22:1142–1152.

Wilson, D. S. 1989. The diversification of single gene pools by density- and frequency-dependent selection. Pages 366–383 in D. Otte, and J. A. Endler, eds. *Speciation and its consequences.* Sinauer: Sunderland, MA.

Wilson, E. O. 1961. The nature of the taxon cycle in the Melanesion ant fauna. *American Naturalist* 95:169–193.

———. 1965. The challenge from related species. Pages 7–24 in H. G. Baker, and G. L. Stebbins, eds. *The genetics of colonizing species.* Academic Press: New York.

———. 1992. *The diversity of life.* Harvard University Press: Cambridge, MA.

———. 2010. *The diversity of life* (2nd edn.). Harvard University Press: Cambridge, MA.

Wilson, E. O., and F. M. Peter. 1988. *Biodiversity.* National Academy of Sciences, Smithsonian Institution: Washington, DC.

Wimberger, P. H. 1994. Trophic polymorphisms, plasticity, and speciation in vertebrates. Pages 19–43 in D. J. Stouder, K. L. Fresh, and R. J. Feller, eds. *Theory and application in fish feeding ecology.* University of South Carolina Press: Columbia, SC.

Windig, J. J., C. G. F. De Kovel, and G. De Jong. 2004. Genetics and mechanics of plasticity. Pages 31–49 in T. J. DeWitt, and S. M. Scheiner, eds. *Phenotypic plasticity.* Oxford University Press: Oxford, UK.

Wirtz, P. 1999. Mother species-father species: Unidirectional hybridization in animals with female choice. *Animal Behaviour* 58:1–12.

Wittkopp, P. J., B. K. Haerum, and A. G. Clark. 2004. Evolutionary divergence of *cis* and *trans* gene regulation. *Nature* 430:85–88.

Wittkopp, P. J., and G. Kalay. 2012. *Cis*-regulatory elements: Molecular mechanisms and evolutionary processes underlying divergence. *Nature Reviews Genetics* 13:59–69.

Wolf, J. B. W., J. Lindell, and N. Backstrom. 2010. Speciation genetics: Current status and evolving approaches. *Philosophical Transactions of the Royal Society B: Biological Sciences* 365:1717–1733.

Wolfe, B. E., V. L. Rodgers, K. A. Stinson, and A. Pringle. 2008. The invasive plant *Alliaria petiolata* (garlic mustard) inhibits ectomycorrhizal fungi in its introduced range. *Journal of Ecology* 96:777–783.

Wood, T. E., N. Takebayashi, M. S. Barker, I. Mayrose, P. B. Greenspoon, and L. H. Rieseberg. 2009. The frequency of polyploid speciation in vascular plants. *Proceedings of the National Academy of Sciences of the United States of America* 106:13875–13879.

Wray, G. A. 2007. The evolutionary significance of *cis*-regulatory mutations. *Nature Reviews Genetics* 8:206–216.

———. 2010. Embryos and evolution: 150 years of reciprocal illumination. Pages 215–239 in M. A. Bell, D. J. Futuyma, W. F. Eanes, and J. S. Levinton, eds. *Evolution since Darwin: The first 150 years.* Sinauer: Sunderland, MA.

Wund, M. A., J. A. Baker, B. Clancy, J. L. Golub, and S. A. Foster. 2008. A test of the "flexible stem" model of evolution: Ancestral plasticity, genetic accommodation, and morphological divergence in the threespine stickleback radiation. *American Naturalist* 172:449–462.

Yoder, J. B., E. Clancey, S. Des Roches, J. M. Eastman, L. Gentry, W. Godsoe, T. J. Hagey et al. 2010. Ecological opportunity and the origin of adaptive radiations. *Journal of Evolutionary Biology* 23:1581–1596.

Yoder, J. B., and S. L. Nuismer. 2010. When does coevolution promote diversification? *American Naturalist* 176:802–817.

Yom-Tov, Y., S. Yom-Tov, and H. Moller. 1999. Competition, coexistence, and adaptation amongst rodent invaders to Pacific and New Zealand islands. *Journal of Biogeography* 26:947–958.

Yoshimura, J., and C. W. Clark. 1994. Population-dynamics of sexual and resource competition. *Theoretical Population Biology* 45:121–131.

Zahavi, A. 1975. Mate selection: A selection for a handicap. *Journal of Theoretical Biology* 53:205–214.

Zuk, M., J. T. Rotenberry, and R. M. Tinghitella. 2006. Silent night: Adaptive disappearance of a sexual signal in a parasitized population of field crickets. *Biology Letters* 2:521–524.

INDEX

Acanthomorpha (spiny-finned fishes), 215f
adaptation. *See also specific topics*
   local, 57, 68, 151, 153, 168–169, 188, 195, 202, 218
   morph-specific, 197, 198f, 200
adaptive divergence, 14, 93, 94, 101
adaptive diversification, 213, 217
adaptive evolution, 46, 59, 90, 93, 102
adaptive landscape, 148. *See also* fitness landscape
adaptive radiation, 206–207, 211, 213–214, 216–217. *See also specific topics*
   causes of, 213–214, 216–217
   character displacement and, 217–219, 221–224, 238, 240
   competition and, 213–214, 216–217
   competitively mediated selection and, 219, 220f
   defined, 213
   different phases of, 220f, 223
   divergent selection and, 221–222
   early-burst model of, 222
   ecological opportunity and, 212f, 213–214, 216–217, 220f, 222–223, 228–229, 235
   examples of, 213
   experimental approach for studying, 211, 212f
   how habitats conducive for character displacement may be crucibles of, 150
   importance of, 213
   key components of, 217, 219, 238
   key innovations and, 214–216, 223
   nature of, 213
   replicated, 211, 212f
   of spiny-finned fishes, 215f
agonistic character displacement, 25
allopatric speciation, 184f, 184–185, 190
alternative phenotypes. *See* phenotypes, alternative
amensalism, 3t
Anole lizards, 42t, 73–75. *See also Anolis* sp.
Anolis sp., 40, 41f, 101–102, 158f, 211, 213. *See also* Anole lizards
   A. cooki, 73–74
   A. cristatellus, 73–74
   character displacement in, 16
   divergence in sexual signals in, 158
   ecomorphs of, 41f
   evidence of adaptive radiations in, 211, 212f, 213
   habitat shifts in, 73–76

Anolis sp. *(continued)*
  phenotypic plasticity in, 42t, 101–102
  resource partitioning in, 40, 41f
  size differences, 16
antagonistic genetic correlations
  defined, 65
  affect on character displacement, 65
apparent competition, 15, 20
arthropods, disparity in modern vs. Cambrian forms, 223
Asellus aquaticus (freshwater isopod), 167
assortative mating by ecomorphs, 127, 199, 201–202
  factors that may preclude evolution of, 199–200

bacteriophage
  φ6, evolution of host-range shift in, 111f
  Lambda, evolution of resource polymorphism in, 118t
barnacles, 31, 34, 134–135
beak morphology, 70, 71f, 72, 75, 86, 102, 163f, 200
  in finches, 6–7, 8f, 9, 11f, 70, 72, 72f, 86, 87f, 102, 113f, 120, 162
beetles, 2f. *See also individual taxa*
  diversity of, 1, 2f
  evidence of ecological character displacement in, 13f
  evidence of reproductive character displacement in, 9f
  giant rhinoceros, 59
  horned, 123, 123f, 124t, 126
  mating polymorphism in, 123, 123f, 124t, 126
benthic ecomorphs. *See under* ecomorphs
biodiversity, 1. *See also* diversity
  competition and, x
  defined, 1
  explaining, 1–2, 2f
  neutral theory of, 143
Blair, W. F., 20–21
body size, 16, 43f, 206f, 209f
  biotic and abiotic factors and, 229
  character displacement and, 20, 22, 23f, 24, 73, 75, 100, 146, 147f, 209f, 226–227, 238, 240, 241
  clades and, 227
  competition and, 3, 32, 146, 226, 227

  evolutionary escalation and, 225–227
  increase over time, xii, 20, 22, 23f, 206f, 209f, 225–227
  oxygen levels and, 241
  reasons extant taxa are not at their maximum, 227
bone morphogenetic protein 4 (*Bmp4*), 86, 87f
Bonner, J. T., 226
bottlenecks, 185
brachiopods, 209f
Brown, W. L., 7, 9–10
Burmeistera, 51, 51f

cadmium-tolerant vs. -intolerant flies, 110
call rates, in anurans, 171–172, 173f
calmodulin (*CaM*), 86, 87f
Calopteryx sp.
  C. splendens, 62f
  C. virgo, 62f
Cambrian explosion, 241–242
Cambrian period, 213, 222, 223, 241–242
canalization, 83f, 100, 101
  genetic, 82–83, 83f, 89, 91, 93–103
  genetically canalized divergence, 82–89, 93–96, 97f
character displacement, x–xii, 2. *See also specific topics*
  agonistic, 25
  alternative manifestations of, 22–24, 23f
  alternative routes to, 58–59, 60f, 61, 63, 64, 138
  as the "best of a bad situation," 148
  asymmetric, 66–67, 67f
  can produce different forms of trait evolution, 22–24, 23f, 234–235
  community-wide, 218, 225, 227
  conflation of process and pattern, 16, 19–21, 24, 70, 110, 175–176, 234
  convergent, 20, 24, 45, 235
  criteria for demonstrating, 16, 17t–18t, 90, 236
  and Darwinian extinction, 146–149
  defined, 24
  detecting, 10–14
  developmental bases of, 82–84, 86–87, 87f, 90–91, 92f
  direct evidence of, 10, 11f
  discovery of, 4–10

292 · INDEX

ecological, xi, 24. *See also* ecological character displacement
ecological opportunity and, 63
evolution of, 10, 11f, 41, 43, 52, 94–98, 236. *See also* plasticity-first hypothesis
forms of, 24–25. *See also* ecological character displacement; reproductive character displacement
in the fossil record, 207,209f
importance of, 4–5, 25–27, 236–237, 238
importance of identifying the mechanisms underlying, 82
intraspecific. *See* intraspecific character displacement
macroevolutionary implications of, 238
observational evidence of, 8f, 9f, 10–14, 13f, 16, 17t–18t
phenomena mistaken for, 14–19
phenotypic plasticity as a mediator of, 89–91, 93–98
and phenotypic similarity between competitors, 37–38, 39f, 209f, 220
and phylogenetic distance between competitors, 144–145, 145f
as process vs. pattern, 16, 19–21, 24, 70, 110, 175–176, 234
as a cause of complexity, xii, 22, 23f, 226t, 228, 238, 240–241
promotes diversification at multiple levels, 236–237
proximate mechanisms mediating, 82–93, 87f, 92f, 239
reinforcement vs. reproductive character displacement, 20–21
reproductive, xi, 24–25. *See also* reproductive character displacement
role in speciation. *See under* speciation
selective barriers that arise as a consequence of, 195, 203
within vs. between species, 129–130
speed/tempo and mode of, 81–82, 93–102
terminology, 21–25
as unifying principle, 25–27
unsolved problems regarding, 238–242
variation in the expression of, 65–69
vs. species sorting. *See under* species sorting
ways it can trigger rapid production of new species, 218–219

character displacement, facilitators of, 58, 68–69, 89, 93–94, 106, 130, 174–175, 237. *See also* ecological and reproductive character displacement; sexual selection and character displacement
 ecological opportunity, 63
 gene flow, 64–65
 initial differences between species, 63–64
 lack of antagonistic genetic correlations, 65
 standing variation, 58–61
 strong selection, 61–62
character release, 8f
 causes of, 109–110
 defined, 108
 examples of, 8f, 108–110
Chthamalus stellatus, 34, 134–135
cichlids, 115, 116f, 120–121, 202, 214
Cicindela sp. (tiger beetles), 13f
clades, 127–129, 128f, 200, 222–223, 227
co-adapted gene complexes, 88
co-evolution, 22, 23f, 32, 149, 224–225, 229, 242
co-mimics, 37
Cocos finch (*Pinaroloxias inornata*), 108
coexistence of species. *See* species coexistence
community organization, 143–146
competition, ix–x, 2. *See also* exploitative competition; interference competition; interspecific competition; intraspecific competition; reproductive competition; resource competition
 as a cause of adaptive radiation, 217
 defined, 2, 3t
 in the fossil record, 207–209, 208f, 209f, 210–211, 210f
competitive ability, 3t, 32
 character displacement and, 22, 23f, 224
 evolutionary escalation of, xii, 3t, 20, 22, 23f, 32, 148–149, 224, 225, 235
competitive displacement, 207, 208f
competitive exclusion, 33–34, 95, 130, 134, 153, 175, 189, 207, 220
 character displacement and, 57, 58, 61, 63, 64, 78, 82, 89, 93, 130, 136, 138, 140, 145, 146, 149–151, 154, 174, 190, 207, 240
 classic examples from natural populations, 34
 classic experimental demonstration of, 33f

competitive exclusion *(continued)*
  competitive displacement and, 207
  defined, 33
  frequency of, 37, 39f, 220, 221f
  individual variation and likelihood of, 139f
  phenotypic plasticity and, 89–90, 93, 95
  phenotypic similarity and, 39f
  phylogenetic distance and, 220, 221f
  sexual selection and, 174
competitive exclusion principle, 6, 35–36, 134
competitive interactions, cost of, 3, 32, 62
competitive mimicry, 45
competitively mediated divergence, 6, 19
competitively mediated escalation, 22
  macroevolutionary trends that may reflect, 224–228, 226t
competitively mediated selection, ix–x, 5–6, 16, 17t, 59, 234–235, 240–241. *See also specific topics*
  character displacement, trait evolution, and, 20, 22, 23f, 38, 41, 43f, 47, 52, 59, 61, 70
  community organization and, 143, 144
  and divergent trait evolution and adaptive radiation, 216–217, 219–220, 220f
  forms of trait evolution generated by, 20, 22, 23f, 234–235
  genetic targets of, 85, 87–88, 90, 91f, 94, 97, 236
  macroevolutionary trends and, 216–217, 219–220, 220f, 225–228, 230, 238
  negative-frequency dependence and, 123
  and niche differences, 133, 135f
  sexual selection and, 157–158, 160–161, 165–167, 174, 237
  standing variation and, 59
  strength of, 65–67
competitively mediated trait evolution, 44, 90. *See also* character displacement
complexity, 228
  arising from character displacement, xii, 22, 23f, 226t, 228, 238, 240–241
  defined, 228
  in reproductive traits, 173, 237
conspecific mates, 77, 84, 85, 161–162, 171–173
  character displacement and, 48–52
  reproductive isolation and, 193–196
conspecific populations, 67–68, 69f, 108, 193–196

conspecifics, 105, 129–130, 146
  growth rate of, 35f
conspicuous male, 202. *See also* male traits
  traits, 166
convergence in trait expression, 43
convergent evolution, 45, 221
convergent trait evolution, 45, 77–78. *See also* character displacement: convergent
Cope's rule, 206f, 225, 226
Court Jester hypothesis, 229
crinoids, 208, 210f
crossbills (*Loxia* spp.), 115, 119t, 121, 122
"cryptic" genetic variation, 76, 93–94

Darwin, C., 4. *See also specific topics*
  development of theory of evolution, ix
  his principle of divergence of character, 4–6, 235
  on importance of competition in evolution, ix–x, 2, 4–6, 34, 219–221
  on resource partitioning, 34, 136, 138
  on sexual selection, 159
  on speciation, 183
Darwinian extinction, 149
  character displacement and, 146–149
  defined, 149
Darwinian vs. pre-Darwinian views of evolution, 4–5, 5f
Darwin's finches, 6, 86, 90, 108, 213
  beak morphology in, 7, 72f, 86, 87f, 102, 113f, 200. *See also* beak morphology: in finches
dimorphism, sexual, 108, 109f, 127
disparity, morphological, 222–223
disruptive selection, 112, 113f, 114f, 115, 117, 123, 181, 187, 202
  competition and, 112, 115, 117, 123, 220f
  defined, 112
divergence. *See also* canalization; ecological divergence; phenotypic plasticity
  between ecomorphs, 196–197, 198f, 202
  causes/mechanisms of, 82–93
  of character, Darwin's principle of, 4–6, 235. *See also* character displacement
  environmentally induced, 89–93
  evolution of, 59, 189–193, 196, 219–224
  exaggerated, between sympatric species as a signature of character displacement, 7–11, 14, 19–20, 23f, 24, 234

how different proximate mechanisms differ in speed of, 81–82, 93–94
how phenotypic plasticity may promote rapid, 93–94
divergent evolution, 219–224
  role of phenotypic plasticity in facilitating, 93–94
divergent selection, 3–4, 186, 193, 200
  adaptive radiation and, 221–222
  forms of, 186–187
divergent trait evolution. *See under* competitively mediated selection
diversification, adaptive, 213, 217. *See also* adaptive radiation
diversity. *See also* biodiversity
  character displacement and alternative perspectives on origins and maintenance of, 25–27
  intraspecific competition and, 126–129
Dobzhansky, T., 191, 192
Dobzhansky-Muller incompatibilities, 182
dragonflies. *See Calopteryx* sp.
Drosophila sp. (fruit flies)
  D. melanogaster, 110
  D. recens, 194
  D. santomea, 192, 193
  D. subquinaria, 194
  D. yakuba, 192–93

early-burst model of adaptive radiation, 222
ecological character displacement
  criteria for demonstrating from observations, 17t–18t
  defined, xi, 9, 24
  detection and evidence of, 10–13, 11f, 12f, 13f
  ecological character displacement, interaction with reproductive character displacement, 235
  facilitating each other, 69–77
  impeding each other, 77–78
  occurring simultaneously, 69–70
  reasons one form is necessary to facilitate the other, 76–77
ecological character displacement, reasons for, 30
  competitive exclusion, 33–34
  resource competition, 29–32
  resource partitioning via character displacement vs. species sorting, 40–48
  species coexistence, 34–39
ecological divergence
  as facilitator of reproductive divergence, 70–75
  reproductive divergence as facilitator of, 75–76
ecological interactions, 2, 3t, 15
ecological niche, 35–37, 134–137
ecological opportunity
  defined, 63, 213
  importance for adaptive radiation, 212f, 213–214, 216–217, 220f, 222–223, 228–229, 235
  importance for character displacement, 63
  importance for intraspecific character displacement, 108–109, 117, 120f, 122–123, 126, 137–138, 150, 153
ecological processes, character displacement and, 240
ecological release. *See* character release
ecological specialization, 20, 22, 112, 114f, 137f, 138, 140, 220f, 223–224, 235
ecological speciation, 186, 187f, 188f
ecological speciation model, 186–187, 187f, 188f
ecomorphs, 12f, 40, 41f, 100–102, 181, 198f, 211, 212f
  A-morph and C-morph in *Timena cristinae* walking insects, 198f
  assortative mating between expected, but not found, 199
  benthic and limnetic in various species of fishes, 117, 119t, 196–197
  benthivore and planktivore in fishes, 11
  carnivore and omnivore in *Spea* spadefoot toad tadpoles, 12f, 42t, 100, 113f
  defined, 40
  ecological segregation between alternative resource-use, 117, 118t–119t, 196–197
  evolution of assortative mating between, 199–200
  phenotypic partitioning and the evolution of, 40
  reed and stonewort in *Asellus aquaticus* isopods, 167
  reproductive isolation between, 118t–119t, 181, 199–201
  resource polymorphism and separation between alternative, 196–197, 198f

ecomorphs *(continued)*
  trunk-ground in *Anolis* lizards, 73–74
  various in *Anolis* lizards, 41f, 101–102, 211, 212f
ecosystem engineering, 216
electric fishes (mormyrid fishes), 141, 216
environmentally induced divergence, 89–93
epigenetic assimilation, 92f, 101
escalation, evolutionary, 224–228
  of competitive ability, xii, 3t, 20, 22, 23f, 32, 148–149, 224, 225, 235
Eucyrtidium sp.
  E. calvertense, 209f
  E. matuyamai, 209f
evolution. *See also specific topics*
  two views of the shape of, 4–5, 5f
evolutionary tape of life, replaying the, 211
experimental evolution, 103, 131, 188, 211, 230
exploitative competition
  as a driver of character displacement, 46
  defined, 31
  prevalence of, 32
extinction. *See also* Darwinian extinction; competitive exclusion; reproductive exclusion
  as alternative to character displacement, 17t, 32, 33f, 54
  causes of, 229
  mass, 206f, 214, 215f, 222, 227, 229
  resource polymorphism and, 129, 239
  role in creating ecological opportunity, 207, 214, 215f, 222
  role of gene flow in rescuing populations from, 151
  role of phenotypic plasticity in shielding populations from, 89, 95
  role of resource polymorphism in shielding species from, 129
  role of specialization in increasing risk of, 138
  size selectivity of, 227
  species richness and, 129

facilitation hypothesis for how positive interactions promote species coexistence, 37
Ficedula sp., *See* flycatchers
female preference polymorphisms, 201, 202
female preferences, 168, 175, 194. *See also* mate choice; mate preferences
  character displacement and, 74, 85, 162, 164
  evolution of, 70, 172

  male traits and, 50–51, 73–76, 160, 163f, 164–166, 169, 172, 194, 202
fibroblast growth factor 8 (*Fgf8*), 86, 87
finches. *See* beak morphology: in finches; Darwin's finches
fitness, hybrid, 65, 182
fitness consequences
  of hybridization, 52, 66
  of sexual selection, 168, 170–174, 201, 237
fitness costs, 3t, 32, 49, 54, 236
fitness effects of ecological interactions between species and populations, 2, 3t
fitness landscape, 107, 112
  as a sphagnum bog metaphor, 107
fitness trade-offs, 121, 148–149, 171–172, 174, 176, 195–196
flight, evolution of, 226t
flycatchers, 85–86
founder effects, 168, 184–185
frequency-dependent character displacement, 61–62, 62f
frequency-dependent selection,
  and evolution of alternative phenotypes, 112, 113–114f, 115, 117, 122, 127, 211, 212f
  competition as a cause of, 107
fruit flies. *See Drosophila* sp.

Gasterosteus aculeatus. *See* stickleback fish
Gause, G., 6, 33, 37, 220
Gause's hypothesis. *See* competitive exclusion principle
gene flow, 14, 58, 61, 64, 78, 88, 93, 151–153, 174–175, 181, 183–186, 189, 192–193, 195–196
  role in facilitating or inhibiting character displacement, 64–65
  role in rescuing populations from extinction, 151
generalists, 12f, 63, 107, 111f, 121–122
genes, regulatory, 86–88
genetic assimilation, 83f, 95–96, 199
  defined, 95
  role in character displacement, 96, 98, 101
genetic canalization. *See under* canalization
genetic correlations, antagonistic, 65, 88
genetic hardening. *See* genetic assimilation
genetic regulatory architecture of phenotypic plasticity, 90, 91f

genetic variation
   cryptic, 76, 93–94
   importance of, for character displacement, 58–59
   origins of, 61
geographic mosaics, species distributions and, 150–151
geographic patterns of trait variation, 19
geographical range, 129, 134, 150–153
Geospiza sp., *See* Darwin's finches
Grant, P. R., 19–20
Gray treefrogs (*Hyla versicolor*), 50
guppies, 168–169

habitat shifts, 73–75, 77, 161, 165–167
habitats, diversity in sexual traits between and within, 169, 170f
Heliconius sp. (butterflies), 85, 200
   H. cydno, 85
   H. pachinus, 85
hoverflies, 44–45
hybrid fitness, reduced, 182
hybrid zygotes, 182
hybridization, xi, 48, 191–192, 194
   and evolution of species-specific traits, 52
   as impediment to character displacement, 65, 88
   as selective basis for character displacement, xi, 52, 64–68, 69f
   as source of new variants, 61
   between *Drosophila* species, 84, 194
   between *S. bombifrons* and *S. multiplicata*, 66, 68, 69f, 171–172
   fitness consequences of, 49, 52, 66
   gene flow and, 64–65
   reinforcement and, 20, 21
   reproductive competition and, 48–49
   reproductive exclusion and, 49–50
Hyla versicolor. *See* Gray treefrogs

incumbent replacement, 210–211
inherited environmental effects, 90, 101, 183
insects. *See also individual taxa*
   host races in, 118, 197, 198f, 199
   mating polymorphism in, 124t
   reproductive partitioning in, 51
   resource polymorphism in, 118t, 198f
   resource partitioning in, 47f, 198f

interchangeability, between genetic and environmental effects in development, 83f, 95–96
interference competition. *See also* exploitative competition
   as a driver of character displacement, 46
   defined, 31
   prevalence of, 32
interspecific competition, 34–35, 43, 90, 129–130, 138, 235
   and divergent character displacement, 137
   gene flow and, 64
   vs. intraspecific competition, 34–35, 105, 129–130, 217
   maternal effects and, 90, 92f, 101
   phenotypic plasticity and, 42t, 44, 90, 98
   and population size, 34–35
   release from, 108, 109, 117
   reproductive partitioning and, 143
   resource partitioning and, 47, 47f
intraguild predation, 15
intraspecific character displacement, 8f, 24, 105, 179
   and character displacement between species, 130
   and speciation, 179, 181f, 196–202
   defined, 105, 196
   ecological opportunity and, 108–109, 117, 120f, 122–123, 126, 137–138, 150, 153
   experimental evidence of, 110–111, 111f
   impacts on diversification, 106
   mechanisms of, 106–108
   observational evidence of, 108–110
   and polymorphism, 123, 126–130
   and species diversity, 126–131, 239–240
intraspecific competition, 112, 115, 117, 129, 138. *See also* intraspecific character displacement
   frequency-dependent, 115
   vs. interspecific competition, 34–35, 105, 129–130, 217
   phenotypic plasticity and, 94, 126
   and polymorphism, 117, 122, 126, 127
   and species diversity, 126–129
isolating mechanisms, 182–183. *See also* reproductive isolation
   evolution of, 183–185, 189, 191–195
isopods, 167

key innovation, 214, 216, 223

Lack, D., 6–7, 8f, 9, 10, 19
learning, 89, 90, 183
   possible role in speciation, 183
   role in environmentally induced divergence, 89, 90
limited resources, defined, 30
limiting similarity, 36, 38f, 50
limnetic ecomorphs. *See* ecomorphs: benthic and limnetic in various species

macroevolution, 205, 228–229
   debates over causes of, 25, 205
   defined, 205
   evidence for competition's role in, 206f, 207–217, 208f, 209f, 210f, 215f
   methods for studying, 211, 213
   microevolution and, xi, 25–26, 205, 230, 238, 241
   patterns of, 26, 226
magic traits, 200
male traits, 146–149
   conspicuous, 166
   divergence in ecological traits and, 190–191
   ecological character displacement and, 73, 162, 164–165, 173
   female preferences and, 50–51, 73–76, 160, 163f, 164–166, 169, 172, 194, 202
   habitat and evolution of, 73–76, 165–166
   novel, 164–166, 175
   predation, parasitism, and, 166
   resource shifts and, 70–73, 160, 162–164
   species divergence in, 51
male-male competition, 162, 201. *See also* sexual selection: intrasexual
Malthus, T., ix
matched filtering, 51
mate choice, 72–74, 85, 146, 159–161, 166, 168, 173–174, 176–177, 193, 237, 239. *See also* female preferences; mate preferences
mate-quality recognition, 171–174, 194
mate preferences, 75, 160, 182, 185, 187–188, 237. *See also* female preferences; mate choice
maternal effects
   defined, 90
   role in character displacement, 90–91, 92f, 101

mating behavior, 50, 122, 159–161, 163f, 164–165, 191–194
mating polymorphism
   defined, 122
   ecological opportunity and expression of, 123, 126
   evolution of, 122–123, 126
   role in character displacement, 127, 130–131
   role in speciation, 201–202
   in various taxa, 123f, 124–125t
mating system, 161, 166–167, 176
Mayr, E., 180–185
microevolution and macroevolution, xi, 25–26, 205, 230, 238, 241
mimicry, 37, 45, 85, 123, 124t–125t
morphs, alternative, 115, 123f, 196, 197, 199, 202, 239. *See also* ecomorphs; phenotypes, alternative
   ecological separation between, 197
   evolution of, 187
   female preferences for, 201
   heterogeneity of, 111–112
   resource polymorphism and, 117, 120, 122, 127, 128
Müllerian mimicry, 37, 85
multicellularity, evolution of, 226t, 228
multiple-allele mechanism, 88
mutation, 110
   role in character displacement, 59, 61, 87–88, 98
   role in speciation, 186, 191
mutation-order model of speciation, 186
mutualism, 3t, 37

natural selection, 74
   and body size, 225
   competition and, 4, 37–38, 39f, 43, 46, 52, 61–62, 107, 115, 116f, 219–221, 220f, 225–228
   and complexity, 228
   Darwin on, 1–2, 4–6
   divergent, 186, 200. *See also* divergent selection
   genetic assimilation and, 96
   sexual selection and, 159, 160, 225
neutral theory of biodiversity, 143
niche
   character displacement and, 35–36, 38f, 136, 137

defined, 35–36
ecological, 35–37
evolution of the, 134–136
fundamental, 134–35, 135f, 137
realized, 134–38, 135f
superior, 146, 147f, 148–149
niche breadth, 135f, 222
niche construction, 216
niche-width expansion, 107–108, 129
novel habitats, 74, 167
novel phenotypes, 59–60, 60f, 61, 63
   as an adaptive response to intraspecific competition, 106f, 110–111
novel preferences and traits, 106f, 110–111. *See also* key innovation; sexual traits: novel
novel resources
   intraspecific competition and, 110–111
   shifts to, displacement onto, 110, 149, 164

On the Origin of Species (Darwin), 4–6
operational sex ratio (OSR), 166–167, 176
overpopulation, ix

parallel character displacement, 14–15, 17t, 26, 98
parallel evolution, 11, 14, 167, 197
parallel speciation, 188
Paramecium, 33, 37, 39f
parasitism, 3t, 15, 36, 146, 166, 197
parental effects, 90. *See also* maternal effects
partitioning, 29, 34. *See also* phenotypic partitioning; reproductive partitioning; resource partitioning; spatial partitioning; temporal partitioning
phenotype shifts, 91
phenotypes, alternative, 121, 130, 201–202
   defined, 111
   evolution of, 111–117
   importance of, 111–112
phenotypic plasticity, 12f, 41, 42t, 51–52. *See also* plasticity-first hypothesis
   as a mechanism of reproductive partitioning, 51–52
   as a mechanism of resource partitioning, 41
   defined, x
   examples in which it reduces competition between species, 42t
   genetic regulatory architecture of, 90, 91f

mating polymorphisms mediated by, 126
resource polymorphisms mediated by, 122
role in character displacement, 10, 44, 45, 81–84, 89–91, 93, 94, 99, 101, 103, 126, 236, 239
role in driving adaptive evolution, 236
role in shielding populations from extinction, 89, 95
why it promotes rapid divergence between species, 93–94
phenotypic partitioning
   stemming from competition for reproduction, 50, 51f
   stemming from competition for resources, 40
phenotypic variation, 58–59, 81, 93, 102, 105–106, 236, 239. *See also* standing variation
Phlox sp. (Texas wildflowers)
   P. cuspidata, 13–14
   P. drummondii, 13–14
phylogenetic approaches, 144
phylogenetic clustering, 145, 145f
phylogenetic distance, 144–145, 145f, 220, 221f
phylogenetic overdispersion, 144, 145, 145f
plants. *See also individual taxa*
   adaptive radiation in, 213
   competition in, 31–32
   competitive displacement in, 208f
   convergent evolution in, 221
   individual variation in, 139, 139f
   key innovations in, 214
   macroevolutionary trends in, 208f, 214, 226t
   phenotypic plasticity in, 42t, 97f
   reproductive character displacement (reinforcement) in, 13–14, 87
   reproductive partitioning in, 48–49, 50–52, 51f
   resource partitioning in, 35f
   transition to land, 226t
   use of different pollinators in, 48–49, 50–52, 51f, 200
plasticity. *See also* phenotypic plasticity
   competitively induced, as character displacement, 44–45
plasticity-first hypothesis, 94–96, 236
   empirical tests of, 98–102
   ramifications for the ways character displacement occurs, 96, 98
pleiotropy, 85, 200

Poecilia. *See* guppies.
polymorphism. *See* mating polymorphism, resource polymorphism
population growth, ix, 49, 240
predation, 3t, 15, 36–37, 146, 166–169, 197, 241–242
   as a possible cause of adaptive radiation, 217
predator-prey interactions, 224–225
pre-existing differences, 46f, 47, 52, 190
preferences. *See* female preferences; mate choice; mate preferences
pterosaurs, 206f

reaction norms, 44, 81, 90, 93–94, 98
recombination, 61, 93, 98
red coloration, 74–75, 165
Red Queen hypothesis and Red Queen effects, 224, 229
regulatory genes, 86–88
reinforcement, 191–193
   defined, 21, 191
   role in character displacement, 20–21
   role in speciation, 191–193
   vs. reproductive character displacement, 20–21
replicated adaptive radiation, 211, 212f
replicated character displacement. *See* parallel character displacement
reproductive character displacement
   criteria for demonstrating from observations, 17t–18t
   defined, xi, 9, 21, 24–25
   detection and evidence of, 9f, 10–14
   vs. reinforcement, 20–21
reproductive character displacement, interaction with ecological character displacement, 235
   facilitating each other, 69–77
   impeding each other, 77–78
   occurring simultaneously, 69–70
   reasons one form is necessary to facilitate the other, 76–77
reproductive character displacement, reasons for, 48
   reproductive competition, 48–49
   reproductive exclusion, 49–50
   reproductive partitioning via character displacement vs. species sorting, 50–53

reproductive competition, 30f, 48–49, 109, 214, 235
reproductive divergence, 70
   ecological divergence as facilitator of, 70–75
   as facilitator of ecological divergence, 75–76
reproductive exclusion, 49–50, 61, 64, 82, 89, 130, 134, 136, 141, 153, 189
   character displacement and, 49–50, 57, 58, 61, 63, 64, 78, 82, 89, 93, 130, 136, 138, 140, 145, 146, 151, 154, 174, 240
   competitive exclusion and, 49, 50
   phenotypic plasticity and, 93, 95
   sexual selection and, 174
reproductive interactions between species, xi, 3
   causes of, 48–49
   consequences of, 49–53
   defined, 3
   importance in ecology and evolution relative to resource competition, 27, 216
   selection and, xi, 27, 216
reproductive interference, 48–49, 53
reproductive isolation, 126–127, 174–175, 180f, 181, 183–196, 202–203, 218f, 219, 239
   between ecomorphs, 199–201
   selection and the evolution of, 185–188
reproductive niche, 134, 141, 148, 219
reproductive partitioning, 50–53, 136–143, 142f
   defined, 50
   how it reduces competition between species, 50–52
   phenotypic plasticity and, 51–52
   proximate mechanisms through which it can arise, 51–52
   resource partitioning and, 51–53
   ways it can manifest, 50–51
   via character displacement vs. species sorting, 50–53
reproductive specialization, 22
reproductive-trait space, 48, 63, 126, 133, 136, 141–143, 153, 213–214, 223, 235
resource competition, 7, 174. *See also* species coexistence
   and character displacement, x, 9, 14, 15, 19, 24, 26–27, 29–32, 57, 70, 71, 71f, 75–77, 89, 100, 196, 234–236
   cost of, 32
   forms of, 31. *See also* exploitative competition; interference competition

reduction of with heterospecifics, 41, 42t
selection stemming from, 53
and trait evolution, xi
resource partitioning, 40–41, 51–53, 136–143, 207, 208, 210f
  defined, 40
  how it reduces competition between species, 47, 47f, 137, 137f
  in the fossil record, 207, 208, 210f
  phenotypic plasticity and, 42t, 44–45
  proximate mechanisms through which it can arise, 40–41, 44–45
  reproductive partitioning and, 51–53
  ways it can manifest, 40
  via character displacement vs. species sorting, 40–47
resource polymorphism
  defined, 117
  ecological opportunity and expression of, 117, 120, 120f, 122
  evolution of, 117, 120–122, 126–129, 196–203
  and extinction, 129, 239
  role in character displacement, 117–122, 127, 130–131
  role in shielding species from extinction, 129
  role in speciation, 127–129, 128f, 196–201
  and species richness, 120, 127–128, 128f, 129
  in various taxa, 118–119t
resources, defined, 30

satellite males, 125t, 201
secondary contact, 185, 192
selection. *See also specific topics*
  direct, 65, 188, 200
  disruptive, 112–117, 123, 181f, 187, 196, 199, 202–203, 220
  and evolution of isolating mechanisms, 184, 186–188
  frequency-dependent, 107, 112–117, 122, 127, 139, 211, 212f
  indirect, 65, 187, 200
  natural. *See* natural selection
  sexual. *See* sexual selection
  and speciation, 184–191, 193–197, 198f, 199–203
Semibalanus balanoides, 34, 134–135
sensory modalities, 141, 166

sensory systems, 73–74, 91, 160, 164
sex ratio, 161. *See also* operational sex ratio
sexual dimorphism, 108, 109f
sexual exclusion. *See* reproductive exclusion
sexual selection, xi, 70, 123, 149, 157–177, 186, 194, 197, 199, 225, 237, 239
  causes/mechanisms of, 159–160
  competitively mediated selection and, 157–158, 160–161, 165–167, 174, 237
  defined, 159–160
  discovery of, 159
  elaborate traits and, 169, 171, 237
  fitness consequences of, 170–174, 201, 237
  importance of identifying factors that influence, 157
  intersexual, 159, 162
  intrasexual, 159, 162, 164
  runaway, 169
sexual selection and character displacement, interaction between, 158–160, 163f, 174–175, 177, 239
  effects of habitat shifts, 165–167
  effects of phenotypic shifts, 161–162, 163f, 164–165
  how character displacement affects sexual selection, 160–162, 163f, 164–167
  how sexual selection affects character displacement, 174–175
  implications of the effects of character displacement on sexual selection, 168–174
  need to determine how they interact, 237
  process vs. pattern, 175–176
sexual traits
  diversity in, between and within habitats, 168–170, 170f
  novel, 162, 164–167, 174–175
single-allele mechanisms, 84, 88
  advantages, 88
single-locus mechanisms, 88
sneaker males, 123, 125t
sneaker morphs, 126, 127
sorting. *See also* species sorting
  of pre-existing variation, 59, 60f, 61, 63–64, 78, 138–140
spadefoot toads. *See Spea* sp.
spatial partitioning
  stemming from competition for reproduction, 50

spatial partitioning *(continued)*
  stemming from competition for resources, 40, 47f
Spea sp. (spadefoot toads)
  evolution of character displacement in, 101
  S. bombifrons, 12f, 39f, 52, 59, 63, 66–68, 69f, 92f, 100, 101, 146, 147f, 150, 171–172, 173f, 192
  S. multiplicata, 12f, 39f, 52, 59, 63, 66–68, 69f, 92f, 100, 101, 146, 147f, 150, 171–172, 173f, 192
specialists, 121–122, 138
specialization, xii, 22, 26, 107, 121, 135, 138, 142, 220, 235. *See also* ecological specialization
  character displacement via increased, 23f, 137f, 138, 140, 154, 223–224
  individual, 107
specialized phenotypic traits, 120–121
speciation, xii, 21, 58, 85, 106, 126–131, 157, 158f, 174, 176, 179–203, 217–219, 237, 239–240
  biogeographical context of, 183–185, 184f
  defined, 179
  ecological, 186–188, 187f, 188f
  how character displacement finalizes, 180f, 190–193
  how character displacement initiates, 180f, 190, 193–196
  how intraspecific character displacement initiates, 181f, 196–197, 198f, 199–202
  isolating mechanisms and. *See* isolating mechanisms
  models of, 186–188
  occurring between populations as an indirect consequence of character displacement between species, 180f, 193–196
  role of character displacement in, 189–203
  secondary contact and, 185, 192
  selection and, 184–191, 193–197, 198f, 199–203
  stages of the three modes of, 184f
speciation cascades, 219
species. *See also specific topics*
  definition and nature of, 179–181
  how their boundaries are maintained, 158f, 182–183
  initial differences between, 58, 63–64, 78
species coexistence, xi, 29, 30f, 34–39, 47–48, 54, 90, 133, 136, 138, 140–146, 150, 153–154, 240
  alternative models of, 34, 36–37
  how differences between species in resource use promote, 34–35, 35f, 39f
  individual variation and, 138–140
species niche, 35–37, 134–136, 135f; *See also* niche; niche breadth
species packing, 138
species proliferation, 217–219
species ranges, 150–154, 240
species recognition, 9, 62f, 73–74, 172f, 173, 191
species richness, 48–49, 128–129, 143, 146, 154, 200–201, 240
  resource polymorphism and, 128–129, 128f, 200–201
species sorting, 15–16, 17t, 37, 53, 90, 136, 140, 143, 144, 145f, 151, 240
  character displacement vs., 29, 38, 53, 136, 143, 144, 145f, 151, 240
  reproductive partitioning via character displacement vs., 50–53
  resource partitioning via character displacement vs., 40–48
  vs. sorting of pre-existing variation, 59
sperm competition, 160, 162
spiny-finned fishes (Acanthomorpha), 215f
standing variation, 58–61, 76, 78, 81, 93–94, 96, 98, 110, 130, 150, 153, 175
  defined, 58
stickleback fish, 74
  divergence in sexual signals in, 74–75, 165, 175–176
  evolution of character displacement in, 74–75, 86, 98–100, 99f, 119t, 175–176
  habitat shifts in, 74–75, 165
  phenotypic plasticity in, 98–100, 99f
  proximate mechanism of character displacement in, 86, 98–100, 99f
stochastic processes, 96, 98, 149, 211, 217
strong selection, 49, 58, 61–62, 78, 160, 164, 191, 224
superior niche, 146, 147f, 148–149, 154, 225
sympatric speciation, 185

temporal partitioning
  stemming from competition for reproduction, 50

stemming from competition for resources, 40, 47f
tiering, 208, 210, 227
tiger beetles (*Cicindela* sp.), 13f
trade-offs, 112, 114f, 121, 149, 171, 173–174, 195–196
trait expression, regulation of, 96, 101
trees, evolution of, 226t
trilobites, 208, 222–223

ultraviolet (UV) light sensitivity, 73–76

variation, individual, 68, 236
and coexistence of species, 138–139, 154
virulence, evolution of, 22

weevil species, 49
Wilson, E. O., 7, 9–10, 19, 94–96

Composition: BookMatters, Berkeley
Text: 9.5/14 Scala
Display: Scala Sans
Printer and Binder: Sheridan Books